The Orientation of Science and Technology

A Japanese View

Shigeru Nakayama

The Collected Papers of
Twentieth-Century Japanese Writers
on Japan

VOLUME 3

Collected Papers

of

SHIGERU NAKAYAMA

The Orientation of Science and Technology

A Japanese View

GLOBAL
ORIENTAL

Series: COLLECTED PAPERS OF TWENTIETH-CENTURY JAPANESE WRITERS ON JAPAN

Volume 3
Shigeru Nakayama: The Orientation of Science and Technology: A Japanese View

First published 2009 by
GLOBAL ORIENTAL LTD
PO Box 219
Folkestone
Kent CT20 2WP
UK

www.globaloriental.co.uk

ISBN 978-1-905246-72-4

British Library Cataloguing in Publication Data
A CIP catalogue entry for this book is available from the British Library

Set in Plantin 10 on 11.5pt by RefineCatch Ltd, Bungay, Suffolk
Printed and bound in England by Antony Rowe Ltd, Chippenham, Wilts

Contents

Preface

When Professor Erich Pauer of Philipps-University Marburg suggested that I publish a book of my collected papers in a Japanese studies series, I initially limited it to works on Japanese science. The scope of the collection has, however, been extended to include technology and geographically it now includes China and the East Asia region. I have arranged the various papers in historical order. Some of the studies show attempts to assimilate Western science, and others emphasize efforts to find alternatives to Western science.

The second stage of my work was marked by my interest in the concept of 'service science' which was applicable not only to Japanese science but science in general. The article where I discuss this had limited citations but William Cummings took up the concept with enthusiasm and even extended it to the idea of a 'Service University'. Interest in the concept eventually died out but I was heartened when, a quarter of a century later at the Siegen symposium, 7–8 December 2004, Berthel Sutter seemed to revive the concept in a paper entitled 'A new kind of societal knowledge creation?'.

I originally hoped that the concept of 'service science' would appeal to those scientists and researchers who weren't merely motivated by profit or intellectual curiosity but sought to work towards social justice. But the idea has currency beyond scientific workers and is also highly relevant if one considers the general title of this book, namely *The Orientation of Science and Technology*. The publisher Mr Paul Norbury of Global Oriental was pleased and encouraged me to address the concept in this collection.

I would like to thank, for their assistance in editing the original papers, Dr Morris Low of the University of Queensland and Professor Nathan Sivin of the University of Pennsylvania. Finally, but not least, I am most grateful for the editorial help provided by Professor Pauer, and Professor Regine Mathias of Ruhr-University Bochum, without whose kind assistance this book would not have been possible.

Shigeru Nakayama
Tokyo, October, 2008

Foreword

Tessa Morris-Suzuki

Professor of Japanese History, Division of Pacific and Asian History, Australian National University

Shigeru Nakayama, whose essays are collected in this volume, has played a unique and far-reaching role in advancing knowledge of the history of Japanese and Chinese science and technology. His prolific writings cover the span from ancient Chinese cosmology to the social, economic and political impact of the Internet (Nakayama 1969, 24–43; Nakayama 2000, 187–258). All are marked by his rigorous scientific training. (He majored in astrophysics before becoming a historian.) But they are also characterized by a profound interest in the way in which evolving social and political structures interact with scientific and technological development.

Nakayama's research has opened up important new theoretical perspectives, helping to introduce key concepts such as 'service science' and 'techno-nationalism', which are now widely used by other scholars. His influence on the history of science and technology has been further enhanced by his work as a teacher and as an enthusiastic participator in, and organizer of, collaborative projects. As one of the first Japanese historians of science and technology to receive a doctorate from a US university, Nakayama has played a key role in introducing the history of East Asian science to English-speaking audiences, working alongside other leading figures in the field such as Joseph Needham in Cambridge and Nathan Sivin in Philadelphia.

Several influences can be seen as having shaped Nakayama's distinctive approach to the past and present. He was born in 1928, and spent the first part of his life in Amagasaki. This port city in Kobe Prefecture was one of the centres of Japan's pre-war industrialization, becoming host to an expanding array of steel, electrical and other industries. During his high-school years Nakayama lived in Hiroshima, and (as he mentions in his introduction to this volume) he was there when the first atomic bomb was dropped on the city on 6 August 1945. Many of his school friends were killed by the blast. Although he has seldom discussed this experience in his writings – perhaps because, more than six decades after the event, it is still too difficult to put into words – there can be no doubt that it had a formative impact on his perception of science, and of its potential both for creation and for destruction.

After graduating from Hiroshima High School in 1948, he entered Japan's most prestigious university, the University of Tokyo, where he majored in astrophysics, and in 1955 he joined the small but highly influential group of outstanding young Japanese scholars who studied in the US under the Fulbright Program. This scheme had in fact been proposed by Senator J. W. Fulbright just a few weeks after the dropping of the atomic bombs 'to plant the seeds of peace through international exchange'.[1] The first scholars to participate in the scheme were selected in 1946, and many of those Japanese students who became Fulbrighters over the two decades that followed went on to become leading figures in their country's post-war scientific, intellectual and political life.

Nakayama's years in the US, as a researcher of the history of science at Harvard Graduate School, and his visit to England to work with Joseph Needham in 1957, had a lasting impact on his view of the history of science and technology. In Harvard he worked closely with Thomas Kuhn, and was later to play a key role in introducing the Kuhnian concept of scientific paradigm change to Japan, and in applying this concept to the Chinese and Japanese context (see for example Nakayama and Ishiyama 1987, 29–32; Nakayama 2000). His meeting with Needham was the start of a collaboration which continued until the latter's death in 2004 – and indeed the link to Needham's legacy has been maintained through Nakayama's ongoing connections with the Needham Institute in Cambridge.

At the same time, however, Nakayama's work was shaped by the distinctive conditions of scientific research in Japan during the immediate post-war decades. The war years had seen the growing militarization of science and technology, as the Japanese government had mobilized scholars to work on war-related projects. After the devastating experience of the atomic bombings and defeat in war, many of Japan's laboratories and research institutes lay in ruins. Japanese scientists not only found it necessary to rebuild their research from scratch: they were also confronted with challenging moral problems about the purposes of that research.

As a reaction to the perceived perversion of science and technology for the purposes of military expansion, many scholars sought our new approaches, which might enable knowledge to serve the aims of peace and social development. Marxism, which had long been suppressed in Japan, now gained considerable influence in intellectual life. Both Marxists and liberal democrats heatedly debated issues of scholarly responsibility, and experimented with new approaches to education and research. The result was a period of ferment and organizational innovation in the Japanese scientific community: an era vividly described by Nakayama in his book *Science, Technology and Society in Postwar Japan* (Nakayama 1991, 14–41), and in Chapter 8 in this volume, 'Grass-Roots Geology – Ijiri Shoji and the Chidanken'.

The radicals and reformers of the post-war years sought a new model of 'democratic science'. This implied a rejection of the old hierarchical research projects, in which the senior scientist occupied an almost god-like position of authority within his laboratory. Democratization of the laboratory went hand-in-hand with new efforts to link scientists to the wider society and to 'bring science to the people'.

As someone who lived through this heady period of experimentation, Nakayama Shigeru was able to see both the value and the limitations of these post-war efforts to democratize science. He was later to criticize the stultifying effect which aspects of Marxist theorizing – particularly as articulated through the

Japanese Communist Party – had on intellectual life. At the same time, however, the post-war democratic aspiration for socially responsible science has remained an enduring feature of Nakayama's thought (Nakayama 1981; Nakayama 1984). This aspiration was later to be expressed in his emphasis on the importance of 'service science' and in his four-sector analysis of science and technology, in which 'the citizenry' plays a crucial role alongside academia, industry and government in influencing the trajectory of research.

In 1960, Nakayama received his Harvard doctorate and took up a teaching position in the University of Tokyo's Faculty of Arts and Sciences [*Kyōyō Gakubu*]. By now, the process of post-war recovery was complete, and Japan was embarking on the period of phenomenal growth which, over the next decade, was to transform it into an economic superpower. In 1960, Japan was still widely regarded as a country whose industry was based on cheap labour and technological imitation, but over the years that followed, this image would be radically revised. As foreign researchers and governments came to recognize the emerging strength of Japanese technology, there was growing interest in understanding the distinctive organizational context of Japanese technological research.

During the 1960s, 1970s and 1980s, Nakayama's publications and conference papers, both in Japanese and in English, helped to meet the international demand for an expert analysis of the role of science and technology in Japanese industrial modernization. However, his work retained characteristics which distinguished it from the growing volume of mainstream writings on Japanese science and technology (see, for example, Nakayama 1978). Within Japan itself, much academic research on the history of science continued to be based on Marxist models of development, which stressed the material foundations of thought and focused on issues such as the transformation from feudalism to capitalism. By contrast, Nakayama, deploying ideas such as Kuhnian notion of paradigm shifts, was able to develop a more flexible approach which casts fresh light on the interaction between social forces and scientific ideas.

English-language studies of Japanese technology, meanwhile, were very often embedded in development theories which presented science and technology in uncritical terms. The aim of these studies was to learn from the Japanese success in converting technological innovation into high economic growth, not to question the social impact of technological innovation itself. Nakayama's work diverged from this approach in that he never ceased to be conscious of the enormous potential of science and technology both to enrich and to harm the quality of human life.

The problem of the social impact of technological innovation became an increasingly pressing one from the second half of the 1960s onwards. It was during this period that the negative effects of high-speed industrialization became increasingly apparent in Japan. A series of notorious cases of environmental pollution – including the Minamata mercury poisoning case and mass outbreaks of asthma in industrial cities such as Yokkaichi – drew worldwide attention to the costs of high growth. Indeed, Nakayama's birthplace of Amagasaki emerged as a focus of some of the most intense environmental disputes of the 1970s and 1980s.

At the same time, the New Left movements of the late 1960s were also producing new critiques of science and technology, and of the very ideas of economic development and social progress. Within Japan as elsewhere, these new challenges produced a variety of responses. A growing number of ecological thinkers developed radical critiques of modern science. In the Japanese context, such critiques were sometimes combined with a rediscovery of premodern ideas and practices, such as those of Edo period (1603–1867) Japan, which were seen as promoting harmony between human beings and nature. Others, on the contrary, argued that the further development of technology itself provided the only realistic means of overcoming the problems of industrial pollution.

Many of Nakayama's writings from the late 1960s onwards take up the problems of environmental destruction and the costs of growth. In the 1990s, together with younger co-researchers, most notably Yoshioka Hitoshi and Morris Low, he examined issues such as the problematic environmental and social implications of the widespread adoption of nuclear power by Japan, and the impact of high-tech consumerism on the environment (see for example Low, Nakayama and Yoshioka 1999). Nakayama's response to such problems was neither a rejection nor an outright endorsement of modern science as a source of human welfare. Rather, drawing on the foundations of his earlier research, he focused on analysing the interaction of science and technology with the political and social realms.

Nakayama has observed how the development of science and technology in Japan throughout much of the modern period has been closely linked to concerns about national security. In particular, the Japanese state has tended to see science and technology as an essential means of compensating for the vulnerability which arises from Japan's lack of natural resources (Nakayama 1986; Nakayama 1991, 201). With the microelectronics revolution of the 1980s, Nakayama noted, the Japanese government had come to place even greater emphasis on establishing Japan's place as a 'country founded on technology' [*gijutsu rikkoku*]. This 'techno-nationalist' policy involved a mobilization of science for national economic aims which could in some ways be likened to the wartime mobilization of science for military aims (see for example Nakayama 1991, 200–201).

Criticizing this approach, Nakayama called for a rethinking of the goals of scientific and technological development which would emphasize the quality of life of citizens, rather than merely focusing on the economic might of the nation. Though he never lost faith in the potential of science to enhance the well-being of ordinary people, he argued that this potential could only be fulfilled through fundamental changes in science and technology policy-making: changes that would give citizens a greater say in the setting of policy goals.

In 1989 Nakayama retired from his position at the University of Tokyo and moved to the Science and Technology Studies Centre at Kanagawa University. This move came at a time when a new set of problems was beginning to beset the world of science and technology in Japan and elsewhere. Emerging neo-liberal economic policies were resulting in reduced government funding and the growing privatization of scientific research. Meanwhile, globalization was leading to increased pressures for collaboration between Japanese and foreign research projects.

Nakayama was one of the first Japanese scholars to provide a conceptual framework for understanding the complex effects of privatization on the production of scientific knowledge (for example, Nakayama 1995a). Although critical of some aspects of contemporary globalization, however, Nakayama remains (as he has always been) a passionate *internationalist*, in the sense of believing deeply in the value of cross-border exchanges of scientific and social ideas.

This belief is expressed in his life as well as his writings. A generous and enthusiastic mentor, who (like all good teachers) is always happy to listen as well as to speak, Nakayama has helped to nurture a new generation of scholars researching the history, politics and social impact of science and technology. One of the enduring legacies of this role is the monumental five-volume publication on the history of Japanese science and technology *Tsūshi: Nihon kagaku gijutsu*, written by a large multi-disciplinary team lead by Nakayama (Nakayama 1995b). (Four volumes of this work have now appeared in English translation.) The *Tsūshi* project, like others which Nakayama has initiated, reflects his enthusiasm for transnational collaboration, involving foreign as well as Japanese researchers.

He has indeed played a particularly great role in nurturing international collaboration between Japan and other countries, both by participation in international meetings and by organizing a range of collaborative projects that bring together Japanese and foreign researchers with shared interests. His collaborative work has extended not only to the United States and the UK, but also to Eastern Europe, Australia, and to many parts of Asia.

Since retirement Nakayama has continued to travel widely and to nurture new projects on emerging areas of technological change. As early as the 1980s he had been among the first to recognize the potential of the new communication technologies of the late-twentieth century. By the mid-1990s, at a time when few academics in the English-speaking world, and even fewer in Japan, had begun to use email on a regular basis, Nakayama was already 'on line', eagerly exploring the possibilities and limits of this latest technological advance.

This enthusiasm, in fact, exemplifies (in a small way) the quality that suffuses all Nakayama's work. His writings are characterized above all by openness, receptiveness to new ideas, and by an urge to communicate across the narrow boundaries of nation and culture. It is that quality which makes his work so accessible to readers from all parts of the world. It is also that quality which has enabled him to bring fresh perspectives to the task of understanding the sources, course and implications of the scientific and technological advances that have transformed the face of Japan, and of the world more generally, over the past century.

REFERENCES

Low, Morris, Nakayama, Shigeru and Yoshioka, Hitoshi. 1999. *Science, Technology and Society in Contemporary Japan*. Cambridge: Cambridge University Press.

Nakayama, Shigeru. 1969. *A History of Japanese Astronomy: Chinese Background and Western Impact*. Cambridge Mass.: Harvard University Press.

Nakayama, Shigeru. 1978. 'Science and Technology in Modern Japanese Development'. In William Beranek and Gustav Ranis (eds). *Science, Technology and Economic Development: A Historical and Comparative Study*. New York: Praeger.

Nakayama, Shigeru. 1981. *Kagaku to Shakai no Gendaishi*. Tokyo: Iwanami Shoten.

Nakayama, Shigeru. 1984. *Shimin no tame no Kagakuron*. Tokyo: Shakai Hyôronsha.

Nakayama, Shigeru (ed.) 1986. *Nihon no Gijutsuryoku*. Tokyo: Asahi Shinbunsha.
Nakayama, Shigeru and Ishiyama, Hiroshi. 1987. *Kagakushi Kenkyû Nyûmon*. Tokyo: Tokyo Daigaku Shuppankai.
Nakayama, Shigeru. 1991. *Science, Technology and Society in Postwar Japan*. London and New York: Kegan Paul International.
Nakayama, Shigeru. 1995a. 'Kagaku no Mineika to sono Inpakuto'. In Kaneko Motohisa ed. *Kinmirai no Daigakuzô*. Tokyo: Tamagawa Daigaku Shuppankai.
Nakayama, Shigeru (ed). 1995b. *Nihon no Kagaku Gijutsu*. Vols. 1–5. Tokyo: Gakuyôshobô
Nakayama, Shigeru. 2000. *20–21 Seiki Kagakushi*. Tokyo: NTT Shuppan.

NOTE

1. See Japan-US Educational Commission/Fulbright Program website. http://www.fulbright.jp/eng/keikaku/overview.html

Introduction

INTRODUCTION AND OVERVIEW OF EAST ASIAN SCIENCE

The title of this book owes much to Joseph Needham, the author of and authority on *Science and Civilisation in China*. When I proposed the publication of a new Japanese journal for the history of science (in Western languages) around 1960, I canvassed him for any ideas regarding an appropriate title. He suggested 'Orientation' as he himself was fascinated by that wording. The 'Orientation of Science' refers not only to the direction of science but also implies a turning to Eastern science. Unfortunately, members of the editorial committee were less enthusiastic. So I have belatedly adopted the title for my collected works, in the context of my life-long efforts to redirect conventional approaches to science and technology.

In order to state my position in the field of East Asian science, I wrote an introductory article on East Asian traditions in science in 'The Empirical Tradition', in Arnold Toynbee (ed.), *Half the World* (Thames and Hudson, 1973), pp. 131–50. Toynbee originally asked Needham to write the article, but Needham in turn asked me. Toynbee found my draft to be 'intriguing'.

For a survey article of my field, turn to (19) 'History of East Asian Science: Needs and Opportunities', *Osiris*, vol. 10 (1995), pp. 80–94. In the article, I note the pronounced difference in the methodology of historians of science who work on pre-modern, traditional science, and those devoted to studying the period of modernization. While the former tend to write more internal histories of science, the latter adopt an external approach to science which foregrounds social factors.

FIRST TURNING POINT: THE DROPPING OF THE ATOMIC BOMB AND THE END OF THE WAR

I first chose to major in science on entering senior high school in 1945, reflecting last-minute attempts by the Japanese to mobilize science in the closing months of the war. I was in Hiroshima when the atomic bomb was dropped. The surrender of Japan soon followed. The impressionable years of my youth thus coincided with a crucial historical turning point, when every value system was turned upside down. Amid the post-war ideological turmoil, science did not seem to save me. Nevertheless, after a long thought travel, I chose to stay in science, for it showed a more reliable and public knowledge than other belief systems did at the time, concealing my existentialism deep in my mind as a private wisdom.

I worked hard in science, focusing on astrophysics at the University of Tokyo, but in 1950 I left science due to the dismal conditions that researchers had to work under, and turned to the history of science, which was largely Marxist-oriented at the time. In 1955, I enrolled in the Harvard Graduate School to work in the history of science, where I met Tom (Thomas) Kuhn, with whom I had a close personal friendship which lasted a lifetime (see 28). My experience in American academia strengthened my belief in science as a social power for economic recovery. Once I had returned to Japan, I decided that my mission was to help instil this in my own country. I strived to promote the development of science for economic recovery during the period of rapid economic growth in the 1960s.

EARLY WORKS: THE HISTORY OF ASTRONOMY

I majored in astrophysics at the University of Tokyo and wrote a graduation thesis in 1950 on the analytical solution of a quantum mechanical problem, mainly based on scientific literature available at the time in Japan, using copies of available journals like *Physical Review*, the most recent copies of which unfortunately dated back to the 1930s. While engaged in this task, a new American journal reached Japan courtesy of the CIE Library in Tokyo. I found, to my dismay, that the problem I was engaged in had already been solved 400 times faster by the 'punch card' method (a forerunner of today's computers). This was the moment when I decided to quit scientific research and move to the history of science where the social condition of scientific research could be better investigated.

For my PhD thesis, I chose the history of astronomy which I knew best. I published a number of articles on this subject early in my career in the history of science but most of them were included in my *A History of Japanese Astronomy: Chinese Background and Western Impact* (Harvard University Press, 1969), 329 pp.

The only exception was (7). I was a member of the Copernicus Committee of the International Congress of the History of Science, which was responsible for a major celebration: 500 years since the birth of Copernicus. The celebrations were held in Toruin, Poland in 1973. I presented this article and it was subsequently published as 'Diffusion of Copernicanism in Japan', *Studia Copernicana*, V (1973), pp. 153–88.

THE HISTORY OF THE UNIVERSITY

My Harvard PhD was conferred by the Program of Higher Degree in the History of Science and Learning (1960). In those days, the history of science was not yet established as an independent department. As an organizational convention, the history of science programme was strengthened by merging it with the history of learning, which was actually the history of universities. This served the useful purpose of encouraging scholars to explore the institutional background of scientific activity and its influence on scientists, and at the same time it helped to fill the gap between studies of the socio-economic foundations of science and scientific infrastructure. I was one of only a few Harvard PhDs with that title. I worked on the history of the university for some time and even created a History of the University Society in Japan as my interest in the history of science began with research into the institutional background of Japanese science.

(6) 'The Role Played by Universities in Scientific and Technological Development in Japan', *Cahiers d'histoire mondiale*, 9, 2 (1965) pp.340–62, was written at the request of the UNESCO Japanese Committee on Japanese Development since the Meiji Period. Though it is short, it is still cited to this day due to the lack of English articles on this topic. (6) and (17).

MASS PARTICIPATORY RESEARCH

(8) 'Grass-roots Geology – Ijiri Shoji and the Chidanken', in S. Nakayama et. al. (eds) *Science and Society in Modern Japan* (MIT & University of Tokyo Press, 1974), pp. 253–69. This paper focuses on an interesting Japanese group that approached geology from a viewpoint of encouraging mass participation along Marxist lines. A somewhat similar group was formed in China during the Cultural Revolution.

SECOND TURNING POINT: CIVILIZATIONAL REVOLUTION

The late 1960s (or more precisely, around 1968) were a turning point for me. I was already in my forties and had established a career as an historian of science. I was not directly influenced by the student protests that had prevailed all over the campuses of Western industrialized society. Nevertheless, they caused me to question the future direction of science and technology, for which I was professionally obliged to answer.

In order to find answers to this question, I wrote (in Japanese) *Kagaku to shakai no gendaishi* (*A Contemporary History of Science and Society*) to explore the various ideas and events in the decade that followed 1968. Most alternative ideas appeared in the first half of the decade, such as technology assessment, the anti-pollution movement, environmental science, anti-war, anti-science, critical science, alternative technology, appropriate technology, and even anti-professionalism in the Chinese Cultural Revolution. These concerns evolved and converged into what can be called an 'ecological' perspective. Out of this research emerged what I now describe as a 'Civilizational Turning Point', in which the *raison d'être* of science and learning and the conventional values in modern industrialized civilization were re-examined.

In the seventies and eighties, I played a role in altering and redirecting the current of civilization as it moved from the industrial to post-industrial stage in Japan, as evidenced by the change in attitude towards nature held by the Japanese people. Among the Japanese, a sharp decline in the notion that humankind should conquer nature, occurred in 1968 and continues to this day, as appeared from the graph overleaf. Since then, a change in my own viewpoint influenced even my professional writings in English. This aspect was beginning to emerge in article No. 6 and can be seen in the following papers in the same vein.

(10) This originally appeared as 'History of Science: A Subject for the

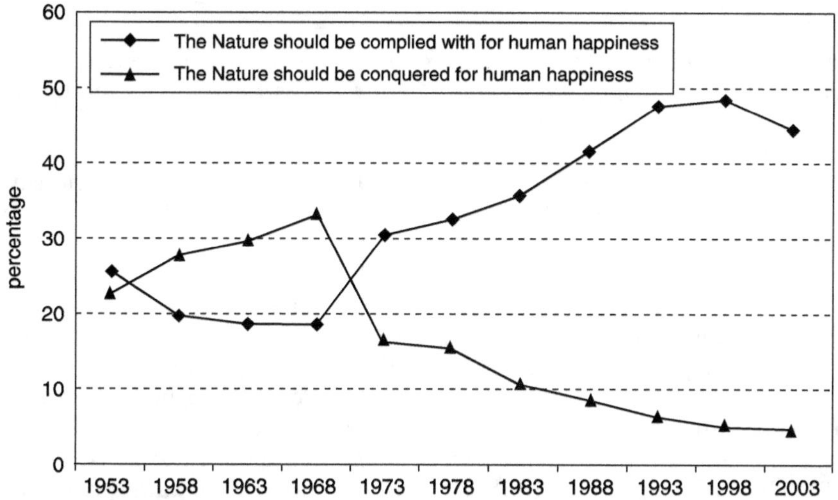

Source: Arranged from the Institute of Statistical Mathematics. *The Japanese National Character; The Eleventh Nationwide Survey (2003)*—English Edition, No. 97 (Jan., 2007)

Frustrated', in S. Nakayama et al. (eds), *Science and Society in Modern Japan* (1974), pp. 3–16. It was prepared for the XIVth International Congress of the History of Science in Tokyo and Kyoto in August 1974, in order to introduce hitherto little-known Japanese activity in the history of science, to international audiences. It appeared again in slightly revised form as 'History of Science: A Subject for the Frustrated. Recent Japanese Experience', in *Boston Studies in the Philosophy of Science,* 15 (1974), pp. 213–24. The text comes from the revised version.

In the late 1960s and early 1970s, historians of science were often involved in the issue of the questioning of old paradigms and reorienting conventional science. While I had written a number of articles in Japanese, this was my first article in English in that direction. When I write for an international audience, I often indicate how Eastern scientific traditions are different from those in the modern West, to show that an alternative approach is possible. Typical of this, although not included in this collection, is 'Alternative Science of the East', E.G. Forbes (ed.) *Human implications of scientific advance* (1977) pp. 36–44 (Proceedings of the XVth International Congress of the History of Science, Edinburgh.) It was read at the symposium on Science and Human Values, organized by Jerome Ravetz. As a final remark, I questioned the approach of Joseph Needham who was a co-panelist. While Needham claimed modern science as a synthesis of Western and Eastern science, I denied the existence of such a predetermined harmony.

(13) 'Japanese Scientific Thought', *Dictionary of Scientific Biography,* Vol. XV (Charles Scribner's, 1978), pp. 728–58 was originally requested by Charles Gillispie, the editor-in-chief, as a monograph. It was later changed into an appendix to the *Dictionary*. This article was specially mentioned in the lead review in *New Yorker* magazine, 16 June 1980, which discussed the completed project of *DSB*.

SERVICE SCIENCE

Apart from the Japanese context, I presented and coined a general concept of 'service science', which differs from academic, industrial and national science, as its objectives are set and assessed by the citizenry. During the peak in activity of the anti-pollution movements in Japan, many public laboratory scientists welcomed the idea of service science and a Japanese local mayor seriously considered organizing a council for service science. I am still hoping to create such a science in future.

(14) 'The Future of Research – A Call for a Service Science', *Fundamenta Scientiae*, vol. 2, no.1 (1981), pp. 85–97. This is the article in which I classified science according to the assessors for which scientific works are addressed: academic peers, public institutions (including defence), private corporations and citizenry. For the latter, I called for a 'service science' to be assessed by and addressed to the interests of citizenry (or humankind). The article was first printed in a new journal, for which I was a member of the editorial board.

(18) 'Human Rights and the Structure of the Scientific Enterprise', in C.G. Weeramantry (ed.), *Human Rights and Scientific and Technological Development* (United Nations University Press, 1990), pp. 137–50. Since 1986, I had been involved in a project on 'Human Rights and Scientific and Technological Development' for the Commission of Human Rights conducted by the United Nations University in Japan. The final paper was read in Geneva. I introduced the concept of 'service science' again and also the new concept of the 'rights of the ignorant.'

WESTERNIZATION AND MODERNIZATION OF JAPANESE SCIENCE AND TECHNOLOGY

Here in this section, I have arranged my papers on Westernization and modernization of Japanese science from the eighteenth century to pre-war times, in roughly chronological order. There is inevitably some overlap between successive papers.

(25) 'Japan in the 18th Century', in Roy Porter (ed.), *The Cambridge History of Science, vol. 4: Eighteenth-century Science* (2003), pp. 698–717. This was commissioned by the Cambridge project in the History of Science. I portrayed eighteenth-century Japan as the century of paradigm shift from Chinese science to Western science.

(15) 'The Transplantation of Modern Science to Japan', *Center for Studies in Higher Education, the University of California Occasional Paper,* no. 23 (1982), 26 p. This was originally the lead article for a Japanese book entitled *Bakumatsu no yogaku* (Western Learning at the End of the Edo Period) which I edited. My paper discusses how the leading disciplines of Western science, from astronomy to medicine and finally military technology, were introduced into Japan over time. This project was headed by me during the years 1978–1981, and funded by research grants-in-aid from the Ministry of Education, Science and Culture. I organized a research team which met monthly and its outcome was the above-mentioned book. This team later developed into the Yogakushi Kenkyukai (Society for the Research of Western Learning). In post-war pacifist Japan, there have been practically no courses on military history at educational institutions. In

order to understand the background of the end of the Tokugawa period, in which national defence became the major focus of Western learning, I participated in the ROTC programme held at the University of California, Berkeley, where I took courses such as military history.

(20) 'The Chinese "Cyclic" View of History vs. Japanese "Progress" ', in A. Burgen, P. McLaughlin and J. Mittelstrass (eds), *The Idea of Progress* (de Gruyter) (1997), pp. 65–75. I was invited to join a conference held in Gargellen, Austria in September 1994. The aim of the meeting was to reconsider the notion of 'progress' in the post-industrial era after the civilizational change in the 1970s. I made the point of how the word 'progress' which was coined in Japan in the imperialistic environment of late-nineteenth century East Asia, differed from that of China at the time.

(9) 'Problems of the Professionalization of Science in Late-Nineteenth Century Japan', XIVth International Congress of the History of Science, Texts of Symposia no.1 (1974), pp. 83–9. For the International Congress of the History of Science held at Tokyo, Japan in 1974, I was an organizer of the symposium 'Professionalization of Science in the Nineteenth Century'. This article was my contribution to the symposium, in which I emphasized the samurai origin of Japanese professional scientists.

(11) 'Science and Technology in Modern Japanese Development', in William Beranek Jr. and Gustav Ranis (eds), *Science, Technology and Economic Development: A Historical and Comparative Study* (Praeger, 1978), pp.202–232. It was a part of the US Bicentennial Symposium, 'The Role of Science and Technology in Economic Development', held in Washington DC, 10–16 October 1976. Development economists arranged seven country papers in linear fashion, one after the other, from Britain to Africa, to extract ready lessons for late-comer nations. Reflecting the spirit of the 1970s, 'intermediate technology', in Shumacherian fashion, was considered. In the case of Japanese science and technology, the public sector effort in the early Meiji period had been so far over-emphasized. But technology transfer could only be successful when it was transplanted and reached the private sector. This, I later called a 'revisionist view', and it appeared in my paper in only a rudimentary way.

FROM THE POST-WAR TO THE FUTURE

From the 1980s, I devoted the rest of my career to editing a social history of Japanese science and technology in the post-war period – the period that I lived through – under the auspices of the Toyota Foundation. The Japanese version of the project was published as *Tsushi: Nihon no kagaku gijutsu*, 5 vols. (Gakuyo Shobo, 1995), with two additional volumes published in 1999. The following paper describes the official methodology adopted by participants in the project.

(17) 'Independence and Choice – Western Impacts on Japanese Higher Education', *Higher Education*, 18 (1989), pp. 31–48, was requested by Philip G. Altbach, who later included it in *From Dependency to Autonomy: The Development of Asian Universities*. His original intention was to show how Asia has developed from colonial dependency to post-war independence, but I showed that the Japanese case was, in contrast, an exception.

(16) 'The American Occupation and the Science Council of Japan' in Everett Mendelsohn (ed.), *Transformation and Tradition in the Sciences* (Cambridge University Press, 1984), pp. 353–69. I had earlier joined the *Senryo Kenkyukai* (Research Group for the Study of the Occupation) to work on the Allied Occupation of Japan in the early post-war period (1945–1952). I used official documents of the GHQ (General Headquarters) preserved in the National Archives and Records Administration (NARA) in Virginia, USA. It was an early attempt to make use of GHQ documents, copies of which are now edited and preserved at the National Diet Library, Tokyo.

(24) 'From PC to Mobile Internet – Overcoming the Digital Divide in Japan', *Asian Journal of Social Science*, 30, 2 (2002), pp. 239–47. This was originally presented as a keynote address at the International Conference of IP Policy held at the National University of Singapore in 2001. I spoke of the mobile Internet as a new phenomena which emerged out of youth culture. The talk was intended to give the message that the mobile Internet cannot be controlled by government. It is a phenomena which gained popularity despite discouragement by school authorities and parents. This talk was in contrast to the keynote speeches given by a government minister of Defence and of Culture in Singapore. It seems that the mobile Internet was not understood by the audience, despite its continued popularity, much more than PC, in Japan.

(26) 'Technology in History, Japan' published by Oxford University Press, 2003 as a digital edition. It dealt with pre-war Japanese military technology, in contrast to post-war market-oriented technology.

SUMMING UP AND FUTURE PROSPECT

Whether under nuclear umbrella or because of peaceful constitution, it was a historical fact that the post-war Japanese had been successfully avoiding the involvement in major trend of the cold-war scientific structure to promote nuclear and space sciences, and concentrated in market-oriented 'science for economic recovery'. The end of war constituted the turning point from military-oriented science to market-oriented. Its success was helped by the post-war democracy, which is characterized by the rejection of militarism and adherence to peace cause and peaceful constitution in comparison of earlier democratization movements in 1880s for parliamentary democracy and 1920s for general election. It is crystal clear for the generation of post-war democracy, in which I belong, the post-war should be preferred in contrast to the gloomy pre-war and wartime recollection.

In the late 1960s and early 1970s when Japanese economic recovery had by then been accomplished, we were, simultaneously with the industrialized Western nations, visited by the second and perhaps more fundamental civilizational turning point from industrial to postindustrial. Signalled by anti-pollution movements, concerned citizenry and critical scientists were keen and acute in recognizing such a turning point. Even though the government and industrial sectors were aware of the turning from an industrial to an ecological perspective rather slowly and not thoroughly, Japanese industry had restructured from pollution-full heavy-large industry to light-small industry of high technology in the 1980s, and

Japan has reached a high point of world acknowledgement in promoting precision technology and the production of quality manufactured goods.

In the 1990s, along with the breakdown of the Cold-War structure, globalization and privatization prevailed. While the monetary slump continued throughout the decade, Japanese science and technology survived and continued, but in the post-Cold-War globalized and privatized world, the government and industrial sectors were preoccupied with the cutthroat competition, especially with United States, in cutting-edge technology without a clear-cut long-range national target. Without fully recognizing the civilizational turning, some parts of government and industrial sectors still continued to hold the ideology of 'science for economic recovery' which in practice does not obtain any longer and has now turned to science 'for competition for its own sake' without knowing where it is heading for and who it would serve.

For Japanese people, however, no more science for economic recovery is needed and competition has nothing to do with their life-style, while the ecological perspective is more engaging. Especially when life science and biotechnology reach into our daily life in twenty-first century society, people would have begun having misgivings that it should not be placed in private profit-making or under bureaucratic control; some sort of democratic control of science and technology, research and development is wanted.

For that reason, we would like to establish a science for the people – a 'Service Science', or however one may wish to describe it, governed and moderated by a people's council. Once the evaluation system, goals and objectives are agreed, it is possible that a new science could emerge, especially in the context of the new digital world of communications and data processing, including the internet, the benefits of which would be shared by all the people. In fact, there are many scientists and engineers around the world who are interested in working along these lines, rather than pursuing careers in the disinterested world of academia or in the service of private sector capital.

Since the 1980s, along with the trend which sees international trade relationships with Asia overtaking those of the USA and the EU, Japanese production technology has been transferred to Asian NIES, ASEAN countries and China through personal contact at shop-floor level. Japanese engineering is still respected in the Third World on account of the contribution it has made in the context of science and technology that has ensured its economic survival, thereby reducing the gap between north-south economic performance.

First published in *Separata de 'Actes du IXe Congrès International d'Histoire des Sciences'*, pp.544-546, Asociacion para la Historia de la Ciencia Española, Universidad de Barcelona, 1959.

1 The First Appearance of Aristotelian Cosmology in Japan, *Kenkon Bensetsu*

On the trail of the Jesuits' cosmological influence in Japan, *Kenkon bensetsu* ([Western] cosmography with critical commentaries, ca. 1650) is an exceedingly interesting example of the confrontation of Eastern and Western ideas. The treatise consists of two parts interwoven phrase by phrase. One is the original text by the apostate missionary Christovaō Ferreira (1580–1650) and the other is the commentary of Mukai Genshō (1609–1677), a Confucian scholar and physician.

The preface relates the circumstances in which this treatise came into being. In 1643, a shipwrecked Western vessel drifted onto an island in the westmost part of Japan. The passengers, all Jesuit missionaries, were arrested and imprisoned for illegal entry under the seclusion policy of the time. One missionary who was well versed in astronomy submitted a treatise to the office. A few years later this work was turned over to Ferreira for translation. He was then serving as a censorial officer for the Japanese government, after having abandoned his Christian faith at the time of the persecution in the preceding time. He could not write Japanese characters, so he spelled out his translation in Latin letters. An interpreter read it aloud and Mukai Genshō wrote it in Japanese.

The subject matter largely coincides with that of *De Caelo, Meteorologica* and *De Generatione et Corruptione*. There is no explicit indication of the original text from which Ferreira worked. After comparison of astronomical data and diagrams, it would be conjectured that most of the *Kenkon bensetsu* is derived from Christopher Clavius's *In Sphaeram Ioannis de Sacro Bosco, Commentarius*. However, its copious and detailed Euclidean demonstrations and its tables of astronomical computations make it greatly superior to *Kenkon bensetsu*, which is an exposition of not much use in practical astronomy. Furthermore, the presence of material on astrology is a contraindication, since the Jesuits were strongly opposed to astrology. Still, it is highly probable that Ferreira consulted Clavius or works based on Clavius.

Whereas Ferreira's text was a straightforward description of Western science of the time, Mukai's commentary was considerably biased by his background, and is thus of utmost interest.

In the opening part, he commented on the characteristics of the learning of Western, Buddhist, Confucian and Shintoist doctrines. For Westerners the heaven is something special, entirely unrelated to the four elements. Their heaven does not share the nature of earthly things. Therefore the heaven cannot be essential to creation. They are ingenious only in devices which deal with appearances and utility, but are ignorant about metaphysical matters and go astray in their theory of heaven and hell. Since they do not comprehend the significance of *li-chi'i* and *yin-yang* (both principles of Chinese *Naturphilosophie*), their theory of material phenomena turns out to be vulgar and unrefined. But this vulgarity appeals all the more to the ignorant populace, and stupefies them. Western scholars are convinced of the superiority of their own learning, and so go abroad to preach it. But their study is utterly erroneous and prejudiced. Their preachments on the past and future worlds are full of fantasticism.

In his phrase-by-phrase commentaries Mukai never deviated from his stern adherence to the Neo-Confucian *li-ch'i* principle. On the whole, he could not reconcile himself to the Aristotelian four elements theory and its characteristically European explication. He accepted the universally valid astronomical measurements of the West, sometimes even freely praising their ingenuity, but here and there he expressed his contempt for the unbalanced emphasis on the phenomena.

Although he was biased, at least he was free of Aristotelian prejudices. He was frankly suspicious of the physical explanation of terrestrial phenomena. For example, he was not convinced that the sphere of water is above the sphere of earth. When water is dropped on a lump of earth, he said, the water penetrates and seeps through the lump; thus the existence of underground water is explained. There is therefore no stratigraphical difference between the natural place of water and that of earth.

Whenever the author mentioned the four elements, the commentator countered with the Chinese five elements, not even bothering with a detailed refutation. He sometimes did not even follow the text, expostulating with a sweeping comment that, because the Westerners do not know the *li-ch'i* theory, they have had to devise a cumbersome materialistic demonstration; once one masters the significance of *li*, he can get the same result without consulting the barbarian's (Westerner's) demonstration.

The commentator clings to the five elements principle with irrational vehemence. After the explanation of terrestrial events in terms of the four elements, the author begins to discuss the heavens. The commentator, pleased, states that since it is obvious that *yin* and *yang* are divided into five elements, even the Westerners, despite their ignorance, had to add the heaven as a fifth element. Even if the heaven is, in their theory, foreign to the other four elements, there can be no doubt that the heaven, in harmony with the four elements, is fundamental to terrestrial phenomena.

The historical situation, conditioned by the anti-Christian sentiment of the first few decades of seclusion, put the commentator in a position of general hostility towards Western learning, as his preface shows. The European view of nature was not his taste. He considered such emphasis on the appearances trivial and vulgar. It seems that his primary interest was his overall world-view, which included not only natural phenomena but also the social order and the nature of man. He would expect to find some socio-ethical issues in a Western astronomical treatise and could never be satisfied without them.

Although Mukai accepted Western astronomical knowledge insofar as it was compatible with the traditional approach, his denunciation was mainly addressed at the physical theory based on the Aristotelian four elements theory. This was the greatest obstacle in Japan – perhaps even greater in China – to the acceptance of Western cosmology, much greater than objections to Copernican heliocentricism.

By the early part of the eighteenth century, Aristotelian cosmology was introduced in a fairly satisfactory manner. It did not, however, replace the traditional cosmology, which was largely incompatible with it. At best, the Aristotelian theory was accepted as a facet of European learning, and merely juxtaposed with the Eastern theory. Yet, it should be remembered that while most of Aristotelian cosmological speculations were shortly overthrown, universally verifiable explanations, such as the cause of eclipses, which were compatible with the traditional East Asian astronomy, have survived to date.

First published in Scientific Papers of the College of General Education, Vol.11, No.1, pp.163-176, University of Tokyo, 1961

2 On the Introduction of the Heliocentric System into Japan

INTRODUCTORY REMARKS

On the subject of the introduction of the Copernican system into East Asia, Boleslaw Szczesniak's two articles 'The Penetration of the Copernican Theory into Feudal Japan'[31] and 'Notes on the Penetration of the Copernican Theory into China from the 17th to the 19th Centuries'[30] are the principal sources hitherto available in a Western language. Since they include misleading points, we shall offer some corrections for the most obvious errors and discuss further the general background of the subject.

Szczesniak said that 'Baba Nobutake [as shown in his *Shogaku Temmon Shinan*, 1706][1] believed implicitly in Copernican astronomy'[31] and that 'Later he [Yoshimune] imposed on Nakane Genkei the task of explaining the principles of European astronomy, and this man is the author of the first book on astronomy written from European sources,* under the title *Temmon-zukwaihakki* (Illustrated Astronomy, 1696).† In 1744, the Shogun founded Japan's first astronomical observatory‡. . . . The observatory was at first situated in the castle of Tokugawa in Edo, and Nakane Genkei was the first director,§ and the first who knew the Copernican system'.[31]

Joseph Needham has made a similar statement. 'When the first modern observatory in Japan was founded about 1725', he writes, 'under the direction of Nakane Genkei, Copernican ideas were fully admitted there. But in China it was not until the early nineteenth century with the contributions of the Protestant missionaries, such as Joseph Edkins, Alex. Wylie and John Fryer, that Copernican views really spread.'[34]

The truth is, however, that these treatises were arranged in the way of the traditional Chinese calendrical astronomy, only with some sketchy knowledge of the Ptolemaic system; there is no evidence whatsoever to suggest any awareness of the Copernican system. Up to the early part of the eighteenth century,

* Genkei had no knowledge of any European language. He translated from Chinese into Japanese a Sino-Jesuit treatise which is mainly based on the Tychonic system.

† It was published posthumously in 1739.

‡ This was neither the first nor of modern European type. A historical record of a Japanese observatory goes back to the seventh century.

§ This is obviously wrong, since Genkei died in 1733.

Japanese astronomy was still dominated by Chinese tradition. We can reasonably say that the notion of the Copernican system in Japanese works appeared only in the last quarter of the eighteenth century.

Compared with the date of the acceptance of the Copernican system in the West, its introduction into Japan was late. To explain this, the following three factors should be examined:

(1) Political action to limit free international communication.
(2) The language barrier for the translation of Western works.
(3) Ideological and technical difficulties in comprehending the heliocentric system.

JESUITS' INFLUENCE

The direct impact of the West finally reached the isolated islands of Japan in 1543, the year in which Copernicus' *De Revolutionibus* was published. Sixteenth-century Japan was subjected to Jesuit evangelism.

Those who have studied the reports of such missionaries to China as Matteo Ricci and Adam Schall tend to project their picture of seventeenth-century Chinese science onto that of Japan in the corresponding period, and often conjecture that Japanese science had also been seriously affected by the early contributions of the Jesuits. There existed, however, a great difference between the Chinese and Japanese situations. While the Jesuits in China generally took a flexible, sometimes conciliatory, attitude towards the elite in Chinese bureaucracy and employed an indirect method to convert them to Christianity through the acknowledgement of the superiority of Western astronomy, the missionaries to Japan never attempted a systematic introduction of Western astronomy, but focused their efforts on intensive direct evangelism. Furthermore, owing to the Japanese government's severe controls over the diffusion of the Christian religion and Western learning in general beginning in the seventeenth century, the Jesuits' impact was relatively short-lived and the seeds planted were almost eradicated. We cannot say, therefore, that the Jesuits contributed as much to Japanese astronomy as Ricci and his successors did to Chinese astronomy.[23]

AVAILABILITY OF WESTERN ASTRONOMY UNDER SECLUSION POLICY

From the late sixteenth century onwards, the Japanese government entertained suspicions against Christians, strictly forbade any belief in Christianity, and finally took steps to expel the foreigners from the country. After 1638, only the Chinese and Dutch were permitted to reside in Nagasaki for the sole purpose of trade. This political action was paralleled with restrictions on the importation of foreign books, which of course included Jesuit teaching. These restrictions naturally affected seriously the course of Japanese astronomy.

It seems that Matteo Ricci was, in the eyes of the government censors, a most dangerous character. Any work by him, or associated with his name, was shut out no matter whether it concerned Christian tenets or not. The government categorically forbade the importation of all Sino-Jesuit treatises. It was unfortunate from the point of view of the subsequent scientific development of the Japanese that the edict of prohibition was issued so shortly after the start of publication of these works.

There was, however, no legal action against the importation of books in Western languages. In the early seventeenth century Japan, the part of the population which could read Western books – and its general level of proficiency – was insignificant compared to that part which could read Chinese books. The prohibition of Christianity, the departure of foreign missionaries from the country, and the limitations on foreign trade left no opportunity for the general public to be trained in European languages.

The only exception was the group of official interpreters at Nagasaki. They were, because of their professional function, officially permitted to study European languages. It is said that towards the close of the seventeenth century, when the ban was most effective, even the interpreters had extremely poor proficiency. Around this time the Portugese language was gradually replaced by the Dutch.

On the whole, the decree banning Christian writings was partly responsible for the predominance of purely traditional astronomy in the Chinese pattern during the first half of the Tokugawa period. Its striking characteristic is the lack of conceptual schemes and cosmological outlook, the final aim being to fit the closest possible algebraic relations to observations like Babylonian astronomy.[23]

RELAXATION OF THE BAN – EARLY EIGHTEENTH CENTURY

While the early Tokugawa period was mainly spent in catching up with traditional Chinese scholarship, the Japanese from the early eighteenth century on began to realize that the Chinese achievement was not enough. In fact, the eighteenth century in China was a time of retreat for Jesuit scientific activity, and a time of reactionary response to the previous Western impact. Under these circumstances, the Japanese were compelled to seek a new source of nourishment.[34]

The relations of Japanese astronomy with that of the West entered a new phase under the Shogun Yoshimune (reigned 1716–45), who relaxed the ban on Sino-Jesuit treatises in 1720. Yoshimune intended to issue a revision of the then current Jōkyō calendar immediately after his appointment. He consulted with astronomers and mathematicians, who are generally believed to have influenced his relaxation policy. Presumably they had read some officially forbidden books, which were preserved only in the Shogunate library, and found them superior to the traditional Chinese works. It seems that they then persuaded the Shogun to collect all Chinese translations and treatises on Western astronomy. One of them was Nakane Genkei (1661?–1733) mentioned before, who himself was allowed to examine imported books, and to use Jesuit works in composing a more precise ephemeris than had formerly existed.

After the promulgation of the relaxation decree, Western knowledge on an advanced and professionally useful level was transmitted to Japan for the first time. It was not the scientific works of the 'first generation of Ricci Corpus' based on Aristotelian and Ptolemaic hypotheses, which exerted a serious influence on

Japanese astronomy. When the ban was lifted, the more detailed works of the 'second generation' of Jesuits, which had adopted the Tychonic system, were already available.

For instance, the important *Ch'ung-chen Li-shu*,[4] by Adam Schall and others, basically followed Tycho Brahe's system, his calculations and observations. For the practical purpose of tracing the apparent courses of celestial bodies, the very goal of Chinese astronomy, whether the scheme used was Copernican or Ptolemaic made no essential difference. Brahe's work was certainly especially useful in this regard. Cosmological discussions did not play a significant role in traditional calendar calculation. 'Let us melt their (Western) materials and cast them into the mould of the (traditional) *Ta t'ung* calendar!' was a celebrated slogan of Hsu Kuang-ch'i (徐光啟), an eminent Chinese collaborator of Matteo Ricci. Their lack of attention to the Copernican heliocentric doctrine, or indeed to any physical model, could not have been primarily due to religious prejudices or intentional suppression[35, 36] as Szczesniak maintained.[30]

Meanwhile, at the time of Yoshimune in Japan, scholars and officials had no source of information on Western astronomy other than Chinese works. While the Sino-Jesuit works were relied upon by Japanese practical astronomers for decades, modern Western astronomical ideas could not have been available until the difficulty of translating original Western treatises was overcome; the works in Chinese were practically useless in this respect. The first mention of Copernicus and a sketchy account of his ideas must have been transmitted to Japan at this period through Sino-Jesuit works, but it is doubted that these short passages had deserved enough notice of the Japanese. The late acceptance of the Copernican system in Japan was mainly due to the exclusive dependence on Chinese sources for external learning up until the early half of the eighteenth century, rather than to ideological obstacles.

BEGINNING OF THE 'DUTCH LEARNING'

One consequence of this new receptivity was that in 1745 the interpreters were officially encouraged to learn to read Dutch books. Prior to this, their proficiency had been almost entirely restricted to interpretation of speech. At this early stage, the task of introducing new ideas was left in the hands of only a few linguistic experts.

Holland, after its golden age in the seventeenth century, could no longer maintain its glory. It contributed little of importance to astronomy during the eighteenth century; its scientific effort was at an ebb.[3] Unlike the Jesuits in China, the handful of Dutchmen at Nagasaki were, first and last, tradesmen. They were not particularly able or willing to spoonfeed curious Japanese. Neverthless, the fortuities of Japan's foreign relations confined Western-oriented Japanese to a peculiar reliance upon Dutch sources.

Under these circumstances a limited number of Japanese pioneers, the 'rangakusha' (蘭学者) (scholars of Dutch learning),* had to undertake prodigious

* The term 'Dutch learning' as commonly used in this connection refers to the Japanese study of works on science and technology written in the Dutch language.

labours with almost no aid from foreigners. A generation after Yoshimune's reign, in the 1770s, a notable expansion of study of the Dutch language and science led to a move for translation of Dutch scientific works – or retranslation of Dutch translations of Western European works.

Choice of the subject matter of 'Dutch learning' depended largely on the government's desire for utilization of practical Western knowledge. In this course of events astronomy was in the vanguard. This is to be accounted for as follows: besides its function as an enterprise of the government, astronomy, like medicine, was an ancient subject familiar to the Japanese, and was thus immediately attractive. Furthermore, the early Sino-Jesuit contribution had already proved the superiority of Western sciences, particularly Western astronomy – a fact which was already apparent from available Chinese astronomical treatises.

Generally, technical aspects of natural sciences were easily accepted for their universality and practicality, while the taking up of Western philosophy and the Western world view – more closely tied to cultural permises – was considerably delayed. While the core of genuine Western science, conceptual scheme and cosmology, did not seem to interest practical astronomers much, they wished to take advantage of Western treatises without much inhibition as to their foreign ideological basis as long as it was directly applicable to their traditional approach to calendar-making, in such as observational data.

MOTOKI RYŌEI'S TRANSLATIONS OF WORKS ON HELIOCENTRICISM

Motoki Ryōei's (1735–94) translations of Dutch works are significant not only as the first Japanese sources on the Copernican heliocentric system, but also as a landmark in the advancement of the study of Western languages in Japan.

Oranda Chikyū Zusetsu

The first translation to refer to the heliocentric theory was drafted in 1772 (the 12th month of the 8th year of Temmei) under the title *Oranda Chikyū Zusetsu* (Dutchmen's illustration of the earth).[18] Its original was the Dutch translation *Atlas van Zeevaert en Koophandel door de geheele weereldt* (1745, Amsterdam) of the French original *Atlas de la Navigation et du Commerce qui se fait dans toutes les parties du monde* (1715) by Louis Renard.

The text is a popular guide to the usage of maps for seamen. As it was not a scholarly account, no mention of the name of Copernicus, or indeed of the name of any scholar, was made in any part of the translation.

The passage referring to the heliocentric hypothesis gives only a short account of the daily rotation and anual revolution and the difference between true (heliocentric) and apparent (geocentric) motions. This is quite far from what is called the introduction of heliocentric Copernican theory.

Tenchi Nikyū Yōhō

The second work by the same translator is *Tenchi Nikyū Yōhō* (The use of the celestial and terrestrial globes) dated 1774.[20] Its Dutch original is *Tweevoudigh Onderwiis van de Hemelsche en Aardsche Globen* (Amsterdam 1666, the first edition is dated 1620).[5]

The book was written and prefaced by Willem Janszoon Blaeu (or Blaaw, 1571–1638) and edited and published by his son Johan. Willem Janszoon Blaeu was a renowned Dutch cartographer and an intimate friend and disciple of Tycho Brahe. He was also one of the early proponents of the Copernican system.[2,28]

TWEEVOUDIGH
ONDERWIIS
van de Hemelſche en Aerdſche
GLOBEN;
Het een
Na de meyning van PTOLEMEUS
met een vaſten Aerdkloot;
Het ander
Na de Natuerlijcke ſtelling van N COPERNICUS
met een loopenden Aerdkloot:
Beſchreven door WILLEM I. BLAEU,
t'AMSTERDAM,
Ter Druckerye van JOAN BLAEU.
MDCLXVI

(1)

EERSTE DEEL
van 't
ONDERWYS
van de
GLOBEN,
Na de oneygen ſtelling van
PTOLEMÆUS,
met een vaſten Aerdkloot.

(2)

TWEEDE DEEL
van 't
ONDERWYS
van de
GLOBEN,
Na de natuerlijcke ſtelling van
N. COPERNICUS,
met een loopenden Aerdkloot.

(3)

Blaeu's original
(1) The title page of the whole treatise
(2) The title page of Volume I (Ptolemaic)
(3) The title page of Volume II (Copernican)

We have no way to prove at what time Blaeu's 1666 edition was brought to Japan. It might be that long before the Japanese translation in 1774 it was imported into Nagasaki and then buried for a long time in obscurity. W. Blaeu is an eminent figure in the seventeenth century, the golden age of the Dutch; his works were translated into French and other languages and diffused throughout the world. Many of his maps and other works had been also brought to Japan. It might not be too absured, therefore, to conjecture that even during the eighteenth century, Dutchmen had favoured their ancestor's glorious works and recommended Ryōei to work on it. After the eighteenth century, we find fewer works by Dutch themselves; instead, Dutch translations of French, German and English works became predominant.

It is manifest that Blaeu wrote his book with the intention of propagating Copernican hypothesis. We may not, however, be able to assume hastily the date of 1774 as the turning point of the history of Japanese astronomy, in which the Copernican theory replaced the older cosmology suddenly and thoroughly. Motoki Ryōei's own preface says: 'The preface of [Blaeu's] book said that there are two theories among astronomers concerning the centre of the heaven and planetary movements. One is that the immobile earth is situated in the middle of the heaven, and seven luminaries and fixed stars go around it in circular orbits.

The other said that the sun is always immobile and that the earth, together with the other five planets, goes around the sun; the heaven of fixed stars is also absolutely immobile. The former is the theory of Hipparchus, Ptolemy and his followers, and is still an established theory now. The latter is that of ancient writers, which had been lost for long. About one hundred years ago, there was a man called Nicolaus Copernicus. Having intimate communications with Tycho Brahe, the biggest figure in astronomical observations, Copernicus investigated this [heliocentric] theory thoroughly and finally reached to enlightment out of cumbersome darkness'.[20]

This is apparently the exposition of Copernican heliocentricism but we have found no more mention of it in all the translation work of Ryōei. While Voor-Reden of Blaeu's original has eight paragraphs, Ryōei translated and put into his own preface only the first four paragraphs. Blaeu's preface states that his intent was to illustrate the Ptolemaic system first, because it is more familiar and more easily comprehensible, and then go on to the true theory of Copernicus. The subsequent paragraphs, 5th, 6th, and 7th, substantially discuss Blaeu's plan of arranging the Copernican system as opposed to Ptolemaic. These are all omitted from Ryōei's preface. Furthermore, he made a historical confusion on the relation between Copernicus and Tycho Brahe.

Blaeu's book consists of two parts as follows:

Volume I: Astronomical principles of celestial and terrestrial globes based on the inadequate hypothesis of Ptolemy.

Volume II: Astronomical principles of globes based on true hypothesis of Copernicus.

Ryōei's translation terminated at the middle of Volume I (the part of Book I, p. 120 out of 163 pp. of the whole volume), namely the part based on Ptolemaic geocentric theory. Apparently Ryōei had no intention to extend his translation work further. It is highly probable that Ryōei deliberately omitted the part of heliocentricism from the main text as well as from his own preface.

It seems unlikely that the Copernican heliocentric theory presented an unsurmountable difficulty for Ryōei to comprehend, even though unfamiliar to the tradition-bounded cosmological outlook. Blaeu's astronomical writing was not particularly advanced; Kepler's contributions did not appear in it. Hence, it would be more plausible to interpret this deliberate omission as being due to Ryōei's inhibition-conscious precaution against any subversive or unorthodox thoughts. We conclude, theorefore, that Ryōei's work in 1774 should not be properly called the full introduction of Copernican theory, but remained only the first oppearance of the name of Copernicus in Japanese works so far discovered.

Shinsei Tenchi Nikyū Yōhō Ki

A more detailed and truely comprehensive account of the Copernican system was translated in part in 1792 and completed the following year. It was titled *Seijutsu Hongen Taiyō Kyūri Ryōkai Shinsei Tenchi Nikyū Yōhō Ki* (The ground of astronomy, newly edited and illustrated; on the use of celestial and terrestrial globes according to the heliocentric system) in seven volumes.[19] The Dutch original has been identified as *Gronden der Starrenkunde, gelegd in het zonnestelzel bevatlijk gemaakt*: *in eene beschrijving van't Maaksel en Gebruik der nieuwe hemel-en aardgloben* (Amsterdam, 1770), 470 pp.[10]

Its author, George Adams, was a maker of mathematical instruments to King George III. He had a world-wide reputation as a maker of celestial and terrestrial globes. The ultimate original is entitled *Describing and Explaining the Construction and Use of New Celestial and Terrestrial Globes* (1766).

Ryōei's translation includes the first 325 of the 360 paragraphs in the Dutch original: only a part on the use of the globes was left untranslated. The work begins with a straightforward description and explanation of the solar system, in which the relation between the apparent and true courses of the planets is expounded on the basis of the heliocentric scheme. It does not seem, however, to be an advanced treatise for professional astronomers, but a textbook for instrument operators.

The arrangement of topics is strikingly different from that of traditional treatises. From the outset, the earth is treated as a member of the solar system; detailed instructions are given for the reduction from geocentric to heliocentric coordinates, and after this preparation the behaviour of the planets and satellites is expounded, whereas the traditional approach starts with an analysis of the apparent motions of the sun and moon, ends with the prediction of eclipses, and the five planets have little significance in the main exposition.

Notwithstanding the significance of his work, Ryōei remained only a faithful translator, not an advocate or original writer on new ideas. At the end of the 1792 edition of his translation of Adams' treatise, he expounded his own view of astronomy, but we find no particular note of enthusiasm about the newly introduced heliocentric system.

In his work he also gave brief summaries of *Philosophische Onderwijzer* and *Beginselen der Natuurkunde*. The ultimate original of the former is Benjamin Martin's *The Philosophical Grammar* (first edition 1735, Dutch edition 1744);[9,13] the latter is *Anfangsgründe der Physik* by Johann Heinrich Winkler (Dutch edition 1768).[83] These works elucidate the Newtonian laws of mechanics, but they were entirely beyond Ryōei's concern or comprehension. He could only compare the Ptolemaic, Tychonian and Copernican systems briefly.

As we have seen from the several instances given above, there must have been not a few Western astronomical treatises available for translation in the last quarter of the eighteenth century in Japan. Hence, at this period, the ban on Western books was not completely operative; to a certain degree, translators had freedom to choose their subject and text.

Ryōei was only a faithful translator, not a specialist in astronomy. The transformation from goecentricism to heliocentricism was not a problem of technical difficulty, unless one has certain religious and philosophical presuppositons. Thus, unlike the comprehension of Newtonianism, the introduction of heliocentric theory never raised difficult questions for him.

Ryōei stated in his 1782 work that there was previously no translation on a professional level, and that hence his predecessors were sceptical about the possibility of even literal translation; this was the reason why they were reluctant to try it. It is not impossible that if he were to fail in his bold attempt to initiate translating, his ability might have been called in question and his inherited position jeopardized.* Thus, the main difficulty in the introduction of heliocentricism lay in the linguistic barrier and in psychological unwillingness to present anything extravagant.

Ryōei's translation works were not printed in his lifetime. His manuscripts were sent to the central or local governments and preserved there, beyond the reach of the general public. His achievement was fairly well known among intellectual circles and his manuscripts were circulated through hand-written copies.

Shiba Kōkan was the first Japanese to popularize Copernican theory, through three printed books, *Chikyū Zenzu Ryakusetsu* (地球全図略説) (An outline world atlas, 1793), *Oranda Tensetsu* (和蘭天説) (Dutch astronomy, 1795), and *Kopperunyu Temmon Zukai* (刻白爾天文図解) (Copernican astronomy illustrated, 1805).[21] Although the depth of his knowledge of Western astronomy did not exceed that of Motoki Ryōei, whose work was his main source, the heliocentric hypothesis was considerably diffused among the populace as a result of his work.

THE RESPONSE OF BUDDHISTS AND OTHERS TO WESTERN COSMOLOGICAL THEORIES

By the middle of the eighteenth century Aristotelian cosmology had been widely diffused through printed editions of *T'ien-ching Huo-wen*.†[37] To refute these ideas, a treatise entitled *Hi Tenkyō Wakumon* (Contra T'ien-ching Huowen), by the learned Buddhist monk Monnō, was published in 1759.[16] Another of his works, *Kusen Hakkai Tōron* (A discussion of the theory of the Nine Mountains and Eight Seas),[17] expounded *Sumeru* cosmology.‡ Apparently Monnō could not tolerate the diffusion of European cosmological ideas, which were incompatible with Buddhist beliefs.

Generally speaking the Buddhists' cosmological outlook enjoyed more freedom in infinity and plurality of worlds than the rigidly constructed Aritstotelian world, but this freedom was due to their anarchistic vagueness and lack of conviction concerning the existence of underlying regularity in the phenomenal world, and the possibility of its discovery.

Some of Monnō's criticisms were just. But one generation later, when more detailed material on Western astronomy became available, the animadversions of the Buddhists were really scandalous.

The creativity of Japanese Buddhism was in eclipse at the beginning of the Tokugawa period, and intellectual leadership shifted to Confucianism, the state orthodoxy. As Western astronomy proved its superiority to Japanese intellectuals and as knowledge of the Copernican system was disseminated through the various popular editions of Shiba Kōkan, the antipathy of the Buddhists was

* It is recorded that in 1791 seven interpreters at Nagasaki were dismissed for having made unfaithful translations of Dutch document.[12,25]

† The Tychonian system, vaguely referred to, does not seem to receive proper notices among Japanese.

‡ It sets at the centre of the earth Mount Sumeru, round which the sun, the moon, and the stars revolved. This idea originated in Jaina cosmography, (ca. fifth century B. C.) and was taken over by the Buddhists.

aroused. While the Confucians were not so strongly opposed to the achievements of Western astronomy, the Buddhists tried to restore their own intellectual glory by scoring a victory for Buddhist cosmology.

Among this group of Buddhists, the most zealous and influential figure was Entsū (1754–1834). His thirty years of labour in defence of Buddhist astronomy culminated in his masterpiece *Bukkoku Rekishō hen* (On the astronomy and calendrical theory of Buddha's country, five volumes, 1810),[6] which was followed by a number of other writings in the same vein.

To defend his own point of view, he conducted amazingly wide researches in Chinese and Japanese materials – Buddhist sutras in Chinese translation, ancient and modern Chinese astronomical treatises, Sino-Jesuit works, and works on Western astronomy written by his Japanese contemporaries.*

From the outset, his single purpose of defending Buddhist doctrine from the invasion of Western scientific ideas was marked. On cosmological questions the main authority for his argument was *Ryūsei Abidonron* (立世阿毘曇論) (Lokaupasthana Abhidharmasastra).† From the beginning to the end, Sumeru cosmology was rigidly and literally maintained. Furthermore, he took advantage of borrowing from the ancient Chinese flat-earth *Kai t'ien* (蓋天) theory, which in some ways resembles Sumeru cosmology. It should be noted, however, that while the *Kai t'ien* theory had had an empirical basis and was entirely free from mythological and anthropomorphic elements, the Buddhist cosmology was full of religious fantasticism and lacked an observational foundation. The only explicit point of similarity lay in their non-spherical model. As to the differences, Entsū explained that whereas the Chinese were skilful at rational investigation, the Indian sages had penetrating 'spiritual eyes' (*tengan* 天眼), which were given only to superior beings.

Entsū's arguments did not always make sense. He often appealed to a supernatural insight. He displayed disgust at the purely materialistic basis of Western theory. Still, whenever scientific proofs could be used to demonstrate the superiority of Indian astronomy, he used them. It is not that Entsū was rabidly anti-scientific; the point is that his object was persuasion at any price. His motivation, after all, was a fear that Christianity would undermine Buddhist teaching in Japan.

Entsū never tired of denouncing Western learning, and of extolling the sacred Buddhist cosmology. His vigorous propaganda certainly created a sensation, and found many adherents among Buddhist monks.

Neverthless, the assiduous efforts of Entsū and his followers never achieved orthodoxy. He attempted to obtain Buddhist authority for the publication of *Bukkoku Rekishō hen*, but the aged commissioner Sen'yo (仙葉) never agreed to sanction it as long as he lived, on the grounds that 'since Entsū was too involved in astronomy, he confused the essentials of Buddhism with astronomical science. His views might cause incalculable damage to genuine Buddhism.'[11]

* Entsū was said to have had some background in 'Dutch learning', but there is no evidence that he was able to read Dutch.

† Particularly 'Nichigetsugyō bon' (日月行 品) (Chapter of the solar and lunar courses).

Other monks also denounced Entsū. If Sumeru cosmology was not really true, would not this cast suspicion on all the teachings of Buddhism? It would be dangerous for Buddhism to commit itself to Sumeru cosmology. Properly speaking, Sumeru theory was not of Buddhist origin, but came from an ancient Indian idea.[11]

Some, like Inō Tadataka (伊能忠敬), bitterly denounced Entsū's misleading unscientific dogma, but astronomers of the top rank generally disregarded and ridiculed Entsū's work.[15] But the sensation caused by Buddhist reaction (which was unusual, for Buddhism was generally apathetic to science) caused a great deal of controversial literature on astronomy to be circulated. Even during the Meiji period, when explosive Westernization had nearly eradicated the roots of the traditionally bound approach, a prolonged effort to defend Buddhist cosmology was made.[26]

The Reaction of the Confucians

Unlike the pronounced violent opposition of the Buddhists, we find relatively little reluctance on the part of the Confucians to accept the Copernican system. Some were outspokenly hostile to everything Western, and others claimed Western appropriation of ancient Chinese ideas. Neo-Confucianism formed its cosmological background as an integral part of unitary principle, and therefore, violations in Western cosmology of this unity between human and physical nature were often criticized; however, the Neo-Confucians did not maintain a detailed religious cosmos like the Buddhists or the medieval Church, and generally were not concerned with the appearances. Therefore there was no clear point of conflict with Western cosmology.

Suga Sazan (1748–1828) commented that 'there is no other use for astronomy than the determination of the correct time; other concerns (viz., cosmology and general astronomy) were merely useless argument and dull speculation'.[29] It would not be far from the truth to assume that his attitude was largely shared by other Confucian scholars, whose pragmatic interest in social and ethical problems excluded a disinterested concern with physical nature. This indifference was almost as important a negative factor as Buddhist hostilily.

The Reaction of the Neo-Shintoists

From the beginning of the nineteenth century, owing to increasing foreign threats, a nationalistic spirit and a desire for independent identity had gradually grown up. Hirata Atsutane (1776–1843) and his followers tried to establish a doctrinal basis for this out of the ancient native mythopoeic tradition, borrowing whatever was profitable from Christian, Confucian and Buddhist tenets. Unlike the Buddhists and Confucians, the Neo-Shintoists did not have quasi-scientific tradition to apologize for. The ambiguity of their mythology invited free interpretation. In other words, being free from historical domination and foreign authority, they could now 'create' their own tradition. Quite cleverly, they utilized the most up-to-date knowledge of Western science available.

The result was a curious amalgamation of primitive myth and modern science. Atsutane and his pupils Satō Nobuhiro (1769–1850) and Tsurumine Shigenobu (1786–1859) all had an unreserved appreciation for modern Western science.

The process of world creation was explained by traditional myths. The Creators, a god and goddess, formed the universe from the primordial chaos, and operated the heliocentric world. Entsū's Sumeru argument was bitterly refuted. Thus, they fully took advantage of Western science in their attempt to systematize primitive mythology into a consistent cosmology.[7,8,27,32]

CONCLUSION

To sum up, the Copernican system did not evoke bitter ideological opposition in Japan except in Buddhist circles. Even the latter could not exert such a commanding reactionary influence as the Renaissance Church had in Europe. The main cause of the delay of the introduction of heliocentric theory into Japan was the seclusion policy of the government, rigidly maintained until the early part of the eighteenth century, and secondly, the linguistic barrier, which remained formidable until the last quarter of that century.

It seems that the question of the reception of Copernicus' theory was more or less resolved into the more general problem of the superiority of Western learning. In this respect, recognition of Western superiority in the domain of 'figure and appearance – the material aspects' had been well established and, in spite of some ideological incompatibility and conflict, Western astronomy had a rather smooth reception during the latter part of the Tokugawa period.

REFERENCES

(1) Baba Nobutake (馬場信武), *Shogaku Temmon Shinan* (初学天文指南) (An elementary introduction to astronomy, 1706).

(2) Bandet, Pierre Henry, *Leven en Werken van Willem Jansz. Blaeu* (1871).

(3) Barnouw A. J. and B. Landheer (eds), *The Contribution of Holland to the Sciences* (New York 1943).

(4) Bernard, Henri, S. J., 'L'Encyclopédie Astronomique du Père Schall' *Monumenta Serica*, (1938).

(5) Blaeu, Willem Janszoon, *Tweevoudigh Onderwiis van de Hemelsche en Aardsche Globen* (Amsterdam, 1666, preserved in Library of Congress).

(6) Entsū (円通), *Bukkoku Rekishō hen* (On the astronomy and calendrical theory of Buddha's country, 1810) 5 vols.

(7) Hirata Atsutane (平田篤胤), *Tama no Mahashira* (霊の真柱) (1812), reprinted in *Hirata Atsutane Zenshū* (平田篤胤全集) (Tokyo, 1911), vol. II.

(8) Hirata Atsutane, *Tenchūki* (天柱記) (ca. 1825), reprinted in *Hirata Atsutane Zenshū*, vol. II.

(9) Imai Itaru (今井溱), 'Maruchin no Kyūrisho' (マルチンの窮理書) (Benjamin Martin in the European Astronomy of Old Japan), *Nihon Temmon Kenkyukai Hobun* (Memoirs of the Japan Astronomical Study Association) 7 145 (1960).

(10) Itazawa Takeo (板沢武雄), 'Edo Jidai ni okeru Chikyū Chidōsetsu no Tenkai to sono Handō' (江戸時代における地球地動説の展開とその反動) (The development of the heliocentric theory and the reaction to it during the Tokugawa period), *Shigaku Zasshi* 68 325.

(11) Kōda Rohan (幸田露伴), *Kagyūan Yobanashi* (蝸牛庵夜譚) (A night tale of Kagyūan) (1907).

(12) Krieger, C. C., *The Infiltration of European Civilization in Japan during the 18th Century* (Leiden, 1940) A partial translation of Ōtsuki Nyoden's *Shinsen yōgaku nenpyō*.
(13) Martin, Benjamin, *The Philosophical Grammar* (London 1738).
(14) Mikami Yoshio, 'A Japanese Buddhist's view of the European astronomy' *Nieuw Archief voor Wiskunde* 11 1.
(15) Mikami Yoshio (三上義夫), 'Nihon Kagaku no Tokushitsu: Temmon' (日本科学の特質: 天文) (Characteristics of Japanese science: astronomy), in *Tōyō Shisō no Tenkai* 東洋思想の展開 (The development of Oriental thought; Tokyo, 1936).
(16) Monnō (文雄), *Hi Tenkyō Wakumon* (非天経或問) (Contra T'ien-ching Huo-wen, 1759).
(17) Monnō, *Kusen Hakkai Tōron* (九山八海嘲輪) (A discussion of the theory of the Nine Mountains and Eight Seas; 1754).
(18) Motoki Ryōei (本木良永), *Oranda Chikyū Zusetsu* (阿蘭陀地球図説) (Dutchmen's illustration of the earth; 1772). A later duplication copy is now extant, preserved in Tenri Library.
(19) Motoki Ryōei, *Seijutsu Hongen Taiyō Kyūri Shinsei Tenchi Nikyū Yōhō ki* (星術本源太陽窮理薪制天地二球用法記) (The ground of astronomy, newly edited and illustrated; on the use of celestial and terrestrial globes according to the heliocentric system; two volumes in 1792; seven volumes in 1793).
(20) Motoki Ryōei, *Tenchi Nikyū Yōhō* (天地二球用法) (The use of the celestial and terrestrial globes, 1774) preserved in Nagasaki Museum.
(21) Nakai Sōtarō (中井宗太郎), *Shiba Kōkan* (司馬江漢) (Tokyo, 1942).
(22) Nakane Genkei (中根元圭), *Temmon Zukai Hakki* (天文図解発揮) (Astronomy illustrated, 1739).
(23) Nakayama, Shigeru, 'An Outline History of Japanese Astronomy, Western Impact. vs. Chinese Background' (1959, Harvard PhD thesis), Chapters XI, XII and XIV.
(24) Needham, Joseph, *Science and Civilisation in China* Volume III, p. 447 (Cambridge, 1959).
(25) Ōtsuki Nyoden (大槻如電), *Shinsen Yōgaku Nenpyō* (新撰洋学年表) (A chronology of Western learning in Japan, Tokyo, 1926).
(26) Sada Kaiseki (佐田介石), *Shijitsu Tōshōgi Shōsetsu* (視実等象儀詳説) (A detailed account of the instrument by which the apparent and real courses of the heavenly bodies are explained; 1880).
(27) Satō Nobuhiro (佐藤信淵), *Yōzō Kaiku ron* (鎔造化育論) (On the creation and formation of the world; ca. 1825), reprinted in *Shinchū Kōgaku Sōsho* (新註皇学叢書) (Series in Nipponology, newly edited; Tokyo 1927), vol. X.
(28) Stevenson, Edward Luther, *Willem Janszoon Blaeu* (New York, 1914).
(29) Suga Sazan (菅茶山), *Fude no Susabi* (筆のすさび) (Writing for amusement's sake), reprinted in *Nihon Zuihitsu Taikei* (日本随筆大系) (A comprehensive collection of Japanese informal essays; Tokyo, 1927), I, 80.
(30) Szczesniak, Boleslaw, 'Notes on the Penetration of the Copernican Theory into China from the 17th to the 19th Centuries', *Journal of the Royal Asiatic Society* p. 30 (1945).
(31) Szczesniak, Boleslaw, 'The Penetration of the Copernican Theory into Feudal Japan', *Journal of the Royal Asiatic Society* p. 52 (1944).
(32) Tsurumine Shigenobu (鶴峰戊申), *Ame no Mihashira* (天のみはしら) (The sacred heavenly pillar; 1821).
(33) Winkler, Johann Heinrich, *Anfangsgründe der Physik* (Leipzig, 1st ed. 1753).

(34) Yabuuchi Kiyoshi (薮内清), 'Edo jidai ni okeru Gairai Kagaku no Yu'nyū' (江戸時代における外来科学の輸入) (The importation of foreign sciences during the Tokugawa period) *Kagakushi Kenkyu* 43, 2.

(35) Yabuuchi Kiyoshi, 'Kinsei Chūgoku ni tsutaerareta Seiyō Temmongaku' (近世中国に伝えられた西洋天文学) (European astronomy introduced into Modern China), *Kagakushi Kenkyu* 32, 16 (1954).

(36) Yabuuchi, Kiyoshi, 'Seiyō Temmongaku no Tōzen–Shindai no Rekihō' (西洋天文学の東漸: 清代の暦法) (Introduction of Western astronomy into the East – calculation of the calendar in the Ch'ing period), *Tōhō Gakuhō, Kyoto,* (1946) **15,** pt. 2, p. 146.

(37) Yu I (遊芸), *T'ien-ching Huo-wen* (天経或問) (Queries on the classics of the heaven, 1675?)

First published in *Japanese Studies in the History of Science*, No.1, pp.14-22, The History of Science Society of Japan, 1962

❸ Japanese Studies in the History of Astronomy

Together with genuine interest in the main current of the development of astronomy in the West, it is natural enough for a Japanese to show particular concern respecting the heritage of East-Asian (mainly Chinese and Japanese) astronomy, where he has the advantages of cultural congeniality and linguistic familiarity. In fact, modern Japanese scholarship has already made essential contributions to the chronology and history of Chinese as well as Japanese astronomy.

There is no clear demarcation line between specialists in Chinese and Japanese subjects. However, for purely descriptive convenience, we shall first describe studies on Chinese topics, and then proceed to Japanese ones which are rather derivatives of the former. In the third section, a brief comment on the general history of astronomy will be made. In the final section, a selected bibliography will be given.

1. CHRONOLOGY AND HISTORY OF THE ASTRONOMY OF CHINA

The dating and chronology of ancient China has, since the early eighteenth century, been the subject of much controversy among European Orientalists. In the second decade of the present century, Japanese scholars also came to be involved in this long-standing debate. One school led by Shinjo Shinzo (1873–1943) claimed that Chinese astronomy developed independently of Western influences.[45] This school may be called the 'Astronomers' School', as its members have all started their careers as astronomers and are well-versed in the modern methods of astronomical chronology. Fully utilizing their background, they have now turned to apply their method, after the manner of P. V. Neugebauer and T. von Oppolzer, to ancient Chinese sources, for the determination of dating as well as the evaluation of ancient Chinese astronomy.

Notable opponents to the 'Astronomers' School' are Iijima Tadao (1875–1954) and Hashimoto Masukichi (1880–1956). Both of them tried to prove the Western derivation of Chinese astronomy. While the 'Astronomers' School' attempted to reconstruct ancient Chinese astronomy from fragments of the classics, the latter scholars (we may call them 'Historians' School') tried mainly to reevaluate the classics with the aid of astronomy. Iijima, with his background in Chinese classics, attempted to settle the dates of Confucian classics and was eventually led to the study of ancient Chinese astronomy. He gathered all the evidence in support of his own thesis that Babylonian (or Greek) influence had been basic in the formulation of Chinese astronomy around the time of the conquest of Alexander the Great. He therefore tried to disprove an independent Chinese development.[10–12] Hashimoto, a historian, did not go to the same lengths as Iijima but developed a view that Western influence had been gradually assimilated into China since sometime before the eighth century B.C.[4–6] Thus, although the views of Iijima and Hashimoto differed considerably, both of them criticized the 'Astronomers' School' for its uncritical reliance on antique sources without meticulous documentation as data for their calculation.

The most immediate pupil of Shinjo is Noda Churyo, who pursued almost the same line of research that Shinjo initiated. He has also published a good work on ancient Chinese cosmology.[32] Whether conclusions of the 'Astronomers' School' are plausible or not, the merit of their astronomical approach is to open entirely new aspects of ancient Chinese astronomy, which a purely historical method could never envisage.

These controversies really contributed to the deepening and widening of knowledge of the subject-matter; however, it seems that in dealing with this obscure subject, they never reached any clearcut conclusion. All of the original disputants, Shinjo, Iijima, and Hashimoto have died; there have been no active discussions in the post-war period on this subject, except that Yabuuti Kiyosi, another pupil of Shinjo, wrote several critical comments on Tung Tso-pin's work on the calendar in the Yin period.

As to the role of Japanese scholars in the study of Chinese chronology, we may say that although their work had been anticipated by Western sinologists, Japanese astronomers and classicists were better provided with general proficiency in language and classical scholarship; on the other hand, they had, to a certain extent, been in advance of Chinese scholarship in introducing modern approaches, a fact well illustrated by the translation into Chinese of Shinjo's two works.

The 'Astronomers' School' has its centre of activities at Kyoto, where research activities in the history of science are now on a permanent basis. In 1929 during Shinjo's residency of Kyoto Imperial University, Toho Gakuin Kyoto Kenkyusho (The Institute of Oriental Culture at Kyoto) was founded (its title was changed to Toho Bunka Kenkyusho in 1938). Noda and Yabuuti, as regular staff members of the calendrical science section, engaged in research into the history of Chinese astronomy and calendrical science with the cooperation of sinologist collegues in the Institute. They have published numerous articles in *Toho Gakuho, Kyoto* (the periodical of the Institute) and monographs in the form of occasional reports of the Institute.[33,51,53–66]

From 1940 to 1944, the Institute undertook as a research project the investigation of East-Asian calendars. The project was headed by Noda with the cooperation of a dozen astronomers and historians. It looked forward to the attainment of precise East Asian astronomical chronology. A Sanscritist, Zemba Amane, who was one of the co-workers on the project, has done original works on ancient Indian astronomy and chronology, chiefly by comparisons of Sanscrit, Tibetan and Chinese texts of Buddhist Sutras.[67–70]

In 1948, the Institute became affiliated with Kyoto University and changed its title again to Jimbun Kagaku Kenkyusho (Research Institute of Humanistic Sciences). The calendrical science section was converted into a research professorship in the history of science. The post is now held by Yabuuti.

The interests of Western sinologists in Chinese astronomy have so far been by and large focused upon the ancient period, in which the Chinese pattern was formed, and on the contributions of the Jesuits, for which abundant Western sources are available. It is regrettable, however, that relatively little is known of the slow evolution of Chinese astronomy in the intervening period; especially Chinese calendar-making has been almost entirely overlooked – probably because of the lack of a Western counterpart, and because of its technical complexity – despite the fact that calendar-calculation had a central position in Chinese exact science. Even Joseph Needham ignored it as 'not of primary scientific importance' and said 'its interest is, we suggest, much more archaeological and historical than scientific. A calendar is only a method of combining days into periods suitable for civil life and religious and cultural observances' (*Science and Civilisation in China* Vol. 3, p.390).

Initially, Japanese interest had also been primarily in the problem of dating and chronology, and only secondarily with the genuine history of science. Shinjo's and even Noda's primary intention was 'to gather and arrange in order every astronomical record serviceable for any academic purpose whatever'. It is Yabuuti, who was instrumental in establishing the subject of Chinese astronomy as an independent discipline of the history of science, and not as a mere aid to classical and historical studies. In the 1930s, he started his life-long programme of tracing all astronomical developments throughout Chinese history. Focusing mainly on the calendrical sections of each dynastic history, he covered nearly every aspect of Chinese astronomical (or indeed scientific) thought. Whereas his predecessors chiefly aimed to apply astronomical methods to historical subjects, Yabuuti devoted his main concern to the current of astronomical thought as a part of intellectual history, using historical methods to deal with the subject of astronomy.

II. HISTORY OF JAPANESE ASTRONOMY

Japanese astronomical records are not so valuable as their Chinese counterparts for antiqueness and authenticity for both historical and astronomical purposes. Furthermore, while traditional Japanese mathematics has enjoyed independent prestige, Japanese astronomy can hardly claim to be original; it was dominated by Chinese astronomy in its earlier phase and it assimilated itself to Western astronomical theories and techniques in the later phase. There is no evidence whatsoever that Japanese astronomy had any influence upon the main currents of world scientific thought; the traffic was one-way.

Nevertheless, the history of Japanese astronomy is interesting not only as a domestic issue, but also as an example of the assimilation of ideas and a scheme of modernization in a non-Western country. Japan was a meeting place of traditional Chinese culture and modern Western ideas. It is of importance, from the viewpoint of the history of ideas, to see how the Japanese, with their Chinese background, reacted to Western astronomy, the Copernican system and Newtonian mechanics.

Unfortunately, however, there is no comprehensive history of Japanese astronomy available in a Western language. Iba Yasuyuki's 'Fragmentary Notes on Astronomy in Japan' is literally fragmentary and proves to be of no great interest to professional students of this subject. Recently Nakayama Shigeru completed his PhD thesis at Harvard University entitled 'An Outline History of Japanese Astronomy; Western Impact versus Chinese Background' (1959). In particular, it clarified various special features of Japanese astronomy by utilizing the approach of comparative history.

In Japanese, Mikami Yoshio's *Nihon Kagaku no Tokushitsu; Temmon* (the characteristics of Japanese science; astronomy; 1936) is a most coherent, well-arranged piece of work, though concise and brief.[25] A recent publication entitled *Meijizen Nihon Temmongakushi* (the history of Japanese astronomy before the Meiji era; 1960, 518pp.) consists of articles contributed by several scholars.[24] Although it is most voluminous, it is far from being a conclusive edition of the history of Japanese astronomy. Most of the articles were written before the end of the Second World War, and revision in the light of post-war researches is needed.

Whereas the Research Institute of Humanistic Sciences at Kyoto is a core of research activities on the history of Chinese astronomy, the Tokyo Astronomical Observatory has provided several outstanding contributors to research in the history of Japanese astronomy. The man who first entered this field with modern training in astronomy is, perhaps, Hirayama Kiyotsugu (1874–1943), who was followed by Ogura Shinkichi, Ogawa Kiyohiko and Kanda Shigeru. At present, Maeyama Jinro succeeds to this tradition, and is now authorized to do historical research as the chief of the calendrical section of the Observatory. The Observatory preserves one of the best collections of materials on the history of Japanese astronomy, bequeathed from the library of the governmental astronomical office of the Tokugawa period. Most works by Observatory personnel appeared, especially in the pre-war period, in *Temmon geppo* (Astronomical Herald), a Japanese journal published by the Observatory.

Before and during the Second World War, Japanese chronology had been left beyond the pale of scientific investigation, since the political authorities backed the nationalistic ideology then prevailing and discouraged derogation from the legendary glory of the ancient Japanese state by purely rational researches, which would have indicated that the length of reign of the Japanese Imperial Household was less ancient than traditional chronology. One of the oldest official Japanese chronicles, the *Nihon shoki*, completed in A.D. 720, has a calendrical index extending back to 660 B.C. Its calendrical notation, however, is not entirely consistent with the Chinese system. This was traditionally accounted for by the postulation of a native calendrical system, developed independently of major Chinese Influence. Ogawa made a significant contribution towards the reinterpretation of this dating system. He concluded that merely careless omissions in copying some calendrical indices into the *Nihon shoki* had led scholars to a mistaken belief in the early existence of an independent calendrical system, and that if these errors were amended, it could be demonstrated that the ancient

Japanese merely followed the Chinese calendar which had been introduced to Japan at that time. This rather sensational result could appear in public only after the War, when thought control had been brought to an end.[36]

Hirayama was one of the pioneers in the history of astronomy in Japan. His research on the *Shou-shih* calendar to reproduce traditional astronomy in terms of modern notation was continued by Yabuuti. Hirayama also tried to give an adequate evaluation to each traditional Japanese astronomer and bequeathed a huge collection of bibliographical cards totalling about 3,000 items, including not only published works but also original manuscripts and handwritten copies. These cards provide annotation rather more astronomical than historical. They will be a most valuable inheritance to his successors.

Kanda carried out assiduous labour to collect every basic source-material concerning Japanese astronomy, both of historical and purely astronomical interest. Subject items include such things as meteorites, local calendars and the reports of astrological portents. In 1935, he published *Nihon temmon shiryo*, which is a comprehensive collection of Japanese astronomical findings as recorded in various documents dating from A.D. 620 to 1600 (with one isolated record of a meteor shower in 15 B.C.).[18,19]

Recalculations and checks of numerous records of ancient eclipses were first conducted by Ogura[37] and later more comprehensively by Suzuki Takanobu.[47] Their investigations proved that most eclipses noted in Japanese records were not observed but calculated only.

Materials on astronomical study during the period from the tenth to the sixteenth centuries are quite meagre. Continuing Ogawa's pioneering work[35] on the *Hsuan-ming* calendar which was in use in Japan, Maeyama with the cooperation of Momo Hiroyuki, Professor of the Historiographical Institute of Tokyo University, is engaged in researching into the calendrical practices of this long 'dark age' as far as extant sources of scientific astronomy permit.

Imai Itaru publishes his own mimeograph journal, *Tenkansho*,[13] in which he contributes original works mostly on the relations between Far Eastern and Western astronomy; for example, his study on the *Genna Kokaisho* (dated 1618) is a penetrating astronomical investigation of the influence of Portugese navigational techniques in the early seventeenth century.

There are several biographical studies of traditional Japanese astronomers. Otani Ryokichi's *Ino Tadataka*[41] and Watanabe Toshio's *Hazama Shigetomi to sono ikka* (Hazama Shigetomi and his family)[52] are most authentic and perhaps conclusive accounts of these early nineteenth century astronomers.

At the present, a research project headed by Hirose Hideo for the investigation of astronomy during the Tokugawa period is planned.

These are the principal works done by people with professional astronomical background, able to handle astronomical chronology and calculations. In addition, there are a considerable number of historians and classicists who, although not technical historians of astronomy, have had considerable influence through their treatment of subjects peripheral to the history of astronomy *per se*.

For instance, for astrology and occult sciences, an old work, *Ocho jidai no on'yodo* by Saito Tsutomu is still consulted as a standard academic work,[43] while there are a number of other less scientific works on this subject. Saito analysed Japanese occult science into three main currents, i.e. Taoist, Buddhist and traditional native (Shintoist).

Many articles on the Buddhist calendar, astrology and cosmology appear in various Buddhist journals. Though most of them are not worth mentioning from the viewpoint of the history of science, Ono Gemmyo's 'Buddhist astronomy' is unique in its reproduction of *sumeru* cosmology as it appeared in various Sutras.[38]

For the Tokugawa period in particular, researches were carried on from several other aspects. Ebizawa Arimichi's *Namban gakuto no kenkyu* (A study of the tradition of Jesuit learning in Japan; 1958) is one example;[3] here he tried to collect every evidence of enduring Jesuit influence on Japanese thought and culture (particularly in astronomy), in opposition to the common belief that this influence was mostly eradicated by the seclusion policy of the succeeding century. Its detailed quotations and notes are useful, although it has a slight overtone of favouritism towards the Jesuits.

Rangaku shiryo kenkyukai (The Institute for the Study of Dutch Books), an organization founded in 1954, is primarily composed of historians; the main object of their study is Dutch influence during the later Tokugawa period, in which astronomy was one of the most important topics. Their works are reported monthly in *Rangaku shiryo kenkyukai hokoku* (The Report of The Institute for the Study of the Netherlands Books).

III. WORKS ON THE HISTORY OF ASTRONOMY IN GENERAL

Compared with works on Far Eastern astronomy, those on the history of astronomy in Greece, the Near East and Europe are disproportionately few. The reason is easily understandable. Japanese historians of astronomy have limited access to first-hand records and materials on these subjects; hence, quite often, the subject of Western astronomy has been treated only for the sake of comparison with the Far Eastern development.

Since the available translations of classical writings in astronomy may provide one indication of the degree of interest in these subjects, we list all of them (except for twentieth century works) here:

1930. Isaac Newton: *Principia Mathematica*
1937–43. Galileo Galilei: *Discorsi e Dimonstrazione Mathematiche Intorno a Due Nuove Scienze* (partial translation)
1943. J. Flamsteed: *Atlas celeste de Flamsteed*
1949–58. Claudius Ptolemy: *Almagest*
1952. Aristotle: *De Caelo*
1953. Nicolaus Copernicus: *De Revolutionibus Orbium Coelestrium* (partial translation)
1954. Immanuel Kant: *Allgemeine Naturgeschichte und Theorie des Himmels*
1959–61. Galileo Galilei: *Dialogo dei Massimi Sistemi*

IV. SELECTED BIBLIOGRAPHY

The following bibliography is not an exhaustive list of all works by Japanese in the field of the history of astronomy and its related subjects. We have attempted, on the contrary, to select only 70 works which are essential in the development of the study of the history of astronomy in Japan, with a brief annotation wherever desirable.

(1) Arisaka Takamichi: 'Kansei ki ni okeru Asada ryu tengakuka no katsudo o megutte' (On the activities of the astronomers of the Asada school during the Kansei era) *Historia* (Osaka, 1955), nos. 11, 13, 16.

(2) Ayuzawa Shintaro: 'Kinsei Nihon ni okeru chikyu setsu chido setsu no tenkai' (Development of the theories of the earth's sphericity and rotation in Tokugawa Japan) *Nihon Rekishi* no. 71, pp.33–55 (1954).

(3) Ebizawa Arimichi: *Namban gakuto no kenkyu* (A study of the tradition of Jesuit learning in Japan) 1958 512pp.

(4) Hashimoto Masukichi: ' "Shokyo no kenkyu' (Researches on the Historical Classic) *Toyo gakuho*, 2:283–31 (1912), 3:331–394 (1913), 4:49–76 369–312 (1914).

(5) Hashimoto Masukichi: ' "Shokyo (Gyoten)" no Shichusei ni tsuite' (On the four stars culminating at dusk at the equinoxes and solstices, recorded in the Yao Tien chapter on the Historical Classic), *Toyo gakuho*, 17 (no.3):303–385 (1928).

(6) Hashimoto Masukichi: *Shina kodai rekiho shi kenkyu* (The study on calendrical sciences in the ancient China) 1943, 590pp. with 15pp. summary in German. (Toyo bunko sosho, no. 29).

Chapter I deals with the origin of the stem-branch system and concludes that the twelve-branch system originated in the Western zodiac. Chapter II treats calendrical science during the Spring and Autumn period. Chapter III is on calenders in the Warring states, Ch'in and early Han periods.

(7) Hirayama Kiyotsugu: 'Shishiza ryusei no kokiroku' (Historical records of the Leonids) *Temmon geppo* 5:61–64(1912).

(8) Hirayama Kiyotsugu: 'Nihon ni okonawaretaru jikokuho' (The time system in Japan) *Temmon geppo*, 5:121–124 135–139 (1913).

(9) Hirayama Kiyotsugu: 'Shokyo no nisshoku' (On the eclipses recorded in the Historical Classic) *Temmon geppo*, 21:3 23 (1928).

(10) Iijima Tadao: *Shina kodai shi ron* (A study of the history of ancient China) 1925.

(11) Iijima Tadao: *Shina kodaishi to temmongaku* (The history of ancient China and astronomy) 1939, 333pp.

(12) Iijima Tadao: *Temmon rekiho to inyo gogyo setsu* (Astronomy, calendrical science and the Yin-yang and five elements principles) 1939, 354pp.

The above two works are collections of the author's Journal articles.

(13) Imai Itaru: *Tenkansho* (Mr Imai's private mimeograph journal). Occasional publications including such as 'Genna kokaisho no temmongaku' (Astronomy in Genna kokaisho) and 'Jujireki kenkyu' (A study of the *Shou-shih* calendar).

(14) Imai Itaru: 'Naracho zengo no rekijitsu' (Calendars about the time of the Nara Period), *Kagakusi kenkyu*, 40:36–40 (1956).

(15) Imai Itaru: 'Clavius to Kenkon bensetsu' (Clavius and Kenkon bensetsu), *Nihon temmon kenkyukai hobun* 4:181–188 (1957).

The author compares the first Japanese Aristotelian treatise with the work of Christopher Clavius.

(16) Imoto Susumu and Hasegawa Ichiro: 'Chugoku, Chosen oyobi Nihon no ryuseiu kokiroku' (Ancient records of Chinese, Korean and Japanese meteor showers), *Kagakusi kenkyu* 37:7–15 (1956).

(17) Itazawa Takeo: 'Edo jidai ni okeru chikyu chidosetsu no tenkai to sono hando' (The development of the heliocentric theory and the reaction to it during the Tokugawa period) *Shigaku zasshi* 52.1:12 (1941).

(18) Kanda Shigeru: *Nihon temmon shiryo soran* (A view of Japanese astronomical records), 1934, 232pp.
Chronological arrangement of the item (19).

(19) Kanda Shigeru: *Nihon temmon shiryo* (Japanese astronomical records) 1935, 760pp.
Classified into the following items: solar eclipses, lunar eclipses, the approaches of the moon and stars, planetary phenomena, day-time observations of stars, comets, meteors and miscellaneous.

(20) Kanda Shigeru: *Toyo temmongakushi shoho* (Brief reports on the history of Far Eastern astronomy) nos. 1–100, September 1944 to December 1951.
Cyanotype sheets distributed to a limited circle. His *Nihon temmon kenkyukai temmongakushi buho* (Reports of the history of astronomy section, Japan astronomy study association) since 1955 and *Wazan rekigakushi noto* (Notes on Japanese mathematics and calendrical science) since 1961 are its continuations.

(21) Kanda Shigeru: *Nihon temmon kisho shiryo* (Japanese astronomical and meteorological records) 1947, 160pp.
The author's collected essays on material-gathering.

(22) Kano Kokichi: 'Shizuki Tadao no seikisetsu' (A theory of stellar fluid by Shizuki Tadao) *Toyo gakugei zasshi* no. 165:294–300 (d895).
The author points out that Shizuki Tadao propounded in his commentary on Keill's work the same hypothesis as Kant and Laplace on the formation of the planetary system.

(23) Maeyama Jinro: 'Kansei rekisho oyobi kansei rekisho zokuroku' ('Compendium of the Kansei calendar' and its 'sequel'), *Temmon geppo* 49.4:2 (1956).

(24) *Meijizen Nihon temmongaku shi* (The history of Japanese astronomy before the Meiji era) 1960, 518pp.
I. 'Far Eastern history of astronomy' by Noda Churyo.
II. 'Influence of Western astronomy' by Yabuuti Kiyoshi.
III. 'Calendrical and time systems' by Noda Churyo.
IV. 'History of astronomical observations' by Kanda Shigeru.
Appendex 'Chronological table' by Ohya Shin'ichi.

(25) Mikami Yoshio: *Nihon kagaku no tokushitsu; temmon* (Characteristics of Japanese science; astronomy) in *Toyo shiso no tenkai* (The development of oriental thought) 1936, 67pp.

(26) Mikami Yoshio: *Nihon sokuryo jutsu shi no kenkyu* (A study of the history of the art of land-surveying in Japan) 2nd ed. 1948.
Partly devoted to 'Nihon boenkyo shi' (History of telescope in Japan) 1936.

(27) Mita Hiroo: 'Coperunicus to revolutio' (Copernicus and revolution) *Kobe daigaku kenkyu*, 1961, pp. 1–36.

(28) Nakayama Shigeru: 'Senseijutsu wa temmongaku no hattatsu ni koken shitaka' (Did astrology contribute to the development of astronomy?), *Temmon geppo* 53.5:96–100 (1960).

(29) Nakayama Shigeru: 'Toyo ni kagaku kakumei wa okorietaka – tokuni temmongaku no baai' (The possibility of a 'Scientific Revolution' in the history of the Far East – particularly in case of astronomy) *Kagaku kakumei*, 1961, pp. 165–187.

(30) Nakayama Shigeru: 'Galileo Newton no uchuseisei ron' (Galileo and Newton's problem of cosmogony) *Kagakusi kenkyu* 56:1–7 (1960).

(31) Nishiuchi Masaru: *Shibukawa Harumi no kenkyu* (A study of Shibukawa Harumi) 1940.

A study of the first Japanese calendar reformer in terms of his astronomical and Shintoist background.

(32) Noda Churyo: *Toyo temmongaku shi ronso* (Collected articles on the history of astronomy in the Far East) 1944, 673pp.

This includes most of his main works.

(33) Noda Churyo and Yabuuti Kiyosi: *Kansho ritsureki shi no kenkyu* (A study of 'Lü-li chin' of Han shu) 1947, 300pp.

(34) Ogawa Kiyohiko: 'Shina seiza kanken' (On the traditional Chinese constellations), *Temmon geppo*, 26:105 123(1933), 27:141 168 185 207 221 (1934)

Revision on I. Kögler and G. Schlegel's works on the identification of Chinese constellations, with the Western ones.

(35) Ogawa Kiyohiko: 'Senmyo reki koyo jidai ni okeru suisan to rekijitsu' (Calendrical calculation during *Shun-ming* calendar period) *Temmon geppo*, 35:79–83 93–100 (1942).

Critical evaluation of old Japanese chronology.

(36) Oguwa Kiyohiko: 'Nihon shoki no rekijitsu ni tsuite' (On the calendrical indices of the Nihon shoki) 1946, printed on *Tenkansho* II.

(37) Ogura Shinkichi: 'Wagakuni kodai no nichigetsushoku kiroku' (Ancient eclipses recorded in Japan) *Temmon geppo*, 9:13–18 25–29 39–42 52–55 62–64 (1916).

Records of eclipses from 628 to 1000 A.D. are examined and compared with the results of modern calculation.

(38) Ono Gemmyo: 'Bukkyo temmongaku' (Buddhist astronomy) *Gendai oukkyo*, Vol. 3, nos. 24, 25, 27, 30, 32, 33 (1926).

(39) Ono Kiyoshi: *Temmon yoran* (An outline of ancient astronomy) 1925 193pp. with 14pp. English summary.

Part I. the celestial enclosures and animal figures of stars observed and delineated by ancient oriental astronomers. Part II. on the twenty-eight constellations and the signs of the zodiac; their origin and history.

(40) Osaki Shoji: 'Rekisho shinsho temmei kyuyakubon no hakken' (The discovery of manuscript translations preliminary to Rekisho shinsho), *Kagakusi kenkyu*, no. 4 and 5 (1943).

(41) Otani Ryokichi: *Ino Tadataka*, 1917, 765pp.

I. Tadataka's career. II. his achievement in surveying. III. his teachers, friends and pupils.

(42) Saegusa Hiroto: 'Kako niseiki Nihon ni atta Newton ni tsuite' (Appreciation of Newton in Japan in the past two centuries) *Yokohama daigaku ronso* 10.1:303–363 (1958).

(43) Saito Tsutomu: *Ocho jidai no in'yodo* (The Yin-yang art during the Ocho era) 1915, 209p.

(44) Shimamura Fukutaro: 'Gendai Nihon kagaku gijutsu shi nempyo – temmongaku' (The chronology of the history of science and technology in modern Japan – astronomy), *Kagakusi kenkyu* 16:40–44, 17:29–44 (1950–51)

Correction and supplement by Kanda Shigeru, *Kagakusi kenkyu* 18:48 (1951).

(45) Shinjo Shinzo: *Toyo temmongaku shi kenkyu* (Researches on the history of astronomy in the Far East) 1928, 671pp.

This single volume includes almost all of his main articles which appeared in various journals. It is chiefly devoted to the development of astronomy in China before Christian era.

(46) Sugimoto Isao: 'Mei-ri no tengaku ni tsuite' (On Mei-ri astronomy) *Nihon daigaku bungaku nenpo* 1954.
Mainly on Nishikawa Joken.

(47) Suzuki Takanobu: 'Honpo kodai no nisshoku ni tsuite' (On the ancient Japanese records of eclipses) *Nihon temmongakukai yoho* vol. 6, no. 4, pp. 143–169.
Extended Ogura's previous researches (37) to 1600 A.D.

(48) Tasaka Okimichi: 'Seiyo rekiho no tozen to kaikai rekiho no ummei' (Introduction of European astronomy in China and the Mohammedan calendars) *Toho gakuho* 31.2 141–180 (1947).

(49) Temmon rekizan kenkyushitsu: 'Taisho kaireki to sono rekiho' (The *T'ai-ch'u* calendrical reform and its calendrical system) *Toho gakuho, Kyoto,* 15:3 (1946).

(50) Uchiyama Moritsune: 'Genna kokaisho no sakubo hyo ni tsuite' (On the table of phases of the moon in Genna kokaisho) *Temmon soho* 10.114:47–48 (1956).

(51) Ueda Yutaka: *Sekishi 'Seikyo' no kenkyu* (Investigations on the Star manual of Shin-shen) *Toyo bunko ronso* no. 12, 1929, 184pp.

(52) Watanabe Toshio: *Hazama Shigetomi to sono ikka* (Hazama Shigetomi and his family) 1943 Kyoto, 478pp.
I. Introduction. II. Linealogy of the Hazama family. III. Astronomical observations done by the Hazama family. IV. Instruments for observations.

(53) Yabuuti Kiyosi: 'Sodai no seishuku' (Descriptions of the constellations in the Sung dynasty) *Toho gakuho, Kyoto* 7:42–89 (1936).

(54) Yabuuti Kiyosi: 'To Kaigen senkyo chu no seikyo' (The star catalogue in the K'ai-yuan chang-ching of the T'ang) *Toho gakuho, Kyoto* 8:56 (1937).

(55) Yabuuti Kiyosi: 'Ryokan rekiho ko' (On the calendar reforms of the Former and Later Han dynasties) *Toho gakuho, Kyoto* 11:327–357 (1939).

(56) Yabuuti Kiyosi: *Shina no temmongaku* (Chinese astronomy) 1943, 271pp.
A popular but informative account of the special features of Chinese astronomy.

(57) Yabuuti Kiyosi: 'Toso rekiho shi' (Calendrical science in the T'ang and Sung dynasties) *Toho gakuho, Kyoto,* 13:491–528, (1943).

(58) Yabuuti Kiyosi: *Zuito rekihoshi no kenkyu* (Researches on the calendrical science of the Sui and T'ang periods) 1943, 260pp.
The author discusses basic astronomical elements, the theories of the sun and moon's orbits and the theory of eclipse prediction. A study of Indian calendar is added. His 'Inshu yori Zui ni itaru Shina rekiho shi' (Calendrical sciences from the Yin and Chou periods to the Sui dynasty) *Toho gakuho Kyoto* 12:99–139 (1941) is appendixed in the same volume.

(59) Yabuuti Kiyosi: 'Genmin rekiho shi' (Calendrical science during the Yuan and Ming dynasties) *Toho gakuho, Kyoto* 13:491 (1944).

(60) Yabuuti Kiyosi: 'Seiyo temmongaku no tozen, Shindai no rekiho' (Introduction of Western astronomy into the East – calendrical science in the Ch'ing period) *Toho gakuho, Kyoto,* 15:133–154 (1946).

(61) Yabuuti Kiyosi: 'Chugoku ni okeru Isuramu temmongaku' (The introduction of Islamic astronomy to China) *Toho gakuho, Kyoto* 19:65 (1950).

(62) Yabuuti Kiyosi: 'Kinsei Chugoku ni tsutaerareta Seiyo temmongaku' (The introduction of Western astronomy into Ch'ing China) *Kagakusi kenkyu* 32 15–18 (1955).

(63) Yabuuti Kiyosi: 'Chugoku temmongaku ni okeru gosei undo ron' (The theories of planetary motions in Chinese astronomy) *Toho gakuho* 26:90–103 (1956).
A provisional summing-up of his research on Chinese planetary theory, hitherto an ignored subject.

(64) Yabuuti Kiyosi: 'Kandai ni okeru kansoku gijutsu to Sekishi Seikei no seiritsu' Observational techniques during the Han dynasty and the formation of the Star manual of Shih-shen) *Toho gakuho* 30:1–38 (1959).

(65) Yabuuti Kiyosi: 'Todai ni okeru Seiho temmongaku ni kansuru nisan no mondai' (Several topics concerning the influence of Western astronomy in the T'ang China) *Tsukamoto hakushi shoju kinen bukkyo shigaku ronso* 1961 pp.883–894.

(66) Yoshida Mitsukuni: 'Kongi to konsho' (Celestial globes and armillary spheres in China) *Kyoto daigaku jimbun kagaku kenkyusho soritsu nijunen kinen ronbun shu* 1954, pp.331–348.

(67) Zemba Amane: 'Nijuhasshuku to Beda seiritsu no nendai' (The twenty-eight mansions and Vedic period) *Toho gakuho, Kyoto*, 13:36 (1943).

(68) Zemba Amane: 'Matoga kyo no temmon rekisu ni tsuite' (On astronomical and calendrical problems, which appeared in Sardulakarnavadana) *Toyogaku ronso* pp.171–213.

(69) Zemba Amane: 'Taishukyo no temmon kiji' (Astronomical records in Ta-chi ching) *Nihon bukkyokai nenpo* No. 22:101–116.

(70) Zemba Amane: 'Butten no temmon rekiho ni tsuite' (Astronomy and calendrical science in Buddhist sutras) *Indogaku bukkyogaku kenkyu* 4.1:18–27 (1956).

First published in *Japanese Studies in the History of Science*, No.3, pp.60-67, The History of Science Society of Japan, 1964

❹ Abhorrence of 'God' in the Introduction of Copernicanism into Japan

In my previous article entitled 'On the Introduction of the Heliocentric System into Japan',[1] I only briefly mentioned the first manuscript translation to refer to the heliocentric theory, the 'Oranda chikyū zusetsu' (1772). Recently, I had opportunities to examine its French and Dutch originals and a variant manuscript copy. Their comparison revealed some interesting aspects worth reporting in the following.

COMPARISON OF TEXTS

In treating the problem, the following four sources are examined:

1. Louis Renard: *Atlas de la navigation et du commerce*, Amsterdam, 1715, 95pp., 35 × 54 cm, abbreviated hereafter 'French edition'.
2. Louis Renard: *Atlas van Zeevaert en Koophandel*, Amsterdam, 1745, 152pp., 36 × 57 cm, abb. 'Dutch edition'.
3. Motoki Ryōei[a] (tr.): 'Oranda chikyū setsu'[b] (Dutchmen's view of the earth), drafted in 1772 (the 12th month of the 8th year of Meiwa), reserved in the Tenri Library under the series title of 'Tenmon hisho'[c] (secret books on astronomy), 22 sheets, abb. 'Tenrī manuscript'.
4. Motoki Ryōei (tr.): 'Oranda chikyū zusetsu'[d] (Dutchmen's view of the earth, illustrated), drafted in 1773 (the 12th month of the first year of An'ei), preserved in the Nagasaki City Museum, 90 sheets divided into three volumes, abb. 'Nagasaki manuscript'.

Both the French and Dutch editions are large-scale marine charts with a guide for seamen. The title page of the Dutch edition can be translated into English as follows:

> Atlas of shipping and trade throughout the entire world showing in separate successive maps, all of its ocean coastliners and harbours; with descriptions of the nature, products, handicrafts, religions, form of government and trade of all regions, etc. Formerly published in the French language by Mr Louis Renard, agent of His Majesty of Great Britain and now in the most exact way, with the help of prominent experts, all maps have been revised in accordance with the latest discoveries, and a large number of places as well as a number of new maps have been included by Reinier and Iosua Ottens. The descriptions have been enlarged by many necessary matters, and

> placed in an up-to-date order with the new maps. Descriptions have been added and enriched with scientific remarks on astronomy, geography, the oceans and winds, etc. by Jan van den Bosch Melchiorsz. At Amsterdam by Reinier and Iosua Ottens. MDCCXLV.

While the original French edition only provides explanations for charts, the Dutch edition incorporates a number of revisions. It also has at its beginning some additional elementary accounts of such matters as the earth, the heliocentric system, constellations, longitude and latitude and other astronomical subjects, wind and geography, the history of astronomy and geography, the use of the compass. These additions are considered to have been contributed by Melchiorsz.

It was Melchiorsz' additions in which the Japanese translator was chiefly interested. The Tenri manuscript has a summary of the 'Preface' (from which could be identified its Dutch original) and a translation of the first five pages. The Nagasaki manuscript, while translating the first sixteen pages, neglected the 'Preface'. The latter translation, being the revision of the former, is in somewhat better Japanese.

A comparison with the Dutch original shows that the translator has omitted certain paragraphs in his translations. The details of omissions are as follows. Both translations omit the opening paragraph of the first page of the Dutch original. This common omission is indicated by (A) in the diagram given below. The Nagasaki manuscript omits material amounting to almost one page on the pp. 8–9 in the Dutch original. This omission is indicated (B) in the diagram.

Dutch edition	'P'	1	5	8 9	16 (page)
Tenri ms.	—	———	———		
Nagasaki ms.	(A)	———	———	(B)	———

Omission (A) is translated as follows:

> The place which God has given people to be inhabited is commonly called the world, the Earth sphere, the inhabited earth; the name world implies the entire created world, the universe; or, to be more precise, everything that is within the circle of the planet Saturn, of which circle the Sun is the centre, and the radius of which extends from the Sun to Saturn. The Earth sphere is composed of earth and ocean, of land and water. It contains many different kinds of materials which God has created for the benefit of man, the quantity of which He has decided upon so that nothing could be added to or taken away from in neither by the human mind nor by natural causes, for the materials by their nature are essentially distinct from one another, indestructible, unincreasable and unchangeable. These are the essentials or main materials which are called single bodies, because no composition is contained therein, such as the pure metals, the water, the sand, the light and many similar ones; they are called single bodies, because no other body is known which goes into their composition; and because they even occur in the formation of compound bodies such as most minerals, stones and other mountain materials, which are not born from seed, but which are formed

> themselves from the composition or combination of fewer or more single bodies; they also contribute to the growth of movable body, such as bodies of people, animals, plants, etc. Those who have been born from seeds in which they have been enclosed in small shape, increase, being nourished, and also are destroyed through decomposition of the parts from which they are made up. These are three kinds of bodies which we meet on the earth; and because of the difference of these principles, the onwards appearance of the earth does change, but because of the immutability of Nature and the number of these principles, God prevents the destruction of the earth; and bodies changes which occured therein; in such a way that the earth although constantly changing, yet always remains same.

The omission is apparently deliberate, motivated by the translator's abhorrence of the account of God's Creation. This is clear from his full and faithful translation of the purely scientific account of the heliocentric hypothesis, which immediately follows the omitted paragraph in the Dutch original. Of course, the Dutch account itself was only a short account of the daily rotation and annual revolution and the difference between true (heliocentric) and apparent (geocentric) motions.

The Dutch original gave a brief history of astronomy up to the time of Copernicus and then proceeds to present the discussion involving Biblical issues caused by Copernican theory. The Nagasaki manuscript translated the historical account up to Copernicus. It was in this connection that the name of Copernicus and the information that Copernicus' heliocentricism is now generally accepted among Western scholars appeared for the first time in Japanese literature.

However, the translator left out all the theological argumentation. This omitted portion is (B) translated below:

> However, lower classes, or too prejudiced (was this the case among only lower classes?), find this feeling strange, which seems to them to be contrary to their sight; apart from the fact that they have peculiar ideas about this rotation. For their benefit, I think it will be neither unpleasant nor useless to bring some clarification in this matter in this shortest possible way, and to give them a certain understanding in this matter by means of comparisons; for which purpose, I shall follow the reasoning of the learned men who has done this in different language. It is an mutable law of nature that we see the objects, the images of which are displaced in our eyes without turning our eyes or head, moving at changing places. It is similarly another law of nature, which is in complete agreement with the previous law, that the objects seem immobile to us when their images remain in our eyes in the same places of our retina without changing. That is why when one is sitting in a boat, one does not see it move. And it seems to remain in one place although it is advancing all the time, because all its parts amongst one and another as well as its relation to us remain in the same configuration and the image of which, as a result, does not move in our eyes. But exactly the opposite takes place with the object which we, while sitting in a boat, meet outside of the boat. Let a yacht take you along a river; ships, buildings, churches, towers, cities, the warehouses of admiralty, trees, which are here and there along the bank of the river, move their images in our eyes, and somehow replace each other from one point to another as the yacht brings us closer to these objects or past them. And so from this movement of the

images of the objects in our eyes, one would necessarily conclude that we consider that the objects coincide with images as being factually in motion; and that they are in reality in motion; and so we see ships, cities, etc. moving past on; the warehouse approaches us; coming to our side and going away from us; whereas it is we who are leaving these places. We are sailing out of the harbour, and land and cities seem to trot away. Let us apply this observation to the entire Nature itself.

Even though it had pleased the Creator of all things to make the Earth and other planets turn in the period of several months around the Sun, and each of them in the period of several hours around their own axis, instead of making the Sun and stars and the immeasurable sky turn with an utterly incomprehensible speed around the Earth for the Earth's benefit, the latter being in companion hardly a single tiny speck, we still would see everything in the same form as we see it now. In the first case the arrangement is very simple and the result so truly perfect. The stars and the Sun, though remaining in the same place and never leaving it, would seem to us to rise, go down and finally to hide themselves. The Earth, in spite of describing daily a good portion of a large circle around the Sun, and apart from it making a complete rotation around its own axis from 24 hours to 24 hours, would appear to us to be immobile. That this should be the case, is beyond any doubt; because all the points, which we observe an Earth, are always in the same configuration, both with relation to one another and to ourselves, and the images thereof which are reproduced on our retina would never be displaced or go over from one point to the other. But, contrariwise, the Sun, the planets and the stars would seem to us constantly to rise and go down, depending on whether their image was formed in the upper or the lower part of the retina. Similarly, the planets, having a special course, whereas our Earth also has its own course, would seem to us to have a special motion, although in fact they have a very uniform motion.

However, we shall not go into it any further in order not to digress from the main subject, considering that this is enough to show that nothing more simple and comprehensive can be understood than this explanation of Copernicus; and that only ignorant prejudices and some passages of the Holy Scriptures have caused this explanation to be discarded as being contrary to common sense. It goes without saying that one should have the highest respect for the Word of the Living God. Never can one think too highly of his work. But He is a God of orderliness and not of confusion; both in the work of creation as in that of re-creation. He himself has taught us through Moses more true science than all the philosophers of all centuries taken together could possibly have done. But in speaking to man. He wanted to be understood (and) the passages with relation to the sky-movement are based on the apparent movements; it was not the purpose in such cases to instruct us in astronomy. How strange it would have sounded if God had said that the Earth is rising or going down; Joshua when praying 'Thou Sun remain in place at Gibeon and the Moon in the valley of Ajalons' prayed for the light of the Sun and the Moon in order to pursue his enemies (and) that he should have this light and not his enemies. He therefore did not pray for the Bodies of the Sun and the Moon to remain in one place, because that would have suited his enemies as well as him, and apart from this it would have caused a disruption in the whole of nature as a good observer may easily understand. One finds again elsewhere an expres-

> sion in accordance with the seeming movement, as the Disciple says 'the land came nearer'. It is, however, certain that it was not the land which came closer to the ship, but the ship to the land. So there is nothing in the Holy Scriptures which could serve as an argument against this explanation of Copernicus.
>
> From the times mentioned, the research on heaven and geography, on winds and low and high tides, on the science of measuring and working of the compass has become lively and became accepted by rulers and people, who became enriched through this. That which in these times caused the perfection of shipping, and as a result, big improvements in the knowledge of geography, was the invention of the compass.

The heliocentric account terminates at this point and then the text goes on to discuss the use of compass. Hereafter, the translator resumed his faithful translation again.

Remarks

From the latter half of the eighteenth century, the attitude of the Japanese government towards the Western knowledge became somewhat relaxed and the study of Western science through Dutch sources began. However, any kind of belief in Christianity was still strictly forbidden and even a mentioning of it was taboo. The translator, who was the best linguist of the day, was an official interpreter for the Dutch trade. He did his translations on official request. Even so, this timid official was reluctant to bring in Christianity in any fashion whatsoever. He was afraid lest his hereditary position be jeopardized. However, he felt safe in translating the purely technical matters of heliocentricism.

Motoki Ryōei's translation works on Copernicanism were three. Namely:

1. 'Oranda chikyū setsu' (1772–3) from *Atlas van Zeevaert* (1745)
2. 'Tenchi nikyū yōhō' (1774) from W. J. Blaeu: *Tweevoudigh Onderwiis van de Hemelsche en Aardsche Globen* (Amsterdam, 1666)[2]
3. 'Shinsei tenchi nikyū yōhō ki' (1792–3) from George Adams; *Gronden der Starrekunde* (Amsterdam, 1770, its ultimate English original: *Describing and Explaining the Construction and Use of New Celestial and Terrestrial Globes*, 1766)[3]

Published in the liberal Netherlands, these three Dutch editions had no prejudice against the Copernican theory. The first one was written in the mid-eighteenth century, when the Copernican theory was already established. Yet, some past theological discussion was added in order to meet popular interest. Most likely, the translator learned from it the intimate relationship between Copernicus and Christian God, and took special pains to introduce only heliocentricism.

At the time of the publication of the second work of Blaeu's book, the first edition of which appeared as early as 1620, the Copernican controversy was still raging; and so naturally it was full of theological arguments. Blaeu, who was one of the early proponents of the Copernican system, divided his work into two volumes with the intention of propagating the new theory. The first volume was devoted to the now inadequate hypothesis of Ptolemy and the second to

Copernicus. The translator, perhaps influenced by his own earlier experience of translation, replaced Blaeu's long theology-filled preface with his own brief summary of purely physical nature and left entirely untranslated the second volume on Copernican system.

The third, Adams' work, was already free of time-honoured religio-cosmological controversy; its straightforwardly scientific arrangement seemed to have appealed to the translator. This work was fully translated and a truly comprehensive account of the Copernican system, for the first time, became available in Japan.

Now at last emancipated from the theological implications and also from the governmental suspicion, the versions of heliocentricism by Shiba Kōkan[e] and other popularizers, based on Motoki's translation of Adams, could be printed without trouble and achieved considerable diffusion among the populace.

All quotations are kindly translated from the Dutch for me by Mrs Helena De Vries.

REFERENCE

(1) *Scientific Papers of the College of General Education, University of Tokyo*, Vol. 11, No. 1, pp. 163–176 (1961)

(2) *Ibid*, pp. 168–170

(3) *Ibid*, pp. 170–171

(a) 本木良永
(b) 和蘭地球説
(c) 天文秘書
(d) 和蘭地球図説
(e) 司馬江漢

First published in *Japanese Studies in the History of Science*, No.3, pp.68-80, The History of Science Society of Japan, 1964

❺ Cyclic Variation of Astronomical Parameters and the Revival of Trepidation in Japan

It is an interesting episode on the history of astronomy that in the late 18th century Japan the idea of cyclic variation of astronomical parameters appeared independently of its Western counterpart and the concept of trepidation was revived to account for it. This is, perhaps, the only original achievement in the entire history of Japanese astronomy, and thus merits critical examination.

CHINESE ACCOUNT FOR THE VARIATION OF TROPICAL YEAR LENGTH

As explained in my previous paper,[1] astronomers at the time of the *Shou-shih*[(a)] calendar (1281) were, in adopting the idea of *hsiao-ch'ang*[(b)] (secular variation of tropical year length), accounting only for the ancient Chinese data. The Chinese and the Jesuits in China did not, so far as is known, challenge *hsiao-ch'ang* concept, although neither the *Ta-t'ung*[(c)] (1368) nor *Shih-hsien*[(d)] (the Jesuit calendar, 1644) calendars adopted it. In fact, the *Ch'ung-chen li-shu*[(e)] (the Jesuit astronomical encyclopedia, 1635) pointed out three possible causes of variation in tropical year length:[2] 1. rotation of the centre of the solar orbit with respect to the earth (perhaps referring to the progressive motion of the solar perigee), 2. variation of the eccentricity of the solar orbit, 3. variable precession (trepidation). They refrained, however, from giving any numerical values because the magnitude of such a minute parameter was not determinable within a single life-span.

During the seventeenth century in China, perhaps stimulated by the Jesuits' geometrical astronomy, the Chinese adherents of *hsiao-ch'ang* began to debate on its theoretical foundation. Wang Hsi-shan[(f)] attributed its cause to a decrease of ecentricity in the solar orbit.[3]

Mei Wen-ting,[(g)] in his *Li-hsueh i-wen*[(h)] (Queries on calendrical science, 1693), made a significant attempt to provide a theoretical account of *hsiao-ch'ang*. Skeptical of the idea of secular diminution in tropical year length, Mei considered it to be subject to periodical change like other celestial parameters. From solstitial gnomon observations, he determined that the length of the tropical year was slightly greater than the value given in the *Shou-shih* calendar.[4] He also noted that at the time of the *Shou-shih* calendar, the solar perigee was nearly coincident with the winter solstice. Since the anomalistic year was not yet distinguished from the sidereal year, he inferred from these two facts that tropical year length undergoes periodical variation in accordance with the precession cycle, with a minimum at the solar perigee.[5]

Mei's argument was not rigorous, but merely based on the inference that variation of tropical year length is caused by the movement of the solar perigee. His conclusion is the opposite of the actual state of affairs; tropical year length reached a maximum in Kuo Shou-ching's time (see my previous paper).[1]

CYCLIC VARIATION OF ASTRONOMICAL PARAMETERS

A Japanese astronomer, Asada Gōryū[(i)] (1734–1799) seems to have been influenced by Mei Wen-ting in the formulation of his own idea of variation of tropical year length. Gōryū introduced the idea of periodic variation of astronomical parameters in a precession cycle of 25,400 years.

Classical Western data, i.e. those listed in the *Almagest* of Ptolemy, became available to Gōryū through the Sino-Jesuit astronomical compendium, *Ch'ung-chen li-shu*. Since Gōryū was not informed of the rapid progress in telescopic observations and celestial mechanics that had been made in the West, he was not overawed by Western authority. Thus, he dared to endeavour to synthesize both Western and Chinese astronomies and to give a numerical account explaining by means of a single principle all the observational data available to him, old or new, Western or Eastern.

It seems that Gōryū did not fully comprehend the epicyclic system which appeared in the Sino-Jesuit works based on Tycho Brahe. The only feature of Western astronomy which interested him was observed data and numerical parameters, which he could utilize for his purely traditional approach, i.e. to obtain an algebraic representation which corresponded to observed phenomena as closely as possible.

Copernicus, who appears in the *Ch'ung-chen li-shu* not as an advocate of heliocentricism but as an observational astronomer and the inventor of the eighth sphere of trepidation,[6] is said in that work to have believed that the ancient tropical year length was longer than the medieval one, which was in turn shorter than the contemporary one.[7] Gōryū was apparently struck by this passage in the *Ch'ung-chen li-shu*, and formulated on this basis his idea of a modified *hsiao-ch'ang*, namely that the ancient tropical year length tended to decrease until it reached its minimum in medieval times and that it has been growing longer ever since. Unlike Mei's semi-theoretical inference, Gōryū's approach was totally empirical; the time of minimum tropical year length was not associated with the solar perigee but arbitrarily chosen in order to fit with recorded data.

He also presumed that the only perpetual constant was the length of the anomalistic (or sidereal) year. Other basic parameters, such as the lengths of the synodic, nodical and anomalistic months, were all subject to variation in a precession cycle of 25,400 years.[8] In the West, the first systematic study of the variation of basic astronomical parameters were carried out by Simon Laplace on the basis of the perturbation theory. Though apparently similar, Gōryū's approach was by no means comparable to Laplace's well-founded theoretical considerations.

To begin with, Gōryū selected three values of the tropical year length at different times, as follows:

value (days)		*years*	
T_o:	365.241 620 438 5	E_o:	133 A.D.
T':	365.242 341 541 48	E':	1787 A.D.
T_2:	365.250 469 717 756	E_2:	720 B.C.

These three values are the basic data supporting his *hsiao-ch'ang* law. Of course it is impossible to obtain, by observation alone, such a high order of accuracy in measuring tropical year length. These values must have been based partly on calculation, but the method is now unknown to us.

Given these data, a contemporary scientist might, first of all, try to connect them with a trigonometric curve. He would not automatically assume that the smallest value given was in fact the minimum value. Gōryū considered, however, the T_o value as definitely the minimum. He was therefore required to deal with the states before and after according to entirely different formulae.

If we assume that the present rate of increase of the tropical year length continues until the half precession cycle t_0 (beginning from E_2) is completed, we may extend the $T_o - T'$ line on fig. 1 to T_1.

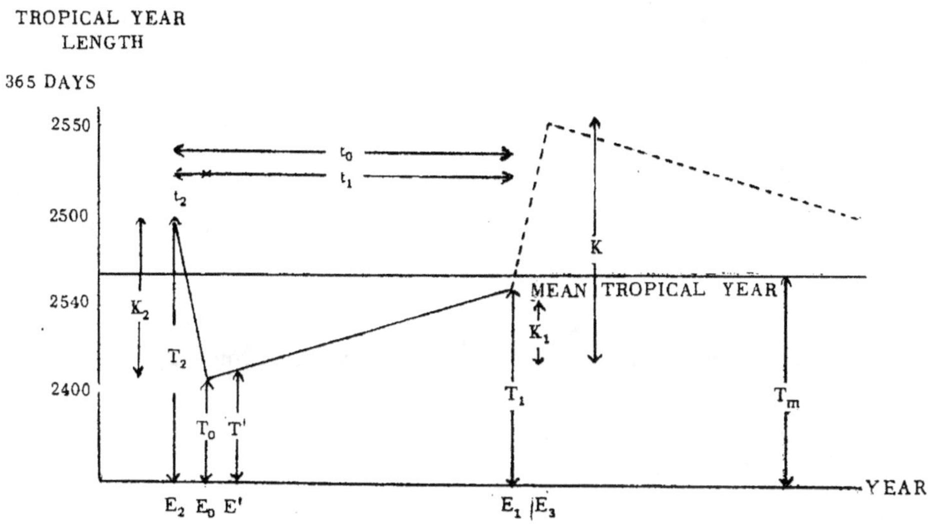

Figure 1

Let us define K_1 and K_2 as follows:

$$T_1 - T_0 = K_1 \qquad (1)$$

$$T_2 - T_0 = K_2 \qquad (2)$$

$$K_1 + K_2 = K \qquad (3)$$

K is the difference between the maximum and minimum values of the tropical year length. The mean tropical year length T_{m} is given:

$$T_0 + \frac{K}{2} = T_m \tag{4}$$

For the convention of calendrical calculation, the Chinese employed, as one of the most fundamental parameters, the accumulated days, P, since epoch (in this case Eo). Then,

$$P = T_m t + aA - bB - C \tag{5}$$

where a and b are constants. Let us define A and B as follows:

$$\left.\begin{aligned} A &= (t_2 + t)\ \{(t_0 - (t_2 + t)\} = (t_2 + t)\ (t - t) \\ B &= t(t_0 - t) \end{aligned}\right\} \tag{6}$$

where t_2 is the number of years elapsed between E_2 and E_0, and t_1 is the number of years between E_2 and E_1. C is another constant, equivalent to at_1t_2.

a and b are defined as follows:

$$a = \frac{\beta - a}{2} \tag{7}$$

$$b = \frac{\beta + a}{2} \tag{8}$$

where

$$a = \frac{K_1}{2t_1} \tag{9}$$

$$\beta = \frac{K_2}{2t_2} \tag{10}$$

When we differentiate P with respect to t, tropical year length T_t in the year t is obtained, as follows:

$$\begin{aligned} T_t = \frac{dP}{dt} &= T_m + a\frac{dA}{dt} - b\frac{dB}{dt} \\ &= T_m + a\ (t_1 - t_2 - 2t) - b\ (t_0 - 2t) \\ &= T_m + \frac{\beta - a}{2}\ (t_1 - t_2) - 2t) - \frac{\beta + a}{2}\ (t_1 + t_2 - 2t) \\ &= T_m + t_1\left(\frac{\beta - a}{2} - \frac{\beta + a}{2}\right) - t_2\left(\frac{\beta - a}{2} + \frac{\beta + a}{2}\right) - 2t\left(\frac{\beta - a}{2} - \frac{\beta + a}{2}\right) \\ &= T_m - 2a_1 - \beta t_2 + 2at \end{aligned}$$

$$= T_m - \frac{K_1 + K_2}{2} + 2at$$

$$= T_0 + 2at \qquad (11)$$

expressed numerically as:

$$T_t = 365.241\,620\,44 + 0.435\,370 \times 10^{(-6)}t \qquad (12)$$

This equation is nothing but a representation of the $T_0 - T_1$ line.

Why, then, did Gōryū introduce a complicated relationship even though the β term is cancelled out in the final formula? The answer must lie in his apparent intent to express all phases of cyclic variation in a single formula.

Equation (12) holds only between E_0 and E_1. In order to extend this formula to other phases of variation, we need some slight modifications.

When we deal with the phase E_2 E_0, we use A' instead of A, which is defined as follows, reversing t_2 and t_1 in formula (6).

$$A' = (t_1 + t)\,(t_2 - t) \qquad (13)$$

When we further adjust the signs of a and b according to the following table, we are able to calculate the tropical year length of all phases on the basis of basic equation (6). In fig. 1, signs are applied as follows:

	E_0~E_2	E_0~E_3	E_3~E_1	E_3~E_4 (E_2)
aA	−	+	+	−
bB	−	−	+	+
	A'	A	A'	A

For the synodic month only the B term, and for the nodical and anomalistic months only the A term, were employed. The obliquity of the ecliptic was also subject to yearly variation.

Applying Gōryū's formula to historical observations, we see in figure 2 how Gōryū's formula reconciles the data represented on fig. 5 in my previous paper[1] After E_0 the formulae of Newcomb, Gōryū and the *Shou-shih* calendar roughly coincide. Before E_0 Gōryū's formula appears as a parabola of deep curvature, which accounts for Greek observations as well as the three ancient Chinese records.

It is apparent, then, that what Gōryū really intended to do was to account for the newly acquired Western data. His basic goal, that of 'saving the ancient records' differs not at all from that of the traditional calendrical astronomy. His consideration of the precession cycle was a mere theoretical disguise.

THE REVIVAL OF TREPIDATION

In spite of conservative resistance from the hereditary official astronomers, Gōryū's pupils finally succeeded in applying Gōryū's variation term to the *Kansei*[j] calendar (1798). Takahashi Yoshitoki[k] (1764–1804) wrote the *Zōshū shōchō* hō" (Hsiao-ch'ang method, revised and augmented, 1798) in order to

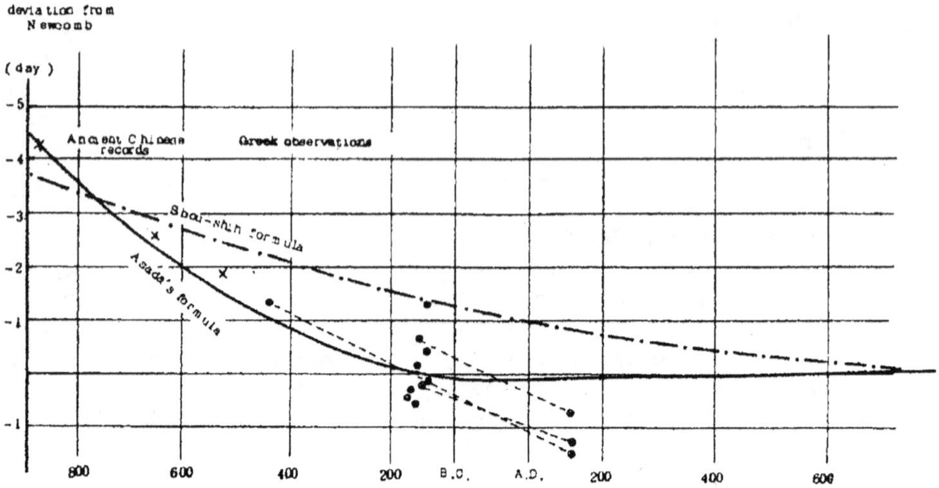

Figure 2

provide a theoretical foundation for his teacher Gōryū's method. Although at that time Yoshitoki probably did not have access to works on Western astronomy unavailable to Gōryū, he had attained a mastery of the contents of the *Ch'ung-chen li-shu*, particularly the theory of spherical geometry and epicycles. Furthermore, rejecting the authority of Tycho, he revived the old idea of trepidation as contained in the *Ch'ung-chen li-shu*, in which trepidation was somewhat vaguely mentioned in order to contrast is to the more accurate view of Tycho. Unlike Alphonsine trepidation, which has a 7,000 year cycle, however, Yoshitoki's cycle of trepidation has the same period as the cycle of precession.

He posited an epicyclic system as illustrated in fig. 3.[9] The deferent, represents precessional motion as a clockwise rotation of the mean equatorial pole about the ecliptic pole K with constant angular velocity (ω = 0° .009 875 per year). In the first and second epicycles representing trepidation, DE and PF have the same radius (*a* = 2°, .847 8) and rotate in opposite senses, the first clockwise and the second counter-clockwise. The angular velocity of the first is ω = 0° .018 75 and that of the second 2ω. P, the true equatorial pole, is situated on the second epicycle. At the epochal time, *P* coincides on the second epicycle with the mean pole at *C*.

The position of the true ecliptic pole is also subject to cyclic variation. The mean ecliptic pole is always at *K*. Thus, the mean obliquity, the distance between the mean equatorial pole and mean ecliptic pole, is arc *CK*, which is 22° .522 34. The true distance between the two poles is accordingly subject to cyclic variation. The deferent and epicycle of the ecliptic pole have equal radii, b = 0° .778 9. The angular velocities of their rotation are ω and 2ω, the first clockwise and the second counter-clockwise. At the epoch, the centre of the epicycle *R* makes an initial angle of deviation $<SKR = \phi = 29°.32595$. The true ecliptic pole is at S on the extension of arc *PK*, making $< SQR = 2\phi$.

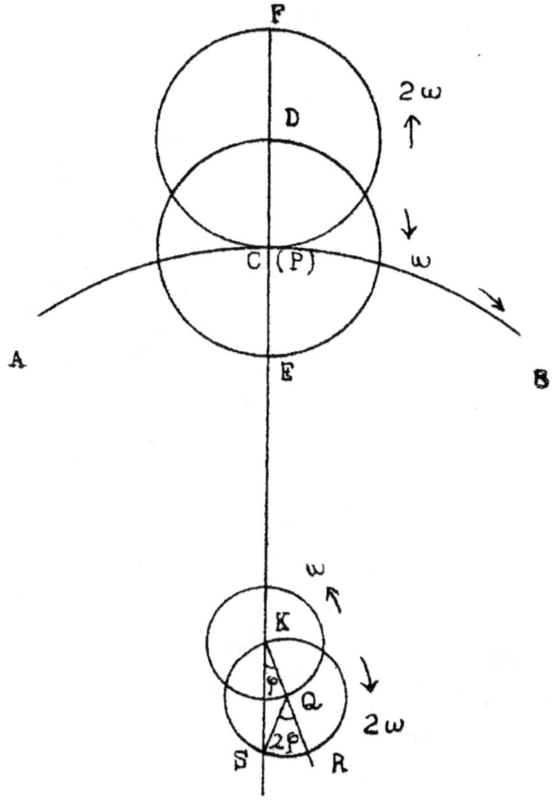

Figure 3

Now, let fig. 4 represent the positions of S and P at a time t after the epoch. C moves to C' and D moves to D'. D' rotates through angle ωt around C' to D''. Equatorial pole P moves to C' and then reaches P' by rotating 2 ωt degrees around D''. While K remains unmoved, Q likewise moves to Q', R to R' and S to S'.

Let us draw a perpendicular from D'' on $D'C'$ and let its foot be Y. Then for spherical triangle $D''YC'$,

$$\frac{\sin Y}{\sin D''C'} = \frac{\sin C'}{\sin D''Y} \tag{14}$$

Namely,

$$\frac{1}{\sin a} = \frac{\sin \omega t}{\sin D''Y} \tag{15}$$

$$\therefore \sin D''Y = \sin \omega t \sin a \tag{16}$$

and

$$arc\ C'\mathrm{P} = 2\ arc\ D''\ \mathrm{Y} \tag{17}$$

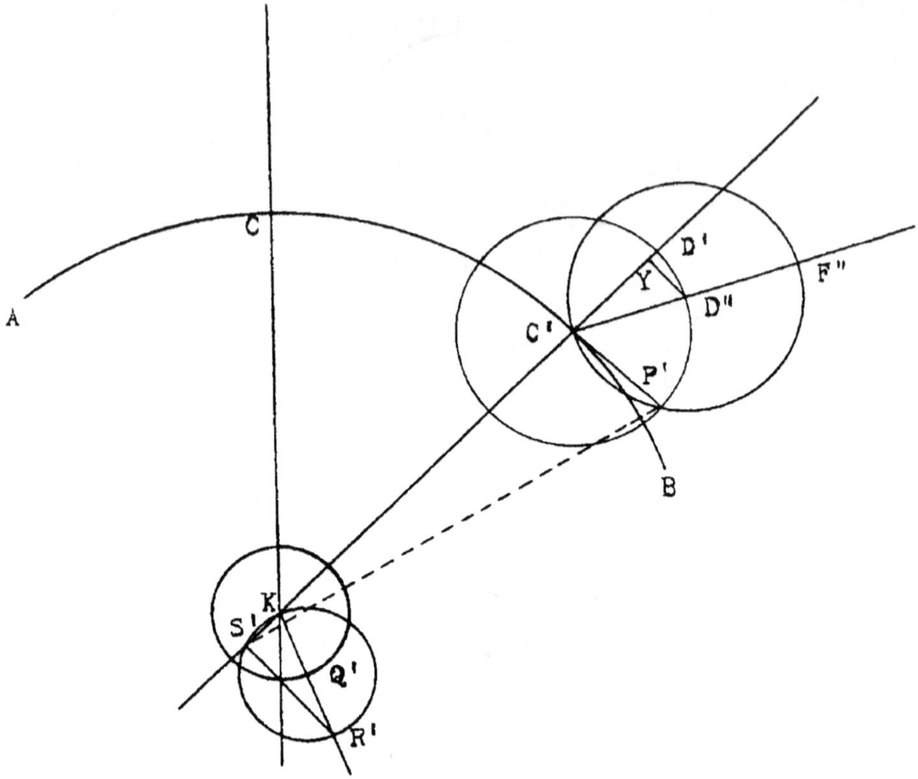

Figure 4

Likewise, for spherical triangle $KR'S'$

$$\tan KS' = \tan KR' \cos K \tag{18}$$

b is small, and hence both *arc* KS' and *arc* KR' are also small. Thus, we may state approximately that:

$$KS' = KR' \cos K \tag{19}$$

$$\therefore\ KS' = 2b \cos(\omega t + \varphi) \tag{20}$$

Next, we consider spherical triangle $C'P'S'$. Then:

$$\tan S' \sin C' = \tan C' P' \tag{21}$$

$$\tan S' = \frac{\tan[2 \sin^{-1}(\sin a \sin \omega t)]}{\sin[i + 2b \cos(\varphi + \omega t)]} \tag{22}$$

Angle S' is the deviation from the mean equatorial pole due to trepidation effect.

Again, for spherical triangle $C'S'P'$, the variation of the obliquity is given by

$$\tan C'P' = \tan C'S' \cos S' \tag{23}$$

S' divided by the mean solar motion (more exactly the true solar motion near the winter solstice) gives the deviation of the date of the true winter solstice from the mean winter solstice (fig. 5). Differentiating further with respect to t, we obtain the difference between true and mean tropical year length due to trepidation effect. This is shown in fig. 6.

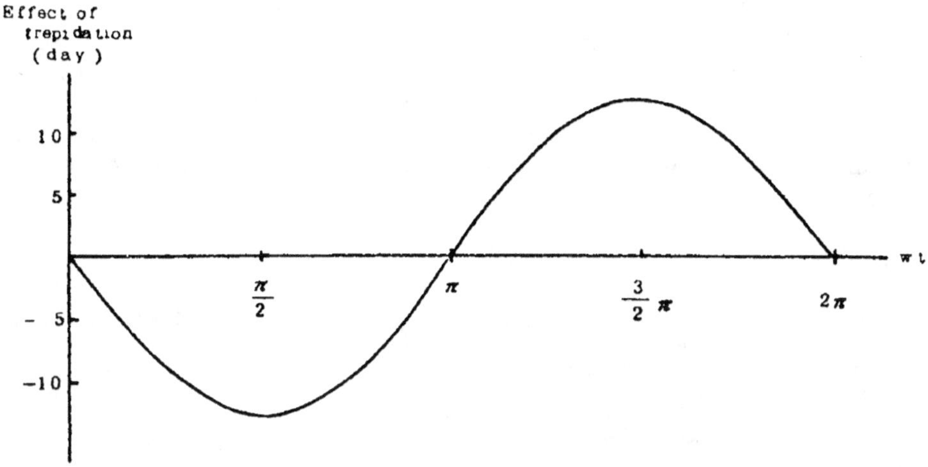

Figure 5

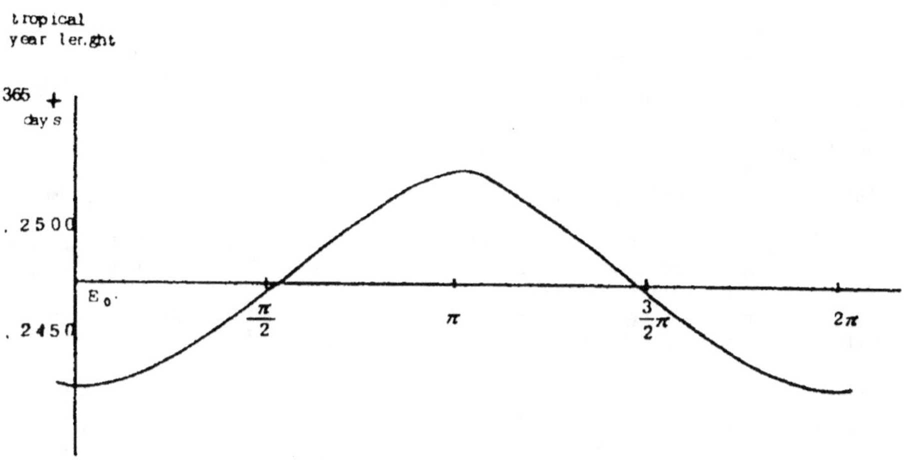

Figure 6

Comparing fig. 1 and fig. 6, we notice a marked dissimilarity. Gōryū's variation formula was empirically induced; Yoshitoki, whose sole purpose was to furnish a theoretical basis for his teacher's idea, did not find it necessary to maintain quantitative agreement with it.

The falsity of Gōryū's variation concept was apparent to Shibukawa Kagesuke,[(m)] Yoshitoki's son, who mastered the rudiments of eighteenth-century

European astronomy through study of Lalande's *Astronomie*. He neglected to adopt Gōryū's idea in the next calendar reform, that of Tenpo[(n)] (1843) and reviewed it critically in compiling the 'Kansei rekisho'[(o)] (Treatise of the Kansei calendar, 1844).[10]

During the Tokugawa period, Japanese astronomers were continually preoccupied by the contrast between Chinese and Western astronomies, and tried to adopt whichever was preferable. While they generally followed Chinese astronomy in the first half of the period, Western astronomy became dominant during the latter half. During the period of transition, there appeared mental attitudes like that of Gōryū who tended to syncretize and synthesize Chinese and Western astronomy. His false originality, however, proved to be rather anachronistic in view of the rapid contemporary development of Western astronomy.[11]

NOTES

1. Shigeru Nakayama 'Accuracy of Pre-Modern Determinations of Tropical Year Length' *Japanese Studies in the History of Science* No. 2, (1963) pp. 101–118.
2. 'Heng-hsing li-chih' 恒星曆指 (Guide to stellar astronomy) in *Ch'ung-chen li-shu* 崇禎曆書 (The Jesuit astronomical encyclopedia, 1635).
3. See, Nakayama Shigeru 中山茂 'Shōchō hō no kenkyū (2)' 消長法の研究 (Variation of tropical year length in the Far Eastern astronomy and its observational basis (II) *Kagakushi kenkyu* 科学史研究 1963: 128–129.
4. See, Mei Wen-ting 梅文鼎 'Sui-chou ti-tu ho-k'ao 歳周地度合考 (On tropical year and latitude) in *Li-suan chuan-shu* 曆算全書 (Complete works on calendar and mathematics, 1723).
5. Mei Wen-ting *Li-hsueh i-wen* 曆学疑問 (Queries on calendrical science, 1693) chapter 2.
6. 'Wu-wei li-chih'. 五緯曆指 (Guide to planetary astronomy) in *Ch'ung-chen li-shu*. Copernicus is not the inventor of the eighth sphere, which goes back to Alphonsine astronomers of the thirteenth century. The erroneous statement of the Jesuit astronomer, Christopher Clavius, attributing the eighth sphere to Copernicus was probably the source of this belief.
7. 'Heng-hsing li-chih'.
8. Asada Gōryū 麻田剛立 'Rekihō shōchō jutsu' 曆法消長術 (The hsiao-ch'ang method in calendar-making, 1788, ms.).
9. Takahashi Yoshitoki 高橋至時 'Zōshū shōchō hō' 増修消長法 (Hsiao-ch'ang method, revised and augmented, 1798, ms.).
10. Shibukawa Kagesuke 渋川景佑 'Kansei rekisho, shōchō hō genri' 寛政曆書消長法原理 (Treatise of the Kansei calendar, principles of Hsiao-ch'ang method, 1844, ms.), 'Saishū shōchō kō' 歳周消長孝 (On the variation of tropical year length, ms.).
11. Nakayama Shigeru 'Shōchō hō no kenkyū (3)' 消長法の研究 (Cyclic variation of astronomical parameters in the 18th century Japan) *Kagakushi kenkyu* 1964: 8–17.

a 授時	b 消長	c 大統	d 時憲
e 崇禎曆書	f 王錫闡	g 梅文鼎	h 曆学疑問
i 麻田剛立	j 寛政	k 高橋至時	l 増修消長法
m 渋川景佑	n 天保	o 寛政曆書	

First published in *Cahiers d'Histoire Mondiale*, Vol.9, No.2, pp.340–362, Editions de la Baconniere, 1965

❻ The Role Played by Universities in Scientific and Technological Development in Japan

1. THE ORIGIN OF JAPANESE UNIVERSITIES

First to modernize among the non-Western countries, Japan undertook the unprecedented task of establishing Western-style universities on an entirely different historical background and of promoting in them scientific research. Throughout the latter part of the nineteenth century, a new system of higher education was sought.

In the Heian period, a system of training imperial officials – created in imitation of a Chinese example – included the fields of mathematics, astronomy and medicine. Down to the Tokugawa period, there were schools sponsored either by the Shogunate itself or by clan governments, but, with rare exceptions, their object was to inculcate upon Samurai boys Confucian ideas and to teach them military skills.

In the Tokugawa period under the national isolation policy, appreciation of the superiority of Western science over traditional Chinese science was popular among intellectuals who eagerly sought knowledge of astronomy and medicine. However, the limit placed on the importation of Western science and the lack of organized opportunities meant that only through individual, voluntary activities could knowledge and information be introduced into Japan. From the beginning of the nineteenth century, the Tokugawa Central Government and some of the feudal clans had indeed adopted systematic policies for the importation of Western science, but these had aimed only at obtaining information necessary for the translation of scientific materials, for the establishment of hospitals and astronomical observatories, or for the modernization of armaments. Importation of educational systems, indispensable to the fostering of scientists, was not yet thought of.

Westernization Policy

After the Meiji Restoration in 1868, the government adopted a new policy and drastically reformed the administrative systems. The introduction of Western science and technology occupied an important position in spite of an ultra-nationalistic or reactionary tendency which was prevalent in most parts of the country. In the 1870s, primary and secondary education were inaugurated. The Kaisei School (1873–1877), a school formerly managed by the Tokugawa

Shogunate and the predecessor of the present University of Tokyo, was a transitional one, and it may be considered as a secondary school in the modern sense of the word. Influenced by the utilitarian climate of the time in which information and knowledge, important for the establishment of a new country, were sought, it was strongly marked by practical and vocational science and technology. Considerable importance was, therefore, attached to subjects in the natural sciences.[1] Teachers were mostly foreigners, and lectures were given in foreign languages. Consequently students had to devote much time to linguistic training, especially at the Kaisei School which adopted English for the three courses of law, science and engineering. This is one of the fundamental reasons why English has attained absolute superiority in the teaching of foreign languages in Japan.[2]

In 1877, the Kaisei School absorbed a medical school and developed into the University of Tokyo, consisting of the four faculties of law, literature, science and medicine. This was the birth of a university-level institution. At the time, the number of foreign teachers in Japan was at its highest, and teaching in the natural sciences was carried out in foreign languages. While foreign teachers could conduct research in Japan's scientific and technological problems (flora and fauna, geological features, earthquakes, traditional industries, etc.) and could introduce their findings abroad. Japanese students could only act as research-workers under the guidance of their foreign teachers. It was a long time before Japan could produce independent researchers.

Foreign teachers were well paid: their salaries amounted to as much as one-third of the budget of the Ministry of Education, and the government planned, therefore, to replace them with Japanese teachers. From the 1870s, at its own expense, the government sent students to Western countries to study in their respective specialities for a period of six years. In 1884 when these students began to return as self-supporting scientists to take the place of foreign teachers, the University of Tokyo was renamed the Imperial University, and a post-graduate course was established. Thus, the Imperial University was Japan's first university in the true sense, where not only education but also scientific research was to be included. The majority of these new scientists who had been educated abroad lectured in foreign languages. Beginning with the twentieth century, however, the situation changed and the balance swung in favour of professors who had been educated solely in Japan; lectures came to be given in Japanese and complete Japanese universities were formed. The disadvantage of scientists thus educated was that they were often poor in foreign languages, especially in writing, and so were handicapped in expressing their opinions on the international stage. This tendency still persists today. From the linguistic point of view, Japanese scientists are more handicapped than are the scientists of developing countries which have just emerged from colonial circumstances.

Social Origins of Japanese Scientists

The new government adopted a national policy of fostering industry, wealth and military power, and emphasized the importance of science and technology as utilitarian means. From a pragmatic point of view, it can be said that the government promoted science and technology. But this is not all: it earnestly wished to revise unequal treaties with the Western Powers and was eager to improve the appearance, not to speak of the substance, of a modern country. Its intention to establish as early as possible a Western-style university resulted from the latter motivation. This is explained by the fact that various branches belonging to pure science occupied the central position among university subjects at an unexpectedly early stage.

In that the universities and the scientific research systems were a part of the national policy, their constitution in Japan was considerably different from that in Western countries. Most of the Japanese scientists who participated in this system belonged to the ex-military class which comprised only a small per centage of the whole population. Generally speaking, the recognition of Western countries on the part of developing countries may be divided into four stages as follows:

(a) Recognition of the military superiority of Western countries;
(b) Recognition of Western military techniques as the basis of military superiority;
(c) Recognition of the necessity to train native people in Western military techniques; and
(d) Recognition of the fact that military techniques are only a part of general Western science and technology, and that in order to reach a parallel development, the importation of pure science and general technology is indispensable.[3]

When compared with China, Japan recognized these stages almost simultaneously: in the former country, a physiocratic bureaucratism was adopted and civil officials held a superior place in the ruling system, whereas in Japan the leading class was composed of samurais who were far more impressed by the threat of the military strength of the Western Powers. But, the majority of the samurais recognized only the first three stages; the fourth stage was reached by a special class.

This special class, which recognized the necessity of the importation of pure science and technology, was composed of three groups: the first consisted of teachers, technical officials and military officers who had worked at the Information Centre of the Shogunate and whose knowledge was useful to the new government in the work of promoting Westernization; the second consisted of physicians and scholars who were different from the samurais as a whole because they knew ways of life other than those of politics and administration and because their views were favourable on the scientific and engineering profession, alienated by the samurais who had a socio-ethical and Confucian system of values; the third was composed of clans which had fought for the Shogunate in the Restoration War. Under the new government, the last group were handicapped in advancing politically and administratively, and they took refuge in science and technology where political influences were less powerful. The boom in Westernization and industrialization which prevailed throughout the country at the beginning of the Meiji era, accounted in part for many members of these groups being attracted by the scientific and technological professions.

These new scientists, endowed as they were with new knowledge, were able to shed the ruling control of the clan government (i.e. the Meiji government), and therefore even if they did not belong to the Satsuma or Chōshū clanship, they could enjoy a kind of academic freedom and autonomy. Another important factor was that Japan was not colonized; it could choose independently what it wished to learn from the Western countries. In the Shogunate days, knowledge was imported from the Netherlands only, but after the Restoration mutual relations with Britain, the United States, France and Germany were promoted.

In the scientific and technological departments of the University of Tokyo there were many foreign teachers from Western countries, but in the medical department, because of the government's 1867 policy for the adoption of German medicine, the foreign teachers were mostly German. In the case of the Engineering College of the Ministry of Engineering, the total number of foreign teachers was 49: its breakdown being 41 English, 7 Italian, and 1 French.[4] Although at the Komaba Agricultural College (the predecessor of the Agricultural Department of the University of Tokyo) intensive agriculture originating in Germany was adopted, at the Sapporo Agricultural College (the predecessor of the present Hokkaido University), an American extensive-farming system prevailed. In 1876, Dr William S. Clark (1826–86), formerly President of Massachusetts College of Agriculture, was elected President of the Sapporo Agricultural College. The College was reconstructed in imitation of that of Massachusetts, and at the beginning lessons were given mainly in the English language by American teachers.

Early in the Meiji era, diplomatic relations were directed towards Britain, France and the United States; but in the academic fields, Germany was respected, perhaps because of the influence of hired foreigners. This tendency is seen in the Draft Regulation of Dispatch of Students Abroad which was prepared in 1870.[5] The consideration for German science by leading academic circles of the world in the mid-nineteenth century, was gradually heightened in Japan. Although the government did not designate the country to which its students were to be sent, the students themselves often chose Germany.

The Meiji government, with the political changes of 1881, adopted a German form of nationalism rather than democracy, and prepared for the coming establishment of the constitution. For the purpose of imitating the Prussian example, it changed literary courses of study at the universities and colleges to conform to the German model, and increased the number of students to be sent to Germany. In the scientific and technological fields also, Japan was consciously bent on imitating the country that was most advanced, and it turned more and more towards Germany. The more the Imperial University developed and acquired its peculiar character, the more it was influenced and coloured by German universities.[6]

Imperial Universities Adopting German Systems

In the 'Imperial University Order' which was promulgated in the nineteenth year of Meiji (1886), German university systems and ideas were strongly reflected, and the objective of universities was defined 'to teach and make a profound study of science and technology which is indispensable to the nation'. Education principles of German universities were different from those of English and American universities; the latter were intended to form ladies and gentlemen, while the former were marked by a nationalistic character and promoted specialized and detailed study in science and technology, the professors being government officials. The faculties of law were above all of a nationalistic character. The Meiji government, with a view to checking the rise of democratic forces by absorbing the men of talent into its own camp, gave graduates from the Faculty of Law of the Imperial University the priority to enter into privileged bureaucratic posts. By a provision in Article 6 of the Imperial University Order that the president of the Imperial University should act also as the dean of the Faculty of Law, the latter was more privileged than any other faculty in the University.[7] The distribution ratio of students between literary faculties and scientific faculties of the former University of Tokyo was 1 to 9, but at the Imperial University the ratio of literary faculties rose higher, and in the twentieth century it was fixed as 6 to 4.

English was taught at secondary education level, but in higher education German was given priority over French, thus occupying the position of the second most important foreign language, the most important being English.

The Japanese university created its own style in the formation of its faculties and was not bound by the Western example of the four faculties of theology, law, medicine and philosophy (or arts and sciences). For example, when establishing the College of Engineering, Mr Henry Dyer, a follower of Dr Rankine, who was commissioned with the organization of the College, gave a theoretical basis to technology. He first set up an educational institution where science and technology were taught in a body, whereas previously technology had been taught only at workmen's schools. Dr Rankine's idea to synthesize technology into engineering to be taught at university level had not yet been realized even in England. Mr Dyer, therefore, referring to the Polytechnical School in Zurich, Switzerland, which was the only synthetic educational institution of engineering at that time, established in the College of Engineering the six faculties of civil engineering, machinery, house-building (architecture), applied chemistry, metallurgy and mining.[8] This College was annexed to the Imperial University, when the latter came into being, as its Faculty of Technology.

The same example is seen in the case of agriculture. In 1890, Tokyo Higher Agricultural School was absorbed into the Imperial University as its Faculty of Agriculture with the three departments of agriculture, forestry and veterinary. There was some opposition against this merger from the University's Council members who said that no Western university had an agricultural faculty within its framework. But, at last, the Council admitted the government's position.[9] The courses of the Faculty resembled those of German-system colleges of agriculture.

With the addition of the Faculty of Technology and the Faculty of Agriculture to that of Science (mathematics, astronomy, chemistry, zoology, botany and geology) and that of Medicine (medicine and pharmacology), the Imperial University came to have four faculties in the scientific field. This system was later imitated by other universities in Japan.

In order to qualify for entrance to the Imperial University, applicants had to pass a very severe examination. The most successful candidates were boys of talent from the upper and the middle classes. Moreover, professors of the University were officials, rather than individual researchers. German universities could keep mobility and flexibility by the system of *Privatdozent*, but the Imperial

University was rigid in structure and far more bureaucratic, if not more so than French universities.

Inactivity of Private Universities

It was the policy of the Meiji government, which adopted a powerful centralized authoritarian rule, to discourage private schools and universities; it set out to intensify the monopolistic position of the University of Tokyo (later the Imperial University). At the beginning of the Meiji era, medical schools dating back to Shogunate days continued their activities, but they were closed with the establishment of the school system in the fifth year of Meiji (1872). In their place, public medical schools were built in such cities as Kyoto, Nagoya, Kanazawa and Niigata.

As for private universities, Keiō, Waseda and Dōshisha universities were established one after another. They had peculiar characteristics reflecting the ideas of their founders in opposing the bureaucracy of the governmental university. Keiō Gijuku, founded by Fukuzawa Yukichi, was characterized by tradesmen's utilitarianism; Waseda University, founded by Okuma Shigenobu, was characterized by non-government politics; and Dōshisha University, founded by Niijima Jō, was influenced by Christianity. While the Imperial University followed a German system, private universities were of British or American style.

In the scientific fields many private schools, with the object of promoting technicians, middle-school teachers of science, or physicians, were established in accordance with the demands of social modernization prior to the governmental school system being set up. These private schools did not develop, however, because of governmental opposition to their obtaining university status and because of the difficulty they had to find ways and means necessary for scientific research and education.

Keiō Gijuku inaugurated a medical department in 1873 in which German-style medicine was taught, but the department was abolished three years later. In the case of Waseda University (which was called at that time the Tokyo Professional School), under the presidency of Hidemaro Okuma, the son of the founder, who had studied in the United States, a department of science was inaugurated in 1882. It was renamed the Department of Civil Engineering in 1884, but it, too, was abolished. To Dōshisha University, the Harris Science School was annexed in 1889 with the resources contributed by Jonathan N. Harris, an American millionaire, but the School was closed in 1896.

Many other private schools in the fields of medicine, veterinary science, dentistry, pharmacology, navigation, agriculture, fishery science, industrial technology, etc., were built to meet the needs of the times and to realize the founders' ideals. Most were abolished or reverted to governmental jurisdiction; some survived and contributed to society by sending out technicians, although it was very difficult for them to obtain a university status before the Second World War. Among the surviving private schools, the most characteristic was the Tokyo School of Physics which was established as the institution for graduates from the Department of Physics of the University of Tokyo to disseminate science. After the war, it was raised to the status of a college, and still continues to-day as the Tokyo College of Science.

In Japan, where higher educational institutions came under the jurisdiction of the Central Government and where priority was given to governmental colleges and universities, the state of affairs was very different to that in China under the Ching Dynasty. For example, universities, especially in scientific fields, run by foreign missionaries did not develop in Japan.

2. MULTIPLICATION OF UNIVERSITIES

The history of Japan as a modern country is influenced by the Russo-Japanese War, the Sino-Japanese War and the First World War, which took place within intervals of ten years. Keeping in step with these historical events, new imperial universities were established which gave fresh impulses to university research systems and which contributed greatly towards the development of scientific research in Japan. Kyoto Imperial University was built in 1897; Tōhoku Imperial University and Kyūshū Imperial University in 1911; Hokkaido Imperial University in 1918; Osaka Imperial University in 1931; and Nagoya Imperial University in 1939.

It would seem that warfare and the establishment of universities is unrelated: in reality, the two were closely related for they had as a common basis the increase of national strength.

In this section the significance of the establishment of each imperial university, and its later development, will be treated.

Kyoto Imperial University

The establishment of Kyoto Imperial University was expressly intended to oppose the monopolistic position of the Imperial University in Tokyo. In February 1890, when the Imperial Diet was considering the bill on the establishment of an imperial university in the Kansai District, Hasegawa Yasushi, a member of the Diet from the Liberal Party, expressed his opinion as follows:

> There is only the Imperial University in Tokyo and, apparently, it seems complete. However, attentive observation shows that because of the lack of competition, the professors have ceased to try to discover new scientific theories, and the students have ceased to pursue their scientific objectives. As a result, the graduates from the Imperial University are not well qualified to be called 'scientists'. It is believed, therefore, that the establishment of an imperial university in the Kansai District is indispensable to the development of education in Japan.

If one university monopolizes higher education it is apt to become ineffective. The government, referring to the examples of Oxford and Cambridge in England, and of Harvard and Yale in the United States, intended to prevent inactivity of scientific research by having two universities stand together and compete with each other.

The project to establish a new imperial university in Kyoto was delayed for some time because of the outbreak of the Sino-Japanese War, but was realized in 1897 when the war was over. The victory brought about Japan's securing an enormous market in Asia and a consequent development of modern industry.

The government-managed heavy industry for military expansion, with the Korean Peninsula and the Chinese Continent in its sight, had started to develop before the Sino-Japanese War. After the war, private light industry, with spinning industry at its centre, was expanded with the perspective of the Asian market. There was a demand for industrial technicians, and the number of applicants to university scientific and technological departments increased so much that the Imperial University in Tokyo alone could not admit them all. Kyoto Imperial University, established in such circumstances, first inaugurated its Faculty of Science and Technology, and in 1897 established the Department of Civil Engineering, absorbing that of the Third Higher School. In 1898, departments of mathematics, physics, pure and applied science, electrical engineering, and mining and metallurgy were added. Although zoological, botanical and geological departments were not created in the early stages – in view of the small number of applicants – the structure of Kyoto Imperial University reflected well the ideal conceived at the inauguration to establish the State on the basis of industry. The Faculties of Law and Medicine were annexed in 1899.

At Tokyo Imperial University, whose inaugural period was over, permanent professional positions were already occupied. The young scientists who had graduated from Tokyo Imperial University positively participated, therefore, in the inaugural programmes of Kyoto Imperial University. But afterwards, the latter, competing with the former, chose professors from among its own graduates. Although Tokyo Imperial University adopted the grade system, Kyoto Imperial University adopted the unit system, still new in Japan, allowing students to stay freely from three to six years. Moreover, Kyoto Imperial University took the lead in adopting the metric system, although the yard-pound system was still used in Japanese scientific circles in general.

Tōhoku Imperial University and Kyūshū Imperial University

After the Russo-Japanese War (1904–05), there was a demand for the establishment of other imperial universities. But, because of financial difficulties and opposition from the Ministry of Education which wanted to preserve the scarcity value of the imperial universities, the government did not act immediately. In addition, Kikuchi Dairoku, Minister of Education, planned to establish higher technological schools rather than universities.

In the 1870s, the government made efforts to foster modern industry by the establishment of government model factories, but after the 1880s, it transferred these to private ownership in order to promote the development of private enterprise. After 1905, Japanese heavy industry, not to speak of light industry, developed considerably and centralization and monopolization of capital was promoted. In the case of Tokyo Imperial University and Kyoto Imperial University, not only the inaugural funds but also the operating funds had been provided by the government; but after the Russo-Japanese War, private enterprises could afford to endow universities. The related local public bodies could make considerable contributions towards the establishment of government universities in the district under their jurisdiction.

The Furukawa family, owners of the Ashio Copper Mine, happened to be the object of public hostility because of the Ashio Mine Pollution Case. In order to mitigate public resentment, the Furukawas accepted the advice of Hara Takashi, Minister for Home Affairs, to contribute towards educational institutions. Their contribution was of great help in the establishment of Tōhoku Imperial University.

Kyūshū Imperial University was inaugurated in 1911, on the basis of the Fukuoka College of Medicine, with the Faculty of Engineering newly created upon the request of the Northern Kyūshū Industries at the centre of which was the government-managed Yawata Iron Works. The Faculty of Agriculture was annexed in 1919, and that of Science in 1939.

The circumstances under which Tōhoku Imperial University was inaugurated are worthy of special mention. As Japan's imperial universities imitated the German system, one single faculty could not constitute an imperial university. Therefore, the Faculty of Science was created at Sendai in 1911; then the Sapporo Agricultural School was converted into the Faculty of Agriculture. Thus, Tōhoku Imperial University started with two faculties. As for its inaugural funds, the Faculty of Science in Sendai received 260,000 yen from the Furukawas and 150,000 yen from the Miyagi Prefectural authorities, and the Faculty of Agriculture in Sapporo received 400,000 yen from the Furukawas and 100,000 yen from the Hokkaido Prefectural authorities. The government did not contribute financially to the establishment of Tōhoku Imperial University.

The Faculty of Science was different from that of the Imperial Universities in Tokyo and Kyoto in that it did not have a predecessor. It was, therefore, free from the restrictions which would otherwise have been placed upon it. Moreover, there was no industrial background which could have made an imprint on its character. Why should a faculty of science have been established without immediate needs to fulfil? Perhaps because the authorities intended to prepare for the future establishment of faculties of medicine and engineering. In fact, the former was created in 1915 and the latter in 1919. When Tōhoku Imperial University was inaugurated, the Faculty of Science and Technology of Kyoto Imperial University emphasized engineering, while at Tokyo Imperial University, departments concerned with pure science were treated in a perfunctory manner. Although the natural sciences in Japan achieved a remarkable development at the beginning of the twentieth century, no age limit was set for professors who continued to be those appointed at the time of inauguration. Most of the newly-graduated scientists were, therefore, anxious about their future prospects. Accordingly, graduates from the Faculty of Science of Tokyo Imperial University were appointed professors of Tōhoku Imperial University and undertook to develop it.

The majority of these young professors were in their thirties, and had studied in Europe for several years. They made efforts to adopt the system of German universities, and consequently created the peculiar character of the University. A most astounding example of the new spirit of the University was its policy of admitting women, hitherto barred from universities in Japan. In addition, the *Tōhoku Sūgaku Zasshi* (the Tōhoku Journal of Mathematics) was issued, in which papers written by top-level scholars of the world were published.

The same can be said about its Faculty of Medicine, established soon after. The first professors sent to Europe before the inauguration intended to study mainly in Germany, but because of the First World War they were obliged to take refuge in France, England and the United States. As a result, on their return to Japan, they were free from the tendency of complete devotion to German medicine, which then prevailed in Japanese medical circles. They appreciated English and American medicine from a fresh point of view, and this was reflected in the initial organization of the Faculty. The *Tōhoku Journal of Medicine* in the English language was a rare achievement at that time in Japan.[10] German medicine was at its height in the 1870s, when Japanese universities decided to adopt it; in the 1890s, the peak was shifting from Germany to the United States. The 'fresh air' which Tōhoku Imperial University introduced into Japanese medical circles was, therefore, most opportune – although traditional devotion to German medicine, with the Faculty of Medicine of Tokyo Imperial University at its centre, continued to prevail until the end of the Second World War, when a definite switch over to American medicine started to be made.

The Faculty of Agriculture of Tōhoku Imperial University continued the tradition of its predecessor, the Sapporo Agricultural School, in adopting the American extensive-farming style and competed with Tokyo Imperial University which followed the German style of intensive farming. But Japan proper being under the influence of the farming style originated by Tokyo Imperial University, the graduates from Tōhoku Imperial University were obliged to establish their influence in Hokkaido, Formosa and Korea. Later, in 1918, this Faculty of Agriculture cut its connections with Tōhoku Imperial University and was reorganized as the Faculty of Agriculture of Hokkaido Imperial University, to which the Faculties of Technology and of Science were annexed in 1924 and in 1930 respectively.

As explained above, Tōhoku Imperial University and Kyūshū Imperial University contributed much towards the development of science, not only by producing scientists but also by introducing new techniques. However, the new outlook did not last long. Among the general public, and especially among the younger generation who wanted to receive higher education, there was a tendency to rate Tokyo and Kyoto Imperial Universities higher than those of Tōhoku and Kyūshū, and these last-named sometimes had difficulty in recruiting students of sufficient promise.

The University Order

At the beginning of the twentieth century there were two opposing opinions as to what a higher educational institution should be.

The professors of imperial universities and the personnel of the Ministry of Education were of the opinion that the Japanese university should keep a level as high as that of the Western universities. Therefore, they observed the principle of the selected few and stood against the increase of universities and the promotion of college status to that of university status. Politically, they were connected with *Seiyūkai*, one of the two largest parties of the time, and were strongly marked by nationalism and authoritarianism. On the other hand, non-official educators and others who were in favour of reforming the educational systems, were of the opinion that in accordance with the development of national industry and in order to meet increasing demands the universities should be increased in number even if by so doing the level was lowered. Moreover, they considered that private universities should have the privilege of awarding academic degrees in the same way as did imperial universities. Politically, they were related to *Kensei-hontō*, the other most powerful party which emphasized democracy and utilitarianism.

In the 1910s, bourgeois democracy was brought about and a violent attack started against the nationalistic authoritarianism of imperial universities. It was in such an atmosphere that, in 1914, the Second Okuma Cabinet set out to reform educational systems, and the following Terauchi Cabinet prepared the draft 'University Order', which was enacted in 1918 and enforced in 1919.

The University Order provided that university status be granted to public or private higher educational institutions and that the university be the place of research as well as of higher education. At the same time, higher educational institutions (governmental, public and private) having only one faculty were granted college status similar to the university status. This provision reflected the educational conditions of the time. Japan was regarded after the First World War as one of the Great Powers, and the number of applicants became so large that government universities alone could not cope. Private schools, if granted university status, would contribute in solving this problem. Upon the enforcement of the University Order, movements started in various parts of Japan to raise the level of schools.

Although in the case of private schools these movements were comparatively successful, government schools whose expenses were met by the national treasury often found it difficult to raise their level. A representative example is seen with the Tokyo Institute of Technology. Originally the Tokyo Workmen's School with the objective to train technicians of middle standing, it was converted into the Tokyo Higher Technical School designed to produce experts for industrialization. It started a movement to raise its status to that of a university and succeeded in overcoming the barrier of the authoritarian imperial universities. Thus, the Tokyo Institute of Technology and the Osaka Institute of Technology were inaugurated in 1929 as the first government colleges in Japan.[11] In the same year, universities of literature and science were established in Tokyo and Hiroshima with a view to training middle school teachers. As for medical colleges, the Kyoto Prefectural Medical College had been established in 1921, and others followed.

The enforcement of the University Order was an epoch-making event: in former days the Japanese university had followed the European pattern, which was maintained for the education of selected youngsters from the upper classes; now, it began to move towards the American pattern which, in keeping with the development of industry and the diffusion of democracy, would produce the scientific and technical experts needed by the society.

Research Institutes Attached to Universities

At the beginning of the Meiji era, each operational government office had annexed an educational and research institute. But, since the birth of the Imperial University, many of these institutes had tended to be incorporated into the latter. For instance, the College of Engineering of the Ministry of Engineering was removed to the Imperial University as its Faculty of Technology, and the former Meteorological Division of the Navy Department and the former Meteorological

Observatory of the Ministry of Home Affairs were merged and attached to the College of Science as its Astronomical Observatory. Scientific undertakings needed in a modernized country, such as research on longitude and latitude and pharmacological examination, which were formerly carried out by various ministries and agencies, were gradually transferred to related research institutions directly attached to universities or to competent faculties.[12]

Before 1891, Tokyo Imperial University had the following research institutes attached to its respective faculties: the Astronomical Observatory (Faculty of Science), the Marine Biological Laboratory (Faculty of Science) and the Experimental Plantation (Faculty of Agriculture). Later, such research institutes as the Fisheries Laboratory (Faculty of Agriculture) and the General Experiment Laboratory (Faculty of Engineering) were added. The Research Institute of Infectious Diseases of the Ministry of Home Affairs was transferred to the Ministry of Education in 1914, and in 1916 was annexed to Tokyo Imperial University.

The Aeronautical Research Institute, attached directly to Tokyo Imperial University, was inaugurated in 1918. Until that time university professors were concurrently engaged in work at research institutions. In the Aeronautical Research Institute, professors and assistant-professors were appointed to be exclusively engaged in research work and had the same status as university professors. The special system of the research professor was thus inaugurated on the recommendation of the University, supported by the Army and can also be considered as an epoch-making event in the history of the Japanese university.[13] A close relationship with the military authorities was ensured by the attendance of Army and Navy officers to the meetings of the Council of the Institute at which the problem of scientific research and military secrecy was first raised. It can be supposed that the military authorities wished to prevent the disclosure of military secrets by directing the professors who were to be engaged exclusively in research work.

The birth of such research institutes meant independence from related faculties and a rise of status to that of faculties. In other words, scientific research came to be looked upon as important a function of a university as education. The example of the Aeronautical Research Institute was followed by the Tokyo Astronomical Observatory of Tokyo Imperial University (1921), the Research Institute for Iron, Steel and Other Metals of Tōhoku Imperial University (1922), the Earthquake Research Institute of Tokyo Imperial University (1925), and the Research Institute for Chemistry of Kyoto Imperial University (1926). From the 1930s on, various ministries and agencies started to re-establish their own research institutes, but their researchers were apt to be hampered by the routine of office work.

Osaka Imperial University and Nagoya Imperial University

Osaka Imperial University, comprising faculties of science and medicine, was established in 1931 with the strong support of local financial circles and the local government, overcoming the resistance of the Diet and the government. It absorbed the Osaka Institute of Technology as its Faculty of Engineering.[14] It established Japan's second nuclear laboratory, next to the Physical and Chemical Research Institute, and had as a lecturer Dr Hideki Yukawa, who initiated the meson hypothesis in 1934 and who is the only Nobel Prize winner in Japan. The

University was thus established as the centre of nucleonics, then in its initial stages in Japan.

Nagoya Imperial University, established in 1939, was also composed of the two faculties of science and medicine, but the outbreak of the Second World War prevented any immediate development. Upon the termination of the war, however, the Department of Physics democratized its management procedures by organizing groups according to the various fields of research, and by setting up department meetings which were attended by the researchers concerned and at which decisions of the highest importance were taken. This example served as a model for post-war reorganization not only of other departments in Nagoya Imperial University but also of those in other universities.

Universities in the Colonies

In Korea, the Keijō (Seoul) Imperial University was inaugurated in 1924. Other colonial universities were Taihoku (Taipeh) Imperial University, established in 1928; the Ryojune (Lushun) College of Engineering, established in 1922; and the Manshū (Manchuria) Medical College (private), established in 1922. These universities were under the jurisdiction of competent colonial governments and not of the Ministry of Education. Keijō Imperial University had no scientific faculties except of medicine (its Faculty of Science and of Technology being inaugurated only in 1941), and Taihoku Imperial University had no physical departments (its Faculty of Science being composed of botanical, zoological, geological, chemical and agricultural departments). This was due to the policy of the Central Government to limit scientific research in the colonies to local sciences necessary for local development, and to entrust basic and pure scientific research to the imperial universities on the main land. Taihoku University had a faculty of medicine and a research institute of tropical medicine. Students of these colonial universities were mostly Japanese.

Private Universities

The 'Private School Order', promulgated in 1899, provided that the establishment, abolishment, change, etc., of a private school should be authorized by the Minister of Education. Thus, the control of the government over private universities was intensified and their development was hampered, the government tending to make much of itself and little of the people. Later, as mentioned above, the promulgation of the University Order made it possible for private universities to be inaugurated. But few comprised faculties in the field of the natural sciences. Ōkuma Shigenobu, the founder of Waseda University, was of the opinion that a university lacking in scientific faculties could not be called a university in the true sense of the word, but, because of the shortage of private capital, he could not put his idea into practice. Other private universities were in the same situation: most of them consisted of literary faculties and departments.[15] The only university-level private institutions authorized in the field of the natural sciences were the Jikei Medical College (established in 1921), the Tokyo University of Agriculture (established in 1925), and the Nippon Medical College (established in 1926). Private universities were hampered by the complex bureaucratic procedures of the scientific circles.

3. THE SECOND WORLD WAR AND UNIVERSITIES IN JAPAN

During the First World War, the mobilization of scientists and technicians for war-time research programmes was not seriously planned and the government regarded scientific research at universities as only an accessory activity in a modern state. However, upon the outbreak of the China Affair, late in the 1930s, it intervened in scientific research at universities, and the outbreak of the Pacific War gave added impetus. Firstly, in 1938, the Scientific Investigation Council was set up, and it dealt with problems of material resources. Secondly, the Cabinet Planning Board created its Science Department and started to elaborate science-mobilization programmes. In the same year, an Investigation Committee for the Advancement of Science was set up as an advisery body to the Minister of Education, and in line with its report a science section was created in the Ministry of Education in 1940. This was expanded into the Science Bureau in 1942, and it undertook the planning of the national structure for scientific research, including the increase of scientific faculties and departments at universities and the expansion of research institutions.

Again, bureaucratic sectionalism hampered science-mobilization programmes and anticipated results were not achieved until the end of the war. For scientists and technical experts, however, this was a happy period, since the significance of their work was recognized by the national authorities.[16] Research funds awarded by the government to university professors were as abundant as they are today, and research projects of the military and the industries benefited university researchers, especially in the field of engineering. Before the war, the ratio of students between literary and scientific faculties was 6 to 4: during the war, it was planned to reverse this ratio to 4 to 6, with the example of Nazi Germany in mind. As the Pacific War became more violent, university students belonging to literary faculties were sent to the front, while those belonging to scientific and technological faculties enjoyed special favours, e.g. postponement of or exemption from military service.

Another noteworthy fact was that, within the framework of a university, the relative importance of research institutes increased. Dr Honda Kōtarō, President of Tōhoku Imperial University, planned to consolidate scientific research facilities, taking advantage of the current situation.[17] He founded the Research Institute of Iron, Steel and Other Metals and other research institutes associated with his university. The number of such institutes created during the war totalled twenty, of which eight were attached to Tōhoku Imperial University. By the end of the war, there were thirty-one attached to the seven imperial universities, of which Tōhoku Imperial University counted ten. Hence, the University being called sometimes a *research institute* university.[18]

Research institutes privileged by war-time measures were nevertheless forced to pursue projects of immediate utility and to play the part of munitions factories, setting aside the more basic, academic projects.[19]

4. OCCUPATION POLICIES AND THE RE-ORGANIZATION OF UNIVERSITIES

For some time after the unconditional surrender of Japan in August 1945, Japanese universities did not know which way to turn. In September, the American occupation forces arrived and started to implement policies, guided by three principles: 1) demilitarization; 2) democratization; and 3) reconstruction of peace-time industries. Renovation of the university system after the American style was included.

Occupied Japan was to be controlled, in principle, by the Allied Powers, but the decisions and execution of the occupation policies were, in fact, taken and carried out by the United States alone. The USSR's proposal to the effect that the policies be decided and carried out by a committee composed of the representatives of the United States, the United Kingdom, the USSR and China was rejected by the United States. Occupied Japan was in a different position from Germany which was controlled by the United States, the United Kingdom, the USSR and France. It is said that the United States pressed the German government to renovate its educational system, according to the American style, but this idea was rejected by Germany. Japan, under the sole control of the United States, was not in a position to reject the programme prepared by the American occupation authorities.

It was a matter of course, since the occupation control was exercised by the Supreme Commander (appointed by the United States government) and, indirectly, through existing Japanese administrative channels, that secondary directives were sometimes withdrawn when strongly opposed by the Japanese authorities. For example, the CIE recommendation for the abolishment of the Ministry of Education, regarded as the citadel of educational centralization, and for the removal of the university lectureship which was apt to promote the 'boss-gang' relationship and sectionalism among researchers, was not realized. The CIE proposed to set up a Board of Trustees as the highest legislative organ of all the government universities; but this proposal was withdrawn in view of the opposition of Japanese authorities who insisted upon the autonomy of the individual universities.[20]

Generally speaking, the Japanese authorities did not understand clearly the necessity of educational renovation. Their educational systems were rather old, and could be renewed only by effacing the ultra-nationalism which had been introduced since the latter part of the 1930s. However, large-scale renovation by the adoption of the grade system and the change in the organization of scientific research did not seem necessary to the Japanese.

The occupation forces sent to Japan in 1946 the First United States Education Mission, which was followed by the Scientific Advisory Group of the United States National Academy of Sciences. On the basis of their recommendations, a renovation programme was drafted. If the Japanese authorities were forced to put it into practice, they were opposed to haste, for they considered that the classical liberalism and the American optimism upon which the programme was built would not suffice to re-establish the educational system in the midst of post-war chaos.[21] In addition, the government was handicapped by the difficulty in finding means to implement the programme.

Nevertheless, new universities, after the American style, were inaugurated: in 1948 privately, and in 1949 under government auspices. The compulsory term of education of six years was prolonged to nine years and the number of higher educational institutions, including universities and colleges, which had been nineteen before the war, was increased to some two hundred. In truth, there had been a remarkable increase in the number of applicants to higher education since the Meiji era so that the existing faculties were insufficient. But while the framework was expanded, the quality did not follow suit, and those who regarded the university as the place where the élite should educate the élite were troubled.

Without the influence of the occupation forces, such a largescale renovation and expansion could never have been realized. Misunderstanding and confusion about the new system lasted approximately ten years, and during this transitional period Japanese educational and scientific circles suffered. Although it is over ten years now since the first graduates from the new-system universities were sent out into the world, it is still too early to make a final appraisal on this renovation. However, it can be said that one of the best results achieved is quantity production of highly qualified experts in science and technology. Secondly, the re-organization of post-graduate research institutions has done much to prevent the level of scientific research from deteriorating.

Most of the new-system institutions created in large quantities did not deserve the name of universities. They could by no means breathe fresh air into the existing scientific circles as the newly-created imperial universities had done earlier. They, indeed, played a rōle in affording working places to those scientific and technical experts thrown out from industries and colonial research institutions, but the majority have never been allowed by the Ministry of Education to have post-graduate courses. Consequently, it is difficult for them to recruit professors from among their own graduates. Again, in this age of science and technology, local universities find it very difficult to recruit able professors.

The post-graduate course under the new system was inaugurated in 1950 in the case of private universities, and in 1953 in that of national universities. Under the old system, it had no relation with doctorates: it was, so to speak, a waiting room for those who were expected to be appointed as assistants or lecturers. Post-graduate students did not follow any special course as in faculties; they pursued individually their own specific research projects. In this sense, the post-graduate course under the old system was similar to that of England. The new-system post-graduate course, formulated after the American style, has a precisely determined study programme and a complete relationship with doctorates. All that may be said at this point is that in industry doctors are often favoured over masters. An academic degree has little to do with the holder's remuneration in the seniority salary scale, which now prevails all over Japan.

Scholarships to cover the living expenses of post-graduate students are still insufficient, although they have considerably increased in number. Graduates from faculties are obliged to seek employment in private firms instead of proceding to a post-graduate course. This, too, creates a problem in the recruitment of the talented into the university professorate. There is also the problem of the appointment of young researchers as assistants to professors. However, in spite of these problems, the training of researchers at faculties is gradually coming to be the function of post-graduate courses.

After the war, loud cries were heard from industries for the increase of university students in the fields of science and engineering. This demand reflected the industrial revival and the technical improvement in post-war Japan. In response, the Ministry of Education inaugurated three-year programmes for the increase of university students in 1957 and in 1961 respectively. Both were successful: the three years following 1957 saw an increase of university students by 8,000; the three years following 1961 an increase by 20,000. Moreover, formalities for the establishment of new faculties and departments in private universities were simplified.

Before the war, it was financially difficult for private universities to maintain scientific faculties, but this became easier as from early in the 1960s when contributions from industries and government subsidies were made available. Some of the faculties at private universities are a transition from scientific colleges established during the war to meet the national demand, while others are allied to 'big business'. Most of the private universities are permitted to create post-graduate courses, the government being under no obligation to share in the costs. If the increase of scientific and engineering faculties and post-graduate courses in private universities have not contributed to the scientific world of Japan to the same extent as have done national and public new-system universities, they have nevertheless opened the door to employment for scientists and engineers and their influence has been favourable to post-war scientific development in Japan. Less favoured are the biological sciences, and, especially, the agricultural sciences owing to the modernization of Japanese agriculture and the conversion of agricultural departments into industrial ones (e.g. agricultural chemistry, food processing, etc.).

The research institutes attached to national universities can be considered as the force behind the mobility of the scientific world, rather than either the national or private universities created after the war. Such institutes as the Cosmic Rays Observatory of the University of Tokyo and the Research Institute for Basic Physics of the Kyoto University (both established in 1953 for joint use to all the scientists concerned) are fulfilling their duties as the place for relations and exchanges among researchers in these fields. In the international scientific world, moreover, an exchange of scientists between Japan and other countries has been promoted. Conferences in diverse fields of science have come to be organized in Japan, and the international academic market has become available to Japanese researchers. However, in most cases, the international scientific exchange of persons still means a one-way outflow from Japan to the United States, where better conditions are attainable.

Recently, research institutes managed by private industries or government organizations have increased in number, but the greater part of scientific researchers in Japan belongs to universities. As of 1956, the ratio of university researchers to the total is 67 per cent, and that of those belonging to research institutes attached to universities is 3,6 per cent.[22]

No voices are heard criticizing the present system but, in the organization of universities, historically-rooted obstructions still remain and hamper the promotion of scientific research. Most of the scientific researchers belong to universities and most of the researchers in the natural sciences belong to national universities. Professors of the latter have a permanent status guaranteed by the government, and the mobility of their researchers is very rare. Such distinguished institutions as the University of Tokyo and Kyoto University are recruiting their professors-to-be from among their own graduates. Thus, a hierarchy is often apt to be established among the university researchers. In the fields where most progress is being made day by day, the depreciation of old scientists comes earlier than before, but how to deal with old-fashioned professors is still only discussed behind the scenes: it has not yet come to the surface.

New-system universities created in post-war Japan are now to be classified into two categories: one to comprise the so-called 'universities for research', with post-graduate courses, which aim primarily at research activities; the other to comprise 'education universities for instruction' whose main function is to instruct students. This may be a natural course for Japanese universities to take, since research in the natural sciences involves expensive equipment and substantial effect could not be expected economically if all universities were to be equipped equally.

NOTES

1. Ōkubo Toshio, *Nippon no Daigaku* (Universities in Japan), 1943, p. 324.
2. Sumeragi Shidō, *Daigaku-seido no Kenkyū* (A Study of University Systems), 1955, p. 335.
3. John K. Fairbank, et al., 'The Influence of Modern Science and Technology on Japan and China', in *Explorations in Entrepreneurial History* (1954).
4. *Kyū Kōbu-Daigakkō Shiryō* (Materials of the Former College of Engineering), 1931, pp. 353–356.
5. Maruyama Kunio, *Waga Kuni ni okeru Doitsugaku no Bokkō* (The Rise of German Science in Japan), 1936, pp. 52–54.
6. Sumeragi Shidō, *Daigaku-seido no Kenkyū* (A Study of University Systems).
7. Ibid.
8. *Kyū Kōbu-Daigakkō Shiryō* (Materials of the Former College of Engineering).
9. *Tokyo Teikoku Daigaku Gojūnenshi* (Fifty Years' History of Tokyo Imperial University), vol. 1, 1932, pp. 972–974.
10. *Tōhoku Daigaku Gojūneshi* (Fifty Years' History of the Tōhoku University), 1960, pp. 113–118.
11. *Tokyo Kōgyō Daigaku Rokujūnenshi* (Sixty Years' History of the Tokyo Institute of Technology), 1940.
12. *Tokyo Daigaku Gakujutsu Taikan: Sōsetsu* (Science in the Tokyo University: its Introduction), 1942, p. 17.
13. *Tokyo Daigaku Gakujutsu Taikan: Kōgakubu; Kōkū-kenkyujo* (Science in the Tokyo University: Faculty of Engineering; Aeronautical Research Institute), p. 397.
14. *Osaka Teikoku Daigaku Sōritsushi* (History of the Establishment of the Osaka Imperial University), 1935.
15. *Waseda Daigaku Hachijūnenshi* (Eighty Years' History of the Waseda University), 1962, pp. 121–125.
16. Yamazaki Toshio, *Gijutsu-shi* (History of Technology), 1961, Chapter VI.
17. *Tōhoku Daigaku Gojūnen-shi* (Fifty Years' History of the Tōhoku University), vol. I, p. 379.
18. *Ibid.*, p. 457.
19. Ibid.
20. *Daigaku Kijun Kyōkai Jūnenshi* (Ten Years' History of the University Standard Society), 1957, pp. 49–57.
21. Ibid., pp. 25–30.
22. 'Daigaku no Gakujutsu Kenkyū-taisei' (Organization of Scientific Research at Universities), *Gakujutsu Geppō*, vol. 10, No. 8, December 1957, p. 447.

First published in *Colloquia Copernicana*, I (V), 1973

7 Diffusion of Copernicanism in Japan

Before we examine closely the diffusion of Copernicanism in Japan, it may be helpful (1) to define briefly what Copernicanism means by analysing it into three conventional elements – astronomical, cosmological and physical – and (2) also to sketch the Far Eastern tradition with respect to each element. This juxtaposition will provide preparation for our main theme, the encounter of Copernicanism with a culture which did not share its historical roots. In so essentially foreign a milieu, it was perhaps inevitable that Copernicus's innovations played a radically different role from that which they played in the Western tradition.

WHAT IS COPERNICANISM?

Astronomical or Observational

Copernicus is, without doubt, primarily a first-rate professional astronomer. His thorough mathematical treatment in *De revolutionibus* from Book II on must have challenged his astronomical successors, such as Erasmus Reinhold and Tycho Brahe. This aspect of Copernicanism may be called 'astronomical Copernicanism' or 'mathematical and quantitative Copernicanism' if not to say 'observational Copernicanism'. As regards practical astronomy, Copernicus was engaged in calendar reform, investigation of long-term variation of planetary positions, and eclipse predictions, which were the serious astronomical topics of his time.[1]

Cosmological or Geometrico-Morphological

On the popular level, Copernicus is, of course best known as a cosmologist. The basic tenet of 'cosmological Copernicanism' is obviously the heliostatic conception. His work on the order and relative distance of planetary orbits was only the outcome of this concept. Man's position in the universe, the taste for harmony, and even the idea of an infinite universe in the later Copernicanism are all closely associated with the cosmological element. We must note that this aspect became weighty in popular thought only from the time of Galileo's recantation, although it must have been what drove Copernicus to the completion of the extensive *De revolutionibus*.

Physical or Mechanical

Finally, 'the moving earth' theory had important consequences for physics, as testified by those critics who denied it, and engaged in a controversy on how that motion, whether rotation or revolution, could possibly take place without causing strong winds on the surface of the earth. This physical argument constituted at the time a vulnerable point of Copernicanism.

THE FAR EASTERN BACKGROUND

Astronomy

Traditional Far Eastern astronomy, which remained centred in China, was oriented towards the special goal of providing the most accurate possible luni-solar calendar, and of predicting lunar and solar eclipses, since these predictions were an excellent means of checking the accuracy of any such calendar.[2]

While calendrical astronomy was at the core of Far Eastern exact science, planetary astronomy, which was not highly developed, held a peripheral position and played its main role in portent astrology.

It is hard to find in the development of Chinese astronomy any notable tendency towards conceptual schemes or mechanistic or geometrical models. The approach of the Chinese official astronomers was to represent numerically the course of the celestial bodies without depending upon a geometrical model. Their final aim was to reduce observations as accurately as possible to algebraic relations. Unlike Ptolemaic astronomy, Chinese astronomy showed no concern for the calculation of dimensions of the universe: for all mensurational purposes the sky was treated two-dimensionally.

Cosmology

During the Han and Six Dynasties period, there had been rivalry between cosmological systems, especially between one school which championed a spherical sky (*hun t'ien*)[a] and another which held that the sky was parallel to a flat or convex earth (*kai t'ien*).[b] Later on, however, the controversy died out and astronomers lost their interest in it, occupying themselves solely with routine observations and calendrical calculations. In the T'ang period the Chinese official astronomers claimed that 'Our business is exclusively calendrical calculations and observations in order to provide the people with the correct time. Whether flat or spherical cosmology is no concern of the astronomers!'[3]

Although without effect on professional astronomical circles, there was a revival of cosmological interest among the Sung philosophers such as Chang Tsai[c] and Chu Hsi.[d] The Neo-Confucians favoured the *hun t'ien* theory. Chu Hsi, the chief figure of the school, argued that the *kai t'ien* theory could not explain how the sky remained in consonance with the earth. The sphericity of sky in the *hun t'ien* cosmology was, however, not paralleled by recognition of the sphericity of the earth. A flat earth, according to Chu Hsi's theory, was located in the middle of the universe, floating on water. Despite considerable developments of eclipse-predicting techniques, it is surprising that the sphericity of the earth was explicitly recognized neither by philosophers nor by astronomers, though the idea was hinted at in the writings of Shen Kua[e] and others.

In the absence of the idea of the sphericity of earth, sun-centredness was simply inconceivable. The sun was accorded a most important role in cosmology, but was not allotted the central position in the geometrical cosmology.

Chu Hsi proposed a nine-layered stratification of the earth, sun, moon, planets and fixed stars, but did not work out in its astronomical details. A rigid and tightly designed universe like the medieval Western cosmos was not to the taste of the Chinese. Their cosmological outlook was not confined inside the shell of *hun t'ien* theory; implicitly or explicitly they admitted the plausibility of an infinite universe, as evidenced in outspoken arguments of the Buddhists.

Physics or Mechanics

The main cosmological concern of the Chinese philosophers was not the shape of the world but its dynamic processes. Their discussions were almost always phrased in terms of the cosmic energetic fluid, *ch'i*,[f] which permeated the universe; meteorological metaphors are more appropriate to it than that of mechanical clockwork. The most important Chinese principle of natural philosophy is the yin-yang[g] principle, which explained all phenomena in the universe in terms of alternating but complementary dynamic phases described by the rhythms of earth and heaven, rest and motion, female and male and so on.

Chinese geocentricism was not based on a sharply defined celestial-terrestrial dichotomy as in Aristotelian cosmology, nor was the cosmology complicated by theological requirements as in medieval Europe. For academic philosophers the earth was simply located at one end of a continuous spectrum running from most rapid (stellar background) to infinitesimally slow (earth) motion. Thus the earth was conceived as not absolutely and self-evidently at rest, but conditioned to be so by external circumstances. There remained a logical option to accept the motion of the earth once those circumstances were no longer accepted.

Contrary to philosophers' view, professional astronomers claimed that the star-bearing sky moves from east to west, and the sun, moon, and planets move from west to east against the stellar background. If to the central earth we extend this notion of eastward turning, the earth must rotate most rapidly against the westward-turning stellar background.

INTRODUCTION OF COPERNICANISM INTO JAPAN

Up to the early part of the eighteenth century, Japanese astronomy was still dominated by Chinese tradition. We can reasonably say that Copernicanism appeared in Japanese works only in the last quarter of the eighteenth century.

Compared with the date of the acceptance of Copernicanism in the West, its introduction into Japan was late. To explain this, the following three factors may be considered:

(1) Political action to limit free international communication,
(2) The language barrier to the translation of Western works,
(3) Ideological and technical difficulties in comprehending Copernicanism.

JESUIT INFLUENCE

The impact of the West finally began to make itself felt on the isolated islands of Japan in about 1543,[4] the year in which *De revolutionibus* was published. In that year, shipwrecked Portuguese introduced firearms into Japan and Jesuit evangelism followed.

Scholars who have studied the reports of such missionaries to China as Matteo Ricci and Johann Adam Schall von Bell tend to project their picture of seventeenth-century Chinese science onto that of Japan in the corresponding period and often conjecture that Japanese science also was substantially affected by the early contributions of the Jesuits. However, circumstances in the two countries differed greatly.

While the Jesuits in China generally took a flexible, sometimes conciliatory, attitude towards the elite in Chinese bureaucracy and employed an indirect method to convert them to Christianity through the demonstration of the superiority of Western astronomy, the missionaries to Japan never attempted a systematic introduction of Western astronomy, but focused their efforts on intensive direct exangelism. From the late sixteenth century onwards, the Japanese government was suspicious of the Christians. Eventually it forbade any belief in Christianity and took steps to expel the Portuguese and Spanish missionaries from the country. Hence, the Jesuits' impact was relatively short-lived and the teachings of the missionaries in Japan were almost eradicated. We cannot say, therefore, that the Jesuits contributed to Japanese astronomy as much as Ricci and his successors did to Chinese astronomy.

AVAILABILITY OF WESTERN KNOWLEDGE UNDER SECLUSION POLICY

After 1638, the Chinese and Dutch were the only foreigners allowed to reside in Japan, and they were restricted to the city of Nagasaki for the pursuit of trade. This political action was paralleled by restrictions on the import of certain Chinese books, which included all works on Christianity and all works by Christian authors.

It seems that Matteo Ricci was, in the eyes of the government censors, a most dangerous character. Any work by him, or associated with his name, was barred whether it concerned Christian tenets or not. The government had categorically forbidden the importation of all Sino-Jesuit (Chinese language) treatises.

The government, however, never took legal action against the importation of Western books. In early seventeenth-century Japan the fraction of the population that could read Western books was insignificant compared to the fraction that could read Chinese books. The prohibition of Christianity, the departure of foreign missionaries from the country, and the limitations on foreign trade left no opportunity for the ordinary intellectual to receive instruction in European languages.

The only exception was the group of official interpreters at Nāgasaki. In view of their professional function, they were officially permitted to study European languages.

On the whole, the decree banning Christian writings was partly responsible for the predominance in Japan during the seventeenth and early eighteenth centuries of purely traditional astronomy in the Chinese pattern.

RELAXATION OF THE BAN – YOSHIMUNE

While the seventeenth century was mainly spent in catching up with traditional Chinese scholarship, the Japanese from the early eighteenth century on began to realize that the Chinese achievements did not suffice. The relations of Japanese astronomy with that of the West entered a new phase under the Shogun Yoshimune,[h] ruler of Japan from 1716 to 1745, who in 1720 relaxed the ban on Sino-Jesuit treatises.

Intending to revise the contemporary Jōkyō[i] calendar immediately after his appointment, Yoshimune consulted astronomers and mathematicians. These men must have read some of the officially forbidden books which were preserved only in the shogunate library and found them superior to the traditional Chinese works, for they apparently persuaded Yoshimune to collect all Chinese translations and treatises on Western astronomy.

After the relaxation, Western knowledge on an advanced and professionally useful level was transmitted to Japan for the first time. Meanwhile, at the time of Yoshimune in Japan, scholars and officials had no source of information on Western astronomy other than Chinese works. While the Sino-Jesuit works were for decades relied upon by Japanese practical astronomers, modern Western astronomical ideas could not have been available until the difficulty of translating original Western treatises was overcome; the works in Chinese were practically useless in this respect. Thus, the late acceptance of Copernicanism in Japan was mainly due to the exclusive dependence until the early half of the eighteenth century on Chinese sources rather than to particular ideological obstacles.

BEGINNING OF THE 'DUTCH LEARNING'

One consequence of this new receptivity was that in 1745 the interpreters were officially encouraged to learn to read Dutch books. Prior to this, their proficiency had been almost entirely restricted to translation of spoken Dutch. At this early stage, the task of introducing new ideas was left in the hands of only a few linguistic experts.

Circumstances of foreign relations confined Western-oriented Japanese to exclusive reliance on Dutch sources. After its golden age in the seventeenth century, Holland could no longer maintain its aureole; in the late eighteenth century its scientific efforts were at an ebb.[5] It was just then that the Japanese depended on Dutch translations of Western European works. In the 1770s, a generation after the reign of Yoshimune, a notable expansion of study of the Dutch langauge and science led to a move for translation of Dutch scientific works – or retranslation of Dutch translations of Western European works.

TWO SOURCES OF COPERNICANISM

Under the circumstances described above, Copernicanism entered Japan via two distinct routes:

(1) Chinese sources, i. e. the writings of Jesuit missionaries in China and their native sympathizers,
(2) Direct contact with Europeans in Japan, or direct translation from European sources.

Chinese Sources

Since the introduction of heliocentricism into China has been exhaustively treated in another paper,[6] it is necessary here only to note by way of summary that the Jesuit monopoly on scientific teaching in China greatly delayed it, and that by the time Copernicus's cosmological convictions were adequately described in Chinese by Michel Benoist in 1760, so many confusing statements had been made about the Polish astronomer's beliefs that the best Chinese astronomers of the time were justifiably little disposed to accept Benoist's description. Benoist's ideas did not enter Japan until they were referred to in 1846 by the Shogunal astronomer Shibukawa Kagesuke[j] (1787–1856) in his 'Shinpō rekisho zokuhen',[7,k] (Sequel to calendrical treatise by the new method). Of course by this time adequate information on heliostatic cosmology had already entered Japan direct from the West.

Prior to that, no substantial Chinese influence in the matter of Copernicanism can be found. Shizuki Tadao[l] (1760–1806) in his *Rekishō shinsho*[m] (new treatise on calendrical phenomena), Part 1 (drafted in 1798) states that 'I have looked at a work entitled probably *Li-ch'i t'u-shuo*[n] (ritual implements, illustrated) and found many instruments made in England. Among them there were some which place the sun at the centre and the planets and the earth outside. Hence, it may be that this (Copernican) theory is now adopted (in China).'[8]

Li-ch'i t'u-shuo is not found in any Chinese bibliography and 'instruments made in England' does not sound genuinely Chinese, but I have located a *Li-ch'i t'u-shih*[o] (prefaced in 1759 and 1766) which contains two illustrations of heliocentric planetaria kept in the Imperial palace. This must have been the work Tadao saw. Again, however, Tadao had been familiar with Copernican theory since his own translation of John Keill's work in the 1780s and hence this vague Chinese source merely strengthened his own conviction of Copernicanism and also his impression of its diffusion in other parts of the world.

Thus we may conclude that despite a tradition of heavy borrowing from Chinese calendrical astronomy until the Jesuits' time, the Japanese were not perceptibly influenced via China in the matter of Copernicanism. They found their own way towards Copernicanism through direct access to Western sources. This is one of the earliest instances of Japanese independence from her historically overwhelming intellectual dependence upon China.[9]

Western Sources

During the Jesuit century in Japan, some works of missionary origin conveyed an Aristotelian kind of cosmology. We have two such sources extant now. One is

Nigi ryakusetsu[p] (Outline theory of celestial and terrestrial globes; basically a translation of *De sphaera* by the Spanish missionary Pedro Gomez) primarily for use in instructing Japanese students at a Jesuit *collegio*. The other is *Kenkon Bensetsu*[q] (Western cosmography with critical commentaries) consisting of an original text by the apostate Jesuit missionary Christovao Ferreira, and indented annotations by a Japanese Confucian commentator.[10] Both of them are merely popular accounts of Aristotelian cosmology. The latter has a passage refuting the concept of the rotation of the earth, saying that if it were so, everything on earth would be whirled out into space. This looks like a physical argument on Copernicanism but there was no mention of the heliocentric system or even of the name of Copernicus. Presumably, the author was not concerned with Copernicanism, since Ferreira left Europe before the Copernican issue came to be noteworthy; he simply repeated such traditional Aristotelian and Ptolemaic arguments as prevailed in the school curricula at the time.[11]

The numerical value adopted in *Kenkon bensetsu* for the period of precession is 25,798 years, closer to the Copernican value than to the Alphonsine-Clavius-Ricci 49,000 years. This is not a traditional Chinese value, either. It may be that some astronomical aspect of Copernicanism had by chance infiltrated.

From that time the ban on Christianity and Western knowledge set in. Even after Yoshimune's relaxation, he and his followers did not seem to have been conversant with cosmological issues. Nishikawa Masayoshi[r] (1694–1756) of Nagasaki, who had access to the residences of Dutch merchants was, on account of a high reputation for knowledge of Western astronomy, appointed Shogunate astronomer, but his writings are still based on a sketchy knowledge of such quasi-Tychonic cosmology as appeared in the *T'ien-ching huo-wen*[s] (Queries on the classics of the heavens, ca. 1675) by Yu I.[t] For the advocacy of Copernicanism, we have to await the emergence of pioneers from a non-astronomical group.

MOTOKI RYŌEI'S TRANSLATION WORKS ON HELIOCENTRICISM – COSMOLOGICAL ASPECT

Translations of Dutch works by Motoki Ryōei[u] (1735–1794) are significant not only as the first Japanese sources on the Copernican heliocentric system, but also as a landmark in the advancement of the study of Western languages in Japan. Ryōei belonged to the third generation of a hereditary family of Nagasaki interpreters. Besides discharging routine duties, he was perhaps the first to receive official requests to translate Dutch works. During his lifetime Ryōei produced many translations, mainly in the fields of astronomy and geography.

He once stated that since there had never been a professional translation of a Dutch work in Japan, his predecessors were sceptical about the possibility of even literal translation and were reluctant to attempt it.[12] He recognized that if he were to fail in his initial translation, his own ability would be brought into question and his inherited position jeopardized.[13] His attitude indicates that the main difficulty in the introduction of heliocentricism lay in the linguistic barrier and the conservative intellectual atmosphere, in which an innovator hesitated to present anything extravagant. Official requests did, of course, encourage Ryōei, although not all his translations originated in this way. The choice of subject matter for translation must have reflected official concern for the material aspects and products of Western culture but the choice of original texts was his own. Ryōei was primarily a faithful translator, but his own favourite subject was heliocentricism.

The Oranda chikyū zusetsu

Ryōei's first translation referring to the heliocentric theory was drafted in 1772 under the title *Oranda chikyū setsu*[14,v] (A Dutchmen's view of the earth.) There is another copy under the title 'Oranda chikyū zusetsu'[15w] (A Dutchmen's view of the earth, illustrated.) These are referred to hereinafter as 'first manuscript' and 'second manuscript' respectively.

Their original was the Dutch translation *Atlas van Zeevaert en Koophandel door de geheele Weereldt* (Amsterdam, 1745, abb. 'Dutch edition') of the *Atlas de la navigation et du commerce qui se fait dans toutes les parties du monde* (Amsterdam, 1715, abb. 'French edition'), by Louis Renard. A comparison reveals some interesting aspects of abhorrence of 'God' in the introduction of Copernicanism into Japan.

Both the French and Dutch editions are big-scale marine charts with a guide for seamen in the margin. While the original French edition provides only explanations for charts, the Dutch edition incorporates a number of revisions. It also has to begin with some additional elementary accounts of such matters as the earth, the heliocentric system, constellations, longitude and latitude and other astronomical themes, wind and geography, the history of astronomy and geography, the use of the compass, and so on. These additions are considered to have been contributed by a Dutch editor, Jan van den Bosch Melchiorsz.

It was Melchiorsz's additions in which the Japanese translator was chiefly interested. The first manuscript has a summary of the 'Preface' (from which could be identified its Dutch original) and a translation of the first five pages of the Dutch edition. The second manuscript, while translating the first sixteen pages, neglected the 'Preface'. The latter translation, revision of the former, is in somewhat better Japanese.

A comparison with the Dutch edition shows that the translator has omitted certain of the original paragraphs. The details of omissions are as follows. Both translations omit the opening paragraph of the first page of the Dutch edition. This common omission is indicated by (A) in the diagram given below. The second manuscript omits material amounting to almost a whole page on pp. 8–9 in the Dutch edition.

This omission is indicated (B) in the diagram.

Dutch edition	'P'	1	5	8	9	16	
First manuscript	______	________					
Second manuscript	(A)____________			(B)	__________		

Omission (A) is translated as follows:

> The place which God has given people for habitation is commonly called the world, the Earth sphere, the inhabited earth; the term world implies the

> entire created world, the universe; or, to be more precise, everything that is within the circle of the planet Saturn, of which circle the Sun is the centre, and the radius of which extends from the Sun to Saturn. The Earth sphere is composed of earth and ocean, of land and water. It contains many different kinds of materials which God has created for the benefit of man, the quantity of which He has decided upon so that nothing could be added to or taken away from either by the human mind nor by natural causes, for the materials by their nature are essentially distinct from one another, indestructible, incapable of increase and unchangeable . . .[16]

The omission is apparently deliberate, motivated by the translator's abhorrence of the account of God's Creation, and what is more, Creation seems to be associated with the sun-centred universe. This is clear in contrast to his full and faithful translation of the purely scientific account of the heliocentric hypothesis, which immediately follows the omitted paragraph in the Dutch edition. Of course, the Dutch account itself was only a short outline of the daily rotation and annual revolution and the difference between true (heliocentric) and apparent (geocentric) motions.

The Dutch edition gave a brief history of astronomy up to the time of Copernicus and then proceeded to present the discussion involving Biblical issues caused by Copernican theory. The second manuscript translated the historical account up to Copernicus. It was in this connection that the name of Copernicus and the information that Copernicus's heliocentricism was now generally accepted among Western scholars appeared for the first time in Japanese literature.

However, the translator left out all the theological argumentation. This omitted portion is (B) translated below.

> However, the lower classes, or the excessively prejudiced (was this the case among lower classes only?), find this feeling strange, which seems to them to be contrary to what they see; apart from the fact that they have peculiar ideas about this rotation . . .
>
> Even though it had pleased the Creator of all things to make the Earth and other planets turn in the period of several months around the Sun, and each of them in the period of several hours around their own axis, instead of making the Sun and stars and the immeasurable sky turn with an utterly incomprehensible speed around the Earth for the Earth's benefit, the latter being in comparison hardly a single tiny speck, we still should see everything in the same form as we see it now . . .
>
> However, we shall not go into this any further in order not to digress from the main subject, considering that this suffices to show that nothing more simple and comprehensive can be understood than this explanation of Copernicus; and that only ignorant prejudice and certain passages of the Holy Scriptures have caused this explanation to be discarded as being contrary to common sense . . . So there is nothing in the Holy Scriptures which could serve as an argument against this explanation of Copernicus . . .[17]

The heliocentric account terminates at this point and then the text goes on to discuss the use of the compass. Hereafter, the translator resumed his faithful translation again.

Published in the liberal Netherlands, the Dutch edition had no prejudice against Copernicanism; it assumed the tone of enlightening ignorant seamen.

After all, it was written in the mid-eighteenth century, when Copernicanism was already established. Yet, some obsolete theological discussion was added in order to meet popular interest. Most likely, the translator learned from it the intimate relationship between Copernicanism and the Christian God, and took special pains to introduce only heliocentricism.[18]

The 'Tenchi nikyū yōhō'

The second work by the same translator was the *Tenchi nikyū yōhō*[x] (the use of celestial and terrestrial globes) dated 1774. Its Dutch original was *Tweevoudigh onderwiis van de hemelsche en aardsche globen* (Amsterdam, 1666,[19] first edition, 1620).

The book was written and prefaced by Willem Janszoon Blaeu (or Blaaw, 1572–1638), and edited and published by his son Johan. Willem Janszoon Blaeu was a renowned Dutch cartographer and an intimate friend and disciple of Tycho Brahe. He was also one of the early proponents of Copernicanism.[20]

We have no way of proving when Blaeu's 1666 edition was brought to Japan. It may be that it was imported into Nagasaki long before the Japanese translation in 1774 and then buried in obscurity. But there may be a hidden reason on the part of the translator why he picked on this particular work for translation.

At the time of the publication of Blaeu's book, the first edition of which appeared as early as 1620, the Copernican controversy was raging furiously; and hence it was naturally full of theological arguments. It is manifest that Blaeu wrote the book with the intention of propagating the Copernican hypothesis.

While Blaeu's preface stated that his intent was to illustrate the Ptolemaic system first, because it was more familiar and more easily comprehensible, and then go on to the true theory of Copernicus, Motoki Ryōei's own preface merely mentions the name of Copernicus as follows:

> About one hundred years ago, there was a man called Nicolaus Copernicus. Being in intimate communications with Tycho Brahe, the biggest figure in astronomical observations, Copernicus investigated this [heliocentric] theory thoroughly and finally out of opaque darkness reached enlightenment.[21]

This extract is apparently an exposition of Copernican heliocentricism, but we have found no further mention of it in this work. While the preface of Blaeu's original had eight paragraphs, Ryōei translated and put into his own preface only the first four paragraphs. Paragraphs 5 through 7 discussed Blaeu's plan of arranging the Copernican system as opposed to the Ptolemaic, but these are all omitted from Ryōei's preface. Furthermore, Ryōei confused the historical relationship between Copernicus and Tycho Brahe. This was the source of confusion among later popularizers, who took Ryōei's writings as a basis.

Blaeu, with the intention of propagating the new theory, divided his work into two volumes.

Volume I: Astronomical principles of celestial and terrestrial globes based on the inadequate hypothesis of Ptolemy.

Volume II: Astronomical principles of globes based on the true hypothesis of Copernicus.

Ryōei's translation terminated near the middle of Volume I (at page 120 of 163 pages) the part based on Ptolemaic geocentric theory. Apparently he had no intention of extending his translation further. It is highly probable that he deliberately omitted the section on heliocentricism from the main text as well as from his own preface.

It seems unlikely that the Copernican heliocentric theory was, despite its unfamiliarity, insurmountably difficult for Ryōei to comprehend. Blaeu's astronomical writing was not particularly advanced; Kepler's contributions did not appear in it. Hence it would be more plausible to interpret this deliberate omission as being caused by Ryōei's precaution against any subversive or unorthodox thoughts. The translator, perhaps influenced by his own earlier experience of translation, replaced Blaeu's long theology-filled preface with his own brief summary of a purely physical nature and left entirely untranslated the second volume on the Copernican system.[22]

The 'Shinsei tenchi nikyū yōhō ki'

A more detailed and truly comprehensive account of the Copernican system was translated in 1792–1793. It was entitled *Seijutsu hongen taiyō kyūri ryōkai shinsei tenchi nikyū yōhō ki* (the basis of astronomy, newly edited and illustrated; on the use of celestial and terrestrial globes according to the heliocentric system), and consisted of seven volumes.[23] The Dutch original has been identified as *Gronden der sterrenkunde, gelegd in het zonnestelzel bevatlijk gemaakt; in eene beschrijving vant'n maaksel en gebruik der nieuwe hemelen aard-globen* (Amsterdam, 1770), 470 pages.[24]

Its author, George Adams the elder (died 1773), was a maker of mathematical instruments under King George III. He had a worldwide reputation as a maker of celestial and terrestrial globes. The ultimate original of Ryōei's translation, Adams' *Treatise describing and explaining the construction and use of new celestial and terrestrial globes* (London, 1766), passed through thirty editions in England and was also printed in America.[25]

Ryōei's translation included the first 325 of the 360 paragraphs of the Dutch original; only a portion on the use of globes was left untranslated. The original began with a straightforward description and explanation of the solar system, in which the relationship between the apparent and true courses of the planets was expounded on the basis of the heliocentric scheme. It was already free of time-honoured religio-cosmological controversy; its purely scientific arrangement seemed to have appealed to the translator. It was translated in full and a truly comprehensive account of the Copernican system, for the first time, became available in Japan. He even ventured to put 'heliocentric system' into the title of his translation.

The original was not an advanced treatise for professional astronomers, but a textbook for navigators. The arrangement of themes was, however, strikingly different from that of traditional treatises of calendrical astronomy. From the outset, the earth was treated as a member of the solar system. First, detailed instructions were given for the reduction from geocentric to heliocentric coordinates; then, the behaviour of the planets and satellites was expounded. But, lack of accurate detail made this treatise of little use to Japanese practical astronomers.

Ryōei also gave brief summaries of *Philosophische onderwijzer* and *Beginselen der Natuurkunde*.[26] The original of the former is Benjamin Martin's *The philosophical grammar* (first edition 1738, Dutch edition 1744); that of the latter is the *Anfangs-gründe der Physik*[27] by Johann Heinrich Winkler (Dutch edition 1768). These works elucidated the Newtonian laws of mechanics, but they were beyond Ryōei's concern or comprehension.[28] He could only briefly compare the Ptolemaic, Tychonic and Copernican systems.

The transformation from geocentricism to heliocentricism did not, in the absence of certain religious and philosophical presuppositions, raise difficult technical problems. Thus, unlike the physical aspect of Newtonianism, the introduction of the cosmological or geometrico-morphological aspect of Copernicanism never raised difficult questions for Ryōei.

SHIZUKI TADAO – THE PHYSICAL ASPECT

While Motoki Ryōei was thoroughly loyal to his official duty and interested in Copernicanism merely as an assignment, his pupil Shizuki Tadao (1760–1806), also born in a family of Nagasaki official interpreters, renounced his hereditary position at the age of eighteen and devoted the rest of his life to the subjects of his own interest – natural philosophy and cosmology.[29]

Immediately after his retirement from office, he undertook the introduction of Newton's doctrines for the first time into Japan. The original version used by Tadao was a Dutch translation by Johan Lulofs, *Inleidinge tot de waare Natuur-en Sterrekunde* (Amsterdam, 1741), of John Keill's (1671–1721) *Introductiones ad veram Physicam et veram Astronomiam* (London, 1739).

Tadao spent more than twenty years on his translation. His preparatory notes were made in three drafts, *Tenmon kanki*[y] (Astronomical collection; 1782), *Dōgaku shinan*[z] (Guide to mechanics; n.d.), and *Kyūshinryoku ron*[aa] (on attraction; 1784).[30] These were revised with substantial amendment into the final monograph, entitled *Rekishō shinsho* (new treatise on calendrical phenomena), which appeared in three volumes (completed 1798, 1800 and 1802).

Tadao's work was not a literal translation at all, but rather a collection of his notes with abundant commentaries of his own.

Tadao was not a mere linguistic expert but the most profound philosophical mind of his day. Although a number of more systematically modernized text-books on natural philosophy, such as B. Martin's, were available to him, he picked up the then outdated work by Keill since Tadao, from his personal propensity for metaphysics, was interested in its polemical manner in early Newtonian days.

Keill's work had the typical post-Newtonian deistic tone; unrestrained praise of God's creation, harmony, order, symmetry, beauty and so forth is lavishly distributed throughout. These embellishments were eliminated from Tadao's translation, except in one place where he rendered 'the idea of God's design' as 'the agency of limitless *prajña* (a Buddhist term, literally 'wisdom').[31]

To Tadao, the location of the sun, whether geocentric or heliocentric, was not a basic issue of Copernicanism; the relativity of location and motion was the most impressive and admirable feature of the heliocentric theory. Motion or rest depended solely on the point of reference, and the two were therefore intrinsically indistinguishable. There was neither absolute rest nor absolute motion. For geographical location, or even for family relations, there was no absolute measure. If the point of observation was the sun, the earth was in motion, and vice versa. Thus there was no reason to prefer either the ancient Chinese or the Western theory concerning the motion of the sun or earth.

From this relativistic notion the plurality of worlds followed easily, since the immediate world had no particular priority. To support this theory, Tadao quoted Buddhist writers and Lieh-tzu,[ab] an ancient and possibly legendary Chinese Taoist thinker, whose grounds for this view were intuitive and metaphorical. Tadao cited a Chinese statement that 'things [fluid] pure and light go up and form the heavens, while the turbid and heavy come down and make the earth,[32]' and argued that every star and planet was composed of the same kind of turbid matter as the earth. Thus, he claimed that the plurality of worlds was known before the Westerners advanced it. (The original content of this statement, which is perhaps a Neo-Confucian saying, had nothing to do with the plurality problem as such.) While he saw that the Western theories of the earth's motion and the plurality of worlds were significant because they were based on properly related reasoning and observation, he was also anxious that the contributions of his own cultural tradition be recognized.

In the appendix to Volume I, Tadao presented his own account of the heliocentric system, entitled *Tentairon*[33ac] (discussions on the heavenly bodies). Tadao did not evince any enthusiasm for the theory he was introducing. His intent was to reconcile modern Western theory with traditional Chinese views.

In the appendix to Volume II, entitled *Fusoku*[ad] (Immensurability), he felt obliged to justify his cosmologic views in terms of Confucian morality:

> However, in everything there always exists a governing centre. For an individual, the heart; for a household, the father; for a province, the government; for the whole country, the imperial court; and for the whole universe, the sun. Therefore, to behave well, to practise filial piety towards one's father, to serve one's lord well, and to respond to the immensurable order of the heavens are the ways to tune one's heart to the heart of the sun. This is the way to admire the sovereign of the universe.[34]

When confronted with the incompatibility of the Copernican theory and the traditional notion, which identified the sky with yang and motion and the earth with yin and rest, Tadao sought to preserve the respectability of ancient Chinese concepts by quoting an ancient passage which, interpreted very freely, referred to the motion of the earth. And when the conservative Chinese attitude towards the new theory was attacked on the grounds that even though the name of Copernicus appeared in the *Li-hsiang k'ao-ch'eng*,[ae] the Chinese still did not adopt the Copernican theory, Tadao again defended the Chinese. In a rather fair apologia, he pointed out that the Chinese were concerned only with observations and predictions of the apparent courses of the heavenly bodies, and not with theory; therefore they had no compelling reason to adopt heliocentricism.

At the end of Volume II, Tadao raised the question as to why all the planets rotate and revolve in the same direction, in planes not greatly inclined to the ecliptic. By way of an answer, he proposed at the very end of the treatise, in a section entitled *Kenkon bunpan zusetsu*[af] (the separation of opposites in the generation of the cosmos illustrated), a hypothesis concerning the formation of the planetary system. He claimed it as his own idea, saying: 'It may be that this theory has already been formulated by some Western scholar, but we have never heard of it.'[35]

Tadao's hypothesis immediately recalls the celebrated hypotheses of Kant and Laplace. Of these Laplace's is scientifically the most advanced. Tadao's argument was not Laplacian abstraction from a cautious synthesis of all relevant observational data, but was somewhat closer to the rationalistic inferences of Kant.

In view of the relative inaccessibility of Western treatises, it is unlikely that Tadao borrowed his idea from anyone else. His hypothesis, considering his background in Neo-Confucian ideas, was not a titanic leap. Many aspects of it were already present in the Neo-Confucian vortex cosmogony, which claims that beginning with primordial chaos the light fluid tends to float to the surface and heavy matter to precipitate at the centre in the course of one-way revolution. Hence, a small portion of the ideas of attraction and centrifugal force provided Tadao with a more elaborate mechanical hypothesis, formulated in accordance with the heliocentric system.

DIFFUSION OF COPERNICANISM

At the time, no work by Motoki Ryōei and Shizuki Tadao was ever published in printed form. For the most part, their works were preserved in manuscript form and circulated as handwritten copies.

It was Shiba Kōkan[ag] (1747?–1818) who popularized the Copernican theory through three printed books:[36] *Chikyū zenzu ryakusetsu*[ah] (An outline world atlas; 1793); *Oranda tensetsu*[ai] (Dutch astronomy: 1795); and *Kopperunyu tenmon zukai*[aj] (Copernican astronomy illustrated; 1805). A literary dilettante and gifted freelance painter in the Western style, Kōkan enjoyed more freedom than did the official interpreters and astronomers. He frankly acknowledged Western achievements and superiority, but never apologized, as Tadao did, for Sino-Japanese tradition. Emancipating himself from the predominant notion that Western knowledge was valuable only from a utilitarian point of view, he incorporated Western astronomical concepts into his thought and used them as one basis of his unique materialistic cosmology, in which fire was the fundamental element.[37] Still the depth of his knowledge of Western astronomy did not exceed that of Motoki Ryōei, whose work was his main source.

Shizuki Tadao's physical ideas were taken over by the freethinker Yamagata Bantō[ak] (1748–1821), the head clerk of an Osaka business firm and an outspoken polemist with conventional beliefs. While Tadao remained introspective and critical to anything extravagant, open-minded Bantō extended Copernicanism freely into his 'great universe' picture, in which plural worlds were arranged in hierarchical order, like Olbers' conception of the universe.[38]

Many other popularizers such as Hoashi Banri[al] and Yoshio Hisasada[am] followed in expounding Copernican and Newtonian thought, but Shizuki Tadao's contribution was so outstanding and penetrating that his work remained unsurpassed until the middle of the nineteenth century.

THE ASADA SCHOOL – ASTRONOMICAL ASPECTS

Among the professional calendrical astronomers, Asada Gōryū[an] (1734–1799) and his school first fully recognized the technical superiority of Western astronomy over the traditional Chinese type, and gave effect to it in the Kansei[ao] calendar reform (1798). This recognition followed upon the importation of such impressively voluminous Sino-Jesuit works as the *Hsi-yang hsin-fa li-shu*[ap] and *Li-hsiang k'aoch'eng*.

Miura Baien[aq] in his *Kizan roku*[ar] (Notes during a trip to Nagasaki) quoted a letter from his close friend Gōryū in 1778 to the effect that: 'later a man called Copernicus appeared. On the nine-layered spheres of the cosmos he superimposed other two heavens to account for the east-west and south-north precessions, building up a total of eleven celestial spheres.'[39] Thus at this time, Gōryū still held an erroneous picture of Copernicus derived from the *Hsi-yang hsin-fa li-shu*. On the other hand, Gōryū must have learned from Baien about the heliocentric hypothesis, the gist of which had been transmitted to the latter by Matsumura Suigai,[as] a collaborator of Motoki Ryōei, during his trip to Nagasaki in 1778.[40]

It is, of course, quite possible that Gōryū was unable to identify the heliocentric Copernicus with the conventional technician mentioned in the Sino-Jesuit treatises. This would have been virtually impossible in view of the limited information provided by the missionaries in China.

Nevertheless, Gōryū and his school committed themselves to Copernican heliocentricism. His pupils claimed for him the honour of having discovered independently, ca. 1796, although he did not publish it, the relationship between the distance of planets from the sun and the periods of their revolution (in other words, Kepler's third law).[41] The independence of his discovery is quite doubtful,[42] but in any case Kepler's third law could not have been reached without starting out from the Copernican heliocentric scheme. Hence, at least Gōryū and such advocates of his discovery as Takahashi Yoshitoki (1764–1804) and Hazama Shigetomi (1756–1816) must have been convinced of the cosmological and physical truth of the Copernican scheme. In fact, Takahashi Yoshitoki in the opening part of his *Zōshūshō chō hō*[43, at] (Hsiao-ch'ang method, revised and augmented, 1798) advocated Copernicanism enthusiastically, although it was completely irrelevant to his *shōchō hō*[au] concept, which deals exclusively with the matter of astronomical parameters. Around the same time, Shigetomi wrote *Tenchi nikyū yōhō hyōsetsu*[av] (A commentary on Motoki Ryōei's translation work on heliocentricism, 1798), attempting to correct the astronomical errors of Ryōei, who knew no professional-level astronomy.

Why was it that calendrical astronomers, who had maintained a professional indifference towards cosmology, developed a keen interest in Copernicanism? It is true that Copernicanism included some computational novelties which calendrical astronomers eagerly adopted, although in general they saw its cosmology as merely a matter of transformation of the geocentric coordinates of Tycho into a heliocentric frame of reference. At least as Ryōei understood and introduced it, it did not show any potential for new solutions to classical Japanese problems, and thus had little reference to the traditionally central issue of calendar reform.

Furthermore, by the late eighteenth century, when Copernicanism reached Japan, the observational and astronomical innovations of Tycho, Kepler and also Cassini, Delahire and others were already available through Sino-Jesuit works. There was no particular reason for the Japanese to pay particular regard to the astronomical aspects of Copernicus's own work.

The answer must involve less obvious factors than simple computational convenience. Perhaps the most important element is that Copernican heliocentricism functioned as the symbol for a novelty-oriented school of calendrical astronomers – that of Gōryū and his followers – who were advocating the superiority of modern Western astronomy over traditional one. The situation was similar to that in medicine around the same period, when men like Sugita Genpaku[aw] (1733–1817) and his associates enthusiastically advocated Western achievements in anatomy although they had no direct application as Japanese medicine was then conceived.

Furthermore, their willingness to diverge from tradition led to a new interest in planetary theory. In China, in the *Li-hsiang k'ao-ch'eng hou-pien*[ax] (1742), the Keplerian ellipse was applied only to the movements of the sun and moon; extending them to the planets would have required Copernican coordinates, which the missionary astronomers hesitated to adopt. Takahashi Yoshitoki, challenged by this lacuna, took upon himself the task of extending the Keplerian ellipse into planetary theory. He thus rejected Tychonic coordinates, which the *Hou-pien* still firmly maintained, in favour of the Copernican frame of reference; otherwise, with the planets turning around the sun, which in turn turns around the earth, the result would have been a complicated and ugly double elliptic scheme, and the aesthetic advantage of adopting the ellipse would have been largely lost. Yoshitoki's scholarly devotion to planetary theory was a remarkable departure from traditional calendrical concerns towards an astronomy based on the concept of the solar system.

Yoshitoki's son, Shibukawa Kagesuke (1787–1856), the last of the great calendrical astronomers, adopted planetary elliptic orbits in the last luni-solar calendar reform in 1843, but even in the final attempt at a traditional official treatise, *Shinpō rekisho*[44, ay] (A calendrical treatise by the new method, 1846), the framework and construction remained traditional; the new cosmology could not be truly integrated in it. Only the last five sections of the 'zokuhen' (Sequel) were devoted to the new Copernican and Newtonian approach, which were thus a mere appendix.

In the preface to the sequel, Kagesuke stated his intention to clarify from the viewpoint of a professional astronomer points of confusion caused by the introduction of heliocentricism. He emphasized that geostatic motion is apparent while heliostatic motion is physically true, but that this is merely a matter of coordinate transformation; the astronomical implications are all the same, and both theories have an equal *logical* claim to verisimilitude. This is the attitude of an astronomical technician towards Copernicanism, not that of a man free to cultivate cosmological or physical interests. Unlike such independent popularizers as Kōkan, or such pioneering enthusiasts as Yoshitoki, the head of the Shogunate's bureau of astronomy, Kagesuke, had to defend traditional bureaucratic responsibilities against modish Western learning, and never accepted any Western theoretical novelty without reservations. But among the professional astronomers of Kagesuke's generation, Copernicanism was accepted as practically self-evident.

IDEOLOGICAL REACTIONS TO WESTERN COSMOLOGICAL THEORIES

The reactions of Buddhists, Confucians and Shintoists to Western cosmologic theories varied widely, resistance to innovation being caused by a conflicting worldview and unyielding commitment to Eastern culture.

The Reactions of the Buddhists

By the middle of the eighteenth century, the Aristotelian type of cosmology had been diffused through the printed editions of *T'ienching huo-wen*. Some Buddhists could not tolerate the diffusion of European cosmological ideas, which were incompatible with Buddhist beliefs. To refute these ideas, a treatise entitled *Hi tenkei wakumon*[az] (Contra *T'ien-ching huo-wen*), by the learned Buddhist monk Monnō,[ba] was published in 1759. Another of his works, *Kusen hakkai tōron*[bb] (A discussion of the theory of the Nine Mountains and Eight Seas), expounded *Sumeru* cosmology, which sets at the centre of the earth Mount Sumeru, round which the sun, the moon, and the stars revolved; this idea originated in Jaina cosmography and was taken over by the Buddhists.[45]

Generally speaking, Buddhist cosmology and astronomy had a much wider scope than corresponding Western disciplines and enjoyed more freedom in the infinity and plurality of worlds than the rigidly constructed Aristotelian world, but this breadth and freedom was due to vagueness and lack of conviction concerning the existence of an underlying regularity in the phenomenal world, and the possibility of discovering one.

Some of Monnō's criticisms were interesting. For instance, according to him, the universe is a limitless void, no bounds being even conceivable. The infinity of the universe is a consequence of the emptiness of the phenomenal world. Attempts to measure the dimensions of the universe are therefore ridiculous. If there are nine spheres within the empyrean, he argued, why not an equal number of spheres outside it?

Monnō's arguments were not very constructive, however. He denounced Aristotelian cosmology, but was unable to replace it with a more consistent system.

The creativity of Japanese Buddhism was in eclipse in the seventeenth century, and intellectual leadership shifted to Confucianism, the state orthodoxy. During the eighteenth century, as Western astronomy proved to Japanese intellectuals its superiority and as knowledge of the Copernican system was disseminated through the various popular editions of Shiba Kōkan, a feeling of anxiety and crisis was aroused among some sects of Buddhists. While the Confucians were not so strongly opposed to the achievements of Western astronomy, the Buddhists tried to restore their own intellectual aura by scoring a victory for Buddhist cosmology and by defending it against Kōkan's ridicule.

Among this group of Buddhists the most zealous and influential figure was

Entsū[bc] (1754–1834). His thirty years of labour in defence of Buddhist astronomy culminated in his masterpiece, the *Bukkoku rekishō hen*[bd] (On the astronomy and calendrical theory of Buddha's country; five volumes, 1810),[46] which was followed by a number of other writings in the same vein.

From the outset, Entsū's single purpose of defending Buddhist doctrine from the invasion of Western scientific ideas was marked. His motivation was a fear that Christianity would undermine Buddhist teaching in Japan. In this respect, he might be following the precedents of Chinese Buddhists' experience under Jesuit influence. This basic attitude was not very different from that of his predecessors such as Monnō, but Entsū's investigations covered an extensive literature, not only in cosmology, but also in the technical aspects of calendar-making. His sources ranged from Buddhist sutras in Chinese translation and ancient and modern Chinese astronomical treatises to Sino-Jesuit works and works on Western astronomy. Although he was said to have had some background in Dutch learning, there is no evidence that he was able to read Dutch. He probably relied on books written by his Japanese contemporaries.

Entsū was not essentially opposed to Western influence in calendrical science, since this traditional activity did not bear on cosmology. He argued, however, that the only worthwhile aspects of calendrical science had originated in India. The Chinese had appropriated the advanced Indian calendar during, for instance, the T'ang period.[47] Citing a superficial Chinese account in the Ming history, he credited India with the origins of Islamic astronomy, and thus also of European astronomy, since the European ephemeris was practically the same as the Islamic. Thus distorting the course of historical development of science, he claimed eventually the Western appropriation of indigenous Indian science.

On cosmological questions, Entsū would yield nothing to the European system, his argument being almost identical to that of his predecessor Monnō, but more amplified and emphatic.[48] From beginning to end, he rigidly and literally supported Sumeru cosmology. Furthermore, he made use of the ancient Chinese flat-earth (*kai-t'ien*) theory, which also propounded a nonspherical model of the universe. The *kai-t'ien* theory differed from Sumeru cosmology in that it had an empirical basis and was virtually free from anthropomorphic and mythological elements, while Buddhist cosmology was replete with religious fantasy and lacked an observational foundation. However, Entsū rationalized the difference by explaining that whereas the Chinese were skilful at rational investigation, the Indian sages had penetrating 'spiritual eyes' which were given only to superior beings.

Entsū was not rabidly anti-scientific; his object was simply to defend the sacred Buddhist cosmology by all available means. Though he found many adherents among Buddhist monks, his biased views never achieved orthodoxy, but were frowned on by more modest people. He attempted to obtain a Buddhist imprimatur for the publication of the *Bukkoku rekishō hen*, but the aged commissioner Sen'yō[be] would not grant it, on the grounds that 'since Entsū was too involved in astronomy, he confused the essentials of Buddhism with astronomical science. His opinions might cause incalculable damage to genuine Buddhism.'[49]

Other monks also denounced Entsū. It would be dangerous for Buddhism to commit itself to Sumeru cosmology. Properly speaking, the Sumeru theory was not of Buddhist origin, but was derived from older Indian ideas.

Some, while in basic agreement with Entsū, considered his views too extreme,[50] and others condemned his attempt to substitute for Sumeru the mathematically elaborate *kai-t'ien* theory.[51] To clarify this point, a number of commentaries on *kai-t'ien* theory appeared at that time.[52] Inō Tadataka,[bf] a prominent surveyor in the Asada school, promptly came out with his *Bukkoku rekishō hen sekimō*[bg] (A refutation of the *Bukkoku rekishō hen*; 1816 or 1817), a bitter denunciation of Entsū's unscientific and misleading dogma. Most top-rank astronomers, however, merely disregarded or ridiculed Entsū's work.[53]

Some of Entsū's writings were suppressed in the 1820s;[54] but other exponents of the same philosophy arose. Even during the Meiji period, when explosive Westernization had almost eradicated traditional attitudes, a prolonged effort to defend Buddhist cosmology was made. The most notable apologist was Sada Kaiseki.[bh] His main work to explain all celestial phenomena in terms of a complicated mechanism, the *Shijitsu tōshōgi shōsetsu*[bi] (A detailed account of an instrument by which the apparent and real courses of the heavenly bodies are explained) appeared as late as 1880.

The sensation caused by these unusually sharp Buddhist reactions to Western cosmology prompted the circulation of a great deal of controversial literature on astronomy. Thus, the Buddhists, by trying to block the advance of Western ideas in Japan, actually stimulated the propagandizing of Copernicanism. The result was a heightened public interest in cosmology.

The Reaction of the Confucians

By contrast with the pronounced, violent opposition of the Buddhists, we find relatively little reluctance on the part of the Confucians to accept the Copernican system. Some were outspokenly hostile to everything Western, and others claimed Western appropriation of ancient Chinese ideas. Neo-Confucianism (Chu Tzuism), the Japanese state orthodoxy, formed its cosmological background as an integral part of an unitary principle, and therefore, violations in Western cosmology of this unity between human and physical nature were often criticized; however, the Neo-Confucians did not, like the Buddhists or the medieval Church, maintain a detailed religious cosmos, and in general were not concerned with the appearance and morphological aspect (shape) of the universe. Therefore, there was no clear point of conflict with Western cosmology.

The Neo-Confucian idea of cosmological unity was challenged by the Ancient Learning (*kogaku*)[bj] school of Itō Jinsai[bk] and Ogiu Sorai.[bl] They accused Neo-Confucians of indulging in fruitless speculation concerning the heavens, when they should be studying only moral and social problems. Regarding astronomy as a technique unrelated to Confucian moral values, they excluded both physics and natural philosophy from their world-view.

Kan Sazan[bm] (1748–1828) commented that 'there is no other use for astronomy than the determination of the correct time; other concerns (viz., cosmology and general astronomy) are merely useless argument and dull speculation'.[55] It would not be far from the truth to assume that his attitude was largely shared by other Confucian scholars, whose pragmatic interest in social and ethical problems excluded a disinterested concern with physical nature.

Influenced by the Ancient Learning school, Neo-Confucians themselves ceased to defend the idea of cosmic unity in the vigorous manner of earlier generations. Thus the Confucian framework of ideas became sufficiently flexible enough to tolerate the reception of Western cosmology without serious ideological difficulty. The acceptance of Western learning was facilitated by two assumptions: that it was historically of Chinese origin, and that as a mere technique it supplemented Eastern values without threatening them.[56]

The belief that current Western scientific theories were originally Chinese was propounded by contemporary Chinese intellectuals, who aimed at apologias for their own tradition and also at the revival of native science, especially mathematics and astronomy. Some of the Japanese Confucians were, as spokesmen for Chinese culture, also in a position to defend the Chinese scientific tradition and its abiding value. Some orthodox Neo-Confucians, Asaka Gonsai[57, bn] and Yasui Sokken,[58, bo] both teachers at the Shōheikō,[bp] the official shogunal school and a stronghold of conservatism, were typical in this respect. They merely imitated the Chinese way of a grandizing classical achievements by extensive investigation of ancient writings.

On the matter of Copernicanism, they welcomed Shizuki Tadao's interpretation. Deeply versed in Chinese classics, they often quoted an ambiguous passage, 'the earth has four displacements', from the *Shang-shu wei k'ao-ling-yao*[bq] (An Investigation of the numinous luminaries; first century B.C.) to support their contention that the heliocentric theory is of Chinese origin.

This classical phrase received the attentions of Chinese philosophers and was interpreted in various ways by later commentators, but it was Japanese scholars who first connected it with Copernicanism, a little earlier than the Chinese. The Chinese priority in the idea of the earth's motion was claimed only on the premise that the truth of Copernicanism was fully recognized. Thus the alleged Western appropriation marks the time when Copernicanism was accepted among intellectuals in general.

This scholarly defence by Japanese Confucians of Chinese culture, though clearly directed against the aggression of Western science, was far less fanatic than the Buddhist reaction. It was also more objective than the contemporary Chinese attitude, which was limited by ethnocentricity. 'It is the vice of the Chinese not to acknowledge the strong points of other countries and always to insist that everything worthwhile comes from China', wrote Ikai Keisho[br] (died 1845).[59]

Rather than merely apologize for Western science, certain private schools of Neo-Confucians acclaimed it outright. Notable was the Kaitokudō[bs] school in Osaka, which produced Yamagata Bantō and Hoashi Banri, famous exponents of Western science. Unencumbered by the responsibilities of public office, they could freely develop their interests and criticism.

The idea that Western learning was a mere technique and as such did not conflict with Confucian values was perhaps insisted upon more and earlier in Japan than in China. Japan thus felt free to choose between Chinese and Western techniques, without being inhibited by cultural ties.

The Reaction of the Shintoists

During the eighteenth century, a group of Neo-Shintoists (*kokugaku*,[bt] literally, 'national learning') gradually became influential. Strongly opposed to the speculations of Buddhists and Confucians, these scholars maintained a more or less positivistic attitude towards scholarly problems and were generous and sympathetic to Western learning.[60]

In 1790 Motoori Norinaga[bu] (1730–1801) wrote a critical essay denouncing Monnō and Sumeru cosmology in favour of the spherical earth theory. He singled out, one by one, the absurdities of Buddhist cosmology and concluded that the Buddhists were so envious of the Western theory that they took unfair advantage of the Chinese *kai-t'ien* theory to strengthen their position. Unlike the tradition-bound Confucians and Buddhists, he plainly acknowledged the advanced state of modern theory, saying: 'The motive for which Western people study astronomy and geography is not merely to succeed in scholastic debate or calendrical work. Their science is of crucial importance for daily use in navigating the oceans; even a small error would result in a grave accident.'[61]

But he was still far from comprehending the view point of Western science. In his *Tenmon zusetsu*[bv] (An illustrated description of astronomy; 1782), he wrote, 'the treatment of the five planets has nothing to do with calendrical science. Astronomers should not regard it as an important concern of theirs.'

By the next generation, general knowledge of Western science, including the Copernican and Newtonian theories, was more widely diffused. At the same time, because of increasing foreign threats, a nationalistic spirit and a desire for independent identity were prevalent. Hirata Atsutane[bw] (1776–1843) and his followers tried to establish a doctrinal basis for this nationalism out of the ancient native mythopoetic tradition, including elements of Christianity, Confucianism, Buddhism and whatever else was available to him. They emphasized native contributions to Japanese thought, which were uncontaminated by Chinese and Buddhist influences, and also attacked current Confucian and Buddhist ideas.

Unlike the Buddhists and Confucians, the Neo-Shintoists did not have a quasi-scientific tradition that required apology. The ambiguity of their mythology invited free interpretation. In the absence of historical domination and foreign authority, they could 'create' their own tradition and include aspects of Western scientific thought.

Atsutane and his pupils, Satō Nobuhiro[bx] (1769–1850) and Tsurumine Shigenobu (1786–1859),[61,by] were all acquainted with fruits of Dutch learning and had an unreserved appreciation for modern Western science. Earlier attempts had been made to amalgamate Neo-Confucian cosmology with the native creation myths, but Atsutane was bolder and quite cleverly utilized the most up-to-date Western theories available. The result was a curious combination of primitive myth and modern science.[62] The process of world creation, ignored in modern science, was explained by traditional myths: the creators, a god and goddess, formed the universe from primordial chaos and gradually modelled the heliocentric system.[63] Thus Atsutane and his followers refuted Entsū's Sumeru argument and took full advantage of Western science in their attempt to systematize primitive mythology into a consistent cosmology.

THE IDENTIFICATION OF COPERNICUS

In China, the Jesuits' distorted account of Copernicus created a confused picture reflected in Juan Yuan's[bz] commentary on Copernicanism. This confusion was still worse in Japan, where the name of Copernicus came to be known not only through the Sino-Jesuits' works but additionally via quite another route, their own translations of Western works.

Motoki Ryōei, a professional interpreter, was not acquainted with the Chinese astronomical writings and hence in his translations he freely employed his own technical terms and transliterations without attempting to identify them with the traditional terminology of calendrical astronomy and standard Chinese transliterations for the names of such Western astronomers as Copernicus and Tycho.

As already indicated, Ryōei, in the preface of his second heliocentric translation, apparently made the error of considering Copernicus a contemporary of Tycho. To his third translation, he added an appendix consisting of summaries from Benjamin Martin's *Philosophical Grammar* and Johann Heinrich Winkler's *Anfangs-grūnde der Physik* (both in Dutch translations), in which three systems, the Ptolemaic, Copernican and Tychonic, were well illustrated. He did not commit any apparent historical error, though passages relating to relationships of the three figures are still somewhat clumsy. He well enough understood the major differences between the Copernican and Tychonic systems. He placed Tycho's scheme after that of Ptolemy, while Copernican was given full credit among 'contemporary' astronomers.

The confusion was increased still further by thoughtless popularizers such as Shiba Kōkan and Honda Rimei.[ca] Kōkan is credited as the first to print an identification of Copernicus as mentioned in Ryōei's translation with the man identified by a Chinese transliteration of his name in the *Li-hsiang k'ao-ch'eng*.[64] But Kōkan confused Copernicus with Kepler taking 'Kopperunyu', the Japanese reading of the Chinese transliteration of 'Kepler', as the equivalent of 'Copernicus'. He even included this erroneous name in the title of his book, *Kopperunyu tenmon zukai* (1805). Honda Rimei in his *Seiiki monogatari* (ca. 1798) wrote that 'Copernicus is a pupil of Tycho Brahe.'[65]

Shizuki Tadao took a much more cautious and scholarly attitude towards this identification problem in the first part of *Rekishō shinsho*[cc] (drafted in 1798), noting that 'someone said that Kopeni (Japanese reading of Chinese Ko-po-ni) spoken of in the *Li-hsiang k'ao-ch'eng* must be identical with the Copernicus spoken of in Western sources'.[66]

Finally the shrewd sense of a professional astronomer settled this identification problem. Takahashi Yoshitoki in a letter to Hazama Shigetomi in 1800 compared the relative distance of the solar perigee (or perihelion) given by various astronomers in the Jesuit treatises with those given in Ryōei's third translation (based on Adams), concluding that the Sino-Jesuits' Copernicus was identical with the Copernicus who appeared in the Dutch sources,[67] despite the fact that the latter writings portrayed his cosmology as heliocentric and the former as geocentric.

'MOVING-EARTH' THEORY

It is interesting to note how and why the present-day Japanese term '*chid setsu*'[cd] and its equivalents in modern Chinese and Korean came to represent Copernicanism or heliocentricism. Literally, *chidō setsu* is 'theory of the earth's motion', but no such set phrase existed in any of the Dutch sources consulted by Motoki Ryōei and Shizuki Tadao. Western sources employed such words as 'Copernican theory', 'heliocentric hypothesis', or 'sun-centred'.

Generally, whenever the word '*chidō*' ('*ti-tung*' in Chinese) is found in the Chinese classics, its interpretation is usually 'earthquake'. Some writers define *ti-tung* as a major earthquake, as distinguished from one of ordinary magnitude *ti-chên*[ce] (earth tremor).[68]

Even if we interpret *ti-tung* liberally as equivalent to *ti-chuan*[cf] (earth-turning), occasionally found in early Chinese cosmological writings, it is not at all clear whether this referred to the rotation or the revolution of the earth. Even in its present usage, the word *chidō* is too vague to be employed as a scientific term. As a matter of fact, a Japanese apologist for Buddhist cosmology, Entsū, attacked the word *chidō* as confusing and susceptible of various divergent interpretations.

Then why was this misleading term adopted to denote such an important concept as the Copernican world picture? Presumably the man who coined the phrase *chidō no setsu* (the theory of the earth's motion) was Shizuki Tadao, who in part I of his *Rekishō shinsho* (drafted in 1798) employed this phrase in his own commentary on the appendix 'tentairon' (on the heavenly bodies);[69] in the preceding straightforward translation of John Keill's work, he simply transliterated the word 'Copernican' to describe the theory.

From the context of 'tentairon', we can deduce two reasons why Tadao coined this phrase:

(1) In order to show that in ancient China, 'there was a theory of the earth's motion, preceding its Western counterpart', and thereby to defend the Asian scientific tradition by demonstrating its priority.
(2) In order to place Copernicanism in the framework of traditional natural philosophy – nemely, the system based on the polar conception of yin and yang. Arguments concerned only with local and morphological relationships, whether heliocentric or geocentric, appeared rather superficial to traditional Asian thinkers. The cosmic polar concepts of motion (to correspond to yang) and rest (to correspond to yin) must have appealed to Shizuki Tadao as more profound and fundamental in their implications. In other words, Copernicanism was comprehensible to him primarily in terms of the physico-dynamical principle of 'motion-rest'. In the traditional list of various dichotomies associated with 'yang-yin', undoubtedly a very important place is occupied by 'motion-rest'; while 'heaven-earth' or 'sun-moon' does form a dichotomy, 'sun-earth' never does.

Tadao's attitude towards Copernicanism was welcomed and shared by Neo-Confucian orthodox philosophers such as Asaka Gonsai and also by the official astronomer, Shibukawa Kagesuke.[70]

We shall now turn to popular treatises in order to estimate the degree of dissemination of the term 'theory of the earth's motion'. In earlier works of the most illustrious Copernican advocate, Shiba Kōkan, such as *Chikyū zenzu ryakusetsu* (1793) and *Oranda tensetsu* (1796), the term 'earth's motion' does not appear explicitly, but *Oranda tsūhaku*[cg] (Dutch oversea activities, 1805), which appeared after the completion of *Rekishō shinsho*, identified the Copernican view by saying that 'this is called the theory of the earth's motion!'.[71] In still later works, however, he did not use 'earth's motion' again. It may be that Kōkan, thoroughly devoted to things Western, might have had a distaste for the chauvinism of Shizuki Tadao.

Perhaps the book most instrumental in disseminating the term 'the theory of the earth's motion' was Yoshio Hisatada's *Ensei kanshō zusetsu*[ch] (An illustrated treatise on Western astronomy, first edition 1823). At the end was appended 'chidō wakumon'[ci] (Queries on the earth's motion), in which he tried to expound further what Tadao had said in *Rekishō shinsho*. This work was reprinted again and again, becoming the most standard textbook of astronomy in mid-nineteenth-century Japan. Thus the term 'earth's motion' entered the popular vocabulary. In the succeeding Meiji period, when the modern educational system was established, the phrase was adopted for mass enlightenment at the elementary textbook level and has remained current since.

On the Chinese scene, 'ti-tung' (earth's motion) appeared in Juan Yuan's *Ch'ou-jen chuan*[cj] (Biographies of Chinese mathematicians and astronomers, 1799) in reference to Michel Benoist's exposition of heliocentricism, though Juan himself did not accept it. In the sequel of that book (prefaced in 1840), Juan tried to identify the origin of heliocentricism in Chang Heng's[ck] *ti-tung*[cl] (earth's motion instrument, second century A.D., presumably a kind of seismometer) and thereby defend his own tradition on the ground that Copernicanism was accepted. The set phrase *ti-tung shuo* (theory of the earth's motion) appears only as late as 1859 in *T'an t'ien*[cm] (Discussion on the heavens), Li Shan-lan's[cn] translation of John Herschel's book. It is impossible to conceive of any influence of Tadao on these Chinese works, but they shared the same traditional mentality with its dichotomy of yin-yang and motion-rest.

In sum, among the various implications of Copernicanism, the physico-dynamic aspect was most attractive, much more so than the geometrical or astronomical aspects, because of the Chinese inclination to use the yin-yang dichotomy and its various equivalents at the most fundamental level of explication of natural phenomena. In this context, 'theory of the earth's motion' received public recognition as a popular as well as professional term.

CONCLUSION

To sum up, the Copernican system did not evoke bitter ideological opposition in Japan except in Buddhist circles. Even the latter could not exert such a commanding reactionary influence as was wielded by the Renaissance Church in Europe. The main cause of the delay in the introduction of heliocentric theory into Japan was the seclusion policy of the government, rigidly maintained until the early part of the eighteenth century and, secondly the linguistic barrier, which remained formidable until the last quarter of that century.

Interest in Western cosmology was initiated not by the camp of traditional astronomers but by linguistic experts, who began with the introduction of the cosmological aspect of Copernicanism. The principal concern with Copernicanism was shown as regards its physico-dynamical aspect, which was interpreted in terms of Eastern *Naturphilosophie*.

It seems that the question of the reception of Copernicanism was more or less used in the more general problem of the superiority of Western learning. Copernicanism never played a pivotal role in overthrowing traditional ideologies or recognizing Western superiority. In this respect, recognition of Western superiority in the astronomical domain had been well established earlier through Sino-Jesuits works, and on the basis of this recognition Copernicanism, in spite of some ideological incompatibility and conflict, had a rather smooth reception in the course of the nineteenth century.

NOTES

1. Cf. Jérôme R. Ravetz; 'The cosmology of Nicolaus Copernicus', *Organon* No. 2, 1965, p. 54, and his *Origin of the Copernican revolution*, 'Nature', March 11, 1961, pp. 859–860.
2. Shigeru Nakayama, 'Characteristics of Chinese calendrical science', *Japanese studies in the history of science*, No. 4, 1965, pp. 124–131.
3. *Li chih* (A) [Treatise on calendar-making] of the *Hsin T'ang shu* (B) [New standard history of the T'ang period], eds. Ou-yang Hsiu (C) and Sung Chi (D) (1060), quoted in Yabuuchi Kiyoshi (E), *Zuitō rekihō shi no kenkyū* [Researches in the history of calendrical science during the Sui and T'ang period; Tokyo, 1944], p. 6.
4. A Portuguese source places the date at 1542. However, according to a Japanese source, the 'Teppōki' (F), written between 1648 and 1651, the year is 1543. See C. R. Boxer, *The Christian century in Japan, 1549–1650* (Berkeley, 1951), p. 27.
5. A. J. Barnouw and B. Lanaheer, eds, *The contribution of Holland to the sciences* (New York, 1943), p. 271.
6. Nathan Sivin's paper, in print.
7. Preserved in the Naikaku Bunko library.
8. *Tentairon* (G) in part one of *Rekishō shinsho*, reprinted in *Nihon tetsugaku shisō zensho* (H) [Source book of Japanese philosophy], ed. Saigusa Hiroto (I) and Shimizu Ikutarō (J), (Tokyo, 1956), vol. 6, p. 142.
9. Hirose Hideo (K), *Kyū Nagasaki tengakuha no gakutō seiritsu ni tsuite* (L), *Rangaku shiryō kenkyūkai kenkyū hōkoku* (M), No. 184 (1966). The manuscript is reprinted with commentaries in *Kinsei kagaku shisō* (N), No. 2, ed. by Hirose Hideo, Nakayama Shigeru (O) and Otsuka Yoshinori (P), (Tokyo, 1971) in *Nihon shisō taikei* (Q) [Source-books of Japanese thought], No. 63.
10. In *Bunmei genryū sōsho* (R) [Series on the origins of civilization; Tokyo, 1914], vol. 2, pp. 1–100.
11. Imai Itaru (S), *Kenkon bensetsu zakki* (T) [Miscellaneous notes on the Kenkon bensetsu], *Tenkansho* (U) [Private journal of Imai Itaru] *22*, 14–16 (mimeographed, 1957).
12. Motoki Ryōei, *Seijutsu hongen taiyō kyūri ryōkai shinsei tenchi nikyū yōhō ki* (V) [The basis of astronomy, newly edited and illustrated, on the use of celestial and terrestrial globes according to the heliocentric system; 1792–1793], vol. 2, reprinted in *Tenmon butsuri gakka no shizenkan* (W) [Japanese astronomers' and physicists' views of nature], in *Nihon tetsugaku shisō zensho* (X) [Source book in Japanese philosophy], ed. Saigusa Hiroto (Tokyo, 1936), vol. 8, p. 342.
13. It is recorded that in 1791 seven interpreters at Nagasaki were dismissed for having made inaccurate translations of Dutch documents. See Otsuki Nyoden (Y), *Shinsen yōgaku nenpyō* (A) [A newly edited chronology of Western learning in Japan; Tokyo, 1926], p. 76. For an English translation, see C. C. Krieger, *The infiltration of European civilization in Japan during the eighteenth century* (Leiden, 1940), pp. 94 ff.
14. Preserved in the Tenri Library under the series title of 'Tenmon hisho' (AA) [Secret books on astronomy], 22 sheets.
15. Preserved in the Nagasaki City Museum, 90 sheets divided into three volumes.

16. A full translation of part (A) is available in Shigeru Nakayama, *Abhorrence of 'God' in the introduction of Copernicanism into Japan*, 'Japanese studies in the history of science', No. 3, 1964, pp. 62–63.
17. A full translation of part (B) in Nakayama, 'Abhorrence', pp. 63–66.
18. Shigeru Nakayama, *Motoki ryōei yaku 'Oranda chikyū setsu' ni tsuite* (AB) [On the 'Oranda chikyū setsu' translated by Motoki Ryōei], *Rangaku shiryō kenkyūkai kenkyū hōkoku*, No. 112 (1962) and No. 162 (1964).
19. Because of the long interval between the date of the original and that of the Japanese translation, Itazawa Takeo (AC) conjectured that the original was published in 1766 instead of 1666 in his *Edo jidai ni okeru chikyū chidōsetsu no tenkai to sono handō* (AD) [The development of the earth's sphericity and motion theory and the reaction to it during the Tokugawa period], *Shigaku zasshi* (AE) 52(1), 12 (1941), but I found the 1666 edition in the Library of Congress.
20. Pierre Henry Bandet, *Leven en Werken van Willem Janszoon Blaeu* (1871); and Edward Luther Stevenson, *Willem Janszoon Blaeu* (New York, 1914), pp. 11–13.
21. The preface is more fully translated in Shigeru Nakayama, *A history of Japanese astronomy; Chinese background and Western Impact* (Cambridge, Mass., 1969), pp. 175–176.
22. Shigeru Nakayama, *Motoki Ryōei no tenmonsho honyaku ni tsuite* (AF) [On Motoki Ryōei's translations on astronomy], *Rangaku shiryō kenkyūkai kenkyū hōkoku*, No. 66 (1960).
23. Vol. I is reprinted in *Nihon tetsugaku zensho*, vol. 8 (1936).
24. Itazawa, *Edo jidai*, p. 10. A first edition and an edition of 1771 were preserved in the shogunate government library. *Yōgaku kotohajime ten, rangaku no shokeifu to Edo bakufu kyūzōbon* (AG) [Catalogue of books, exhibited at the 'Yōgaku kotohajime ten', on earlier phases of Western civilization in Japan], ed. Okubo Toshiaki [AH], pp. 17–18.
25. *Dictionary of national biography*, s. v.
26. See Johannes van Abkoude, *Naamregister van de Bekendset en meest in Gebruik Zynde Nederduitsche Boeken* (Amsterdam, 1787), p. 579.
27. First edition published in Leipzig in 1753. An English translation was published in London in 1757.
28. Kuwaki Ayao (AI), *Reimeiki no Nihon kagaku* (AJ) [Japanese science at the dawn; Tokyo, 1947], pp. 105–107.
29. Watanabe Kurasuke (AK), *Oranda tsūji Shizukishi jiryaku* (AL) [Outline biographies of the Shizuki family, hereditary Dutch interpreters; Nagasaki, 1957], pp. 32–34.
30. Ōsaki Shōji (AM), *Rekishō shinsho tenmei kyūyakubon no hakken*, (AN) [The discovery of manuscript translations preliminary to the Rekishō shinsho], *Kagakushi kenkyū* (AC), Nos. 4 und 5 (1943), p. 101.
31. In *Rekishō shinsho*, reprinted in *Nihon tetsugaku shisō zensho*, vol. 6, p. 117.
32. *Nihon tetsugaku shisō zensho*, vol. 6, p. 141.
33. *Nihon tetsugaku shisō zensho*, vol. 6, pp. 135–142.
34. *Nihon tetsugaku shisō zensho*, vol. 6, p. 246.
35. *Nihon tetsugaku shisō zensho*, vol. 6, p. 288.
36. Kōkan's works are reprinted in Nakai Sōtarō (AP), *Shiba Kōkan* (Tokyo, 1942).
37. Muraoka Tsunetsugu (AQ), *Zoku Nihon shisōshi kenkyū* (AR) [Studies in the history of Japanese thought, sequel; Tokyo, 1939], pp. 239 ff.
38. Takamichi Arisaka (AS), *Chidōsetsu no denrai to shin uchūron no shutsugen* (AT) [An introduction to the theory of earth's motion and the emergence of a new cosmology], *Nihonshi no kenkyū* (AU) [Researches on Japanese history; Kyoto, 1970], pp. 281–301.
39. Volume II. Reprinted in *Nihon tetsugaku zensho*, vol. 8, p. 185.
40. *Nihon tetsugaku zensho*, vol. 8, p. 174
41. Takahashi Yoshitoki (AV), *Shinshū goseihō zusetsu; furoku* (AW) [Appendix to the illustrated planetary theory, newly revised, ca. 1802, preserved in Inō Museum]; Hazama Jūshin (AX), *Senkō Taigyō sensei jiseki ryakki* (AY) [Outline of the work of my father Hazama Shigetomi, n. d.], reprinted in Watanabe Toshio (AZ), *Hazama Shigetomi to sono ikka* (BA) [Hazama Shigetomi and his family, 1943], p. 456.
42. Shigeru Nakayama, *On the alleged independent discovery of Kepler's third law by Asada Gōryū*, 'Japanese studies in the history of science', No. 7 (1968), pp. 55–59.
43. Preserved in Tokyo Astronomical Observatory.
44. Preserved in Naikaku Bunko Library.
45. Cf. Sukumar Ranjan Das, *The Jaina school of astronomy*, 'Indian historical quarterly' 8, 30–42, 565–570 (1933).
46. A good summary of this work in English is Yoshio Micami's *A Japanese Buddist's view of the*

European astronomy, 'Nieuw-Archief voor Wiskunde' 11, 1–11 (1912). In it the name 'Entsū' appears as 'Yentsū'.
47. *Bukkoku rekisho hen*, vol. 1.
48. *Bukkoku rekisho hen*, vol. 2.
49. Kōda Rohan (BB), *Kagyūan yatan* (BC), [A night tale of Kagyūan; Tokyo, 1907], pp. 67–76.
50. For instance, Kanchū (BD), *Shiji idō ben* (BE) [The four seasons compared; 1843].
51. An example is found in Kojima Tōzan (BF), *Bukkoku rekishō benmō* (BG) [On the absurdity of the *Bukkoku rekishō hen*; 1818].
52. For instance, Ishii Kandō (BH), *Shūhi sankei seikaizu* (BI) [True illustration of the *Chou-pi suan-ching*; 1813] and Shinohara Yoshitomi (BJ), *Shūhi sankei kokujikai* (BK) [A Japanese annotated edition of the *Chou-pi suan-ching*; 1819].
53. Mikami Yoshio (BL), *Nihon kagaku no tokųshitsu: tenmon* (BM) [The Characteristics of Japanese science: astronomy], in *Tōyō shichō no tenkai* (BN) [The development of Oriental thought; Tokyo, 1936], pp. 60–63.
54. Miyatake Gaikotsu (BO), *Hikkashi* (BP) [History of the suppression of literature in Japan, Tokyo, 1911], p. 69.
55. Kan Sazan, *Fude no susabi* (BQ) [Writing for amusement's sake], reprinted in *Nihon zuihitsutaikei* (BR) [A comprehensive collection of Japanese informal essays, Tokyo, 1927], I, 80.
56. Nakayama Shigeru, *Edo jidai ni okeru jusha no kagakukan* (BS) [Confucian views of science during the Tokugawa period], *Kagakushi kenkyū*, No. 72, 157–168 (1964). See also George H. C. Wong, 'China's opposition to Western Science during the late Ming and early Ch'ing', *Isis* 54, 29–49 (1963) and his China's opposition to Western religion and science during late Ming and early Ch'ing (University Microfilms, Inc., Ann Arbor, Mich. L. C. Card no. Mic 58–7381; 1958); and N. Sivin, 'On China's opposition to Western science during late Ming and early Ch'ing', *Isis* 56: 201–205, (1965).
57. Asaka Gonsai, *Nanka yohen* (BT), vol. I (circa, 1837), reprinted in *Nihon jurin sōsho* (BU) (Source books of Japanese Confucianism], vol. 2 (1927).
58. Yasui Sokken, *Suiyo manpitsu* (BV), in *Nihon jurin sōsho*, vol. 2.
59. *Ikai Keisho sensei shokan shū* (BW). Undated correspondence; reprinted in *Nihon jurin sōsho*, vol. 3.
60. Muraoka Tsunetsugu, *Nihon shisōshi kenkyū* (Tokyo, 1930), pp. 297 ff.
61. *Shamon Monnō ga kusen hakkai tōron no ben* (BX) [A confutation of the monk Monno's argument of the nine mountains and eight seas], reprinted in *Zōho Motoori Norinaga zenshū* (BY) [Complete works of Motoori Norinaga, revised edition, Tokyo, 1926], vol. 10, pp. 131–136.
62. Fujiwara Noboru (BZ); *Edo jidai ni okeru kagakuteki shizenkan no kenkyū* (CA) [A study of scientific view of nature in Edo period; Tokyo, 1966], p. 77 ff.
63. Hirata Atsutane, *Tama no mahashira* (CB) (1812), reprinted in *Hirata Atsutannshūe ze* (CC) [Complete works of Hirata Atsutane], vol. 2 (Tokyo, 1911); Sato Nobuhiro, *Yōzō kaiku ron* (CD) [On the creation and formation of the world; ca. 1825], reprinted in *Shinchū kōgaku sōsho* (CE) [series in Nipponology, re-edited; Tokyo, 1927], vol. 10; Tsurumine Shigenobu, *Ame no mihashira* (CF) [The sacred heavenly pillar; 1821].
64. Though I suspect that Tadao identified them still earlier in his unpublished manuscripts.
65. Part I, reprinted in *Nihon shisō taikei* 44 (Tokyo, 1970), p. 95.
66. *Nihon tetsugaku shisō zensho* 6, p. 141.
67. *Seigaku shukan* (CG) [Notes and correspondence on astronomy; ed. by Shibukawa Kagesuke] in *Nihon shisō taikei* 63, p. 213.
68. I am indebted for this remark to Professor Jeon Sang-woon (CH), a historian of Korean science.
69. *Nihon tetsugaku shisō zensho* 6, p. 136 ff.
70. *Shinpō rekisho zokuhen*.
71. Volume I, in Nakai, *Shiba Kōkan*, p. 116.

First published in S.Nakayama et al. (eds), *Science and Society in Modern Japan*, pp.253-269, MIT and University of Tokyo Press, 1973

8 Grass-roots Geology - Ijiri Shōji and the Chidanken*

Of the various societies embraced by the Association of Democratic Scientists (Minshu-shugi kagakusha kyōkai, abbreviated as Minka), the only one still quite active is the Society for Corporate Research in Earth Science (Chigaku dantai kenkyūkai, or Chidanken) founded in 1947. The Historical Section of Minka, a similar organization with close relations to Chidanken, also remains in existence by issuing its journal *Rekishi hyōron* (Historical review). The reason why these two particular organizations have so far survived is found in an element common to both of them, namely, that they consist of local field researchers. This is particularly true of Chidanken, which serves as a meeting place for 'grass-roots geologists', the typical member being a provincial teacher's college graduate now teaching in a primary or secondary school.

There are, of course, many other organizations which promote nationwide cooperation for localized studies. Good examples are the ethnographic study group inspired by the late Yanagida Kunio and a group of amateur astronomers organized by Yamamoto Issei. In the history of science we often encounter the birth of a new research paradigm in some marginal group outside any established academic structure, but not all such groups are necessarily as radical and progressively-minded as Chidanken. Many local historical societies are marked, rather, by an unsophisticated conservatism. The reason why Chidanken still maintains a powerful progressive orientation is due largely to the personal character of its able organizer. The man who has sat in the seat of charisma for two decades since the end of the Second World War is Ijiri Shōji.

Chidanken is not merely a scientific society but a crusading body out to propagate its ideology and methodology. In such a group, not only charisma but also a 'bible' is indispensable equipment. To meet this need, Ijiri wrote his *Koseibutsugaku* (Paleontology, 1949), which was reprinted in 1954 under the title *Kagakuron – koseibutsugaku o chūshin to shite* (On science – centring on paleontology) and later reissued without the subtitle. Readers were immediately struck with the distinctive individuality of the author and, at the same time, could find the cornerstone of his organization strategically placed in the treatise.

* Published in *Shisō no kagaku* (The science of thought), Vol. 5, No. 50 (1966), pp. 100–106.

PHILOSOPHY OF SCIENCE BASED ON GEOLOGY AS A PARADIGM

The major topics of the philosophy of science during the early half of the 20th century have focused mainly on new conceptions generated from problems in modern physical science, such as causality, complementarity, and a relativistic interpretation of space-time. Ijiri's philosophy of science is, however, unique to the extent that it is heavily coloured by the real experiences of a mountaineer geologist, as distinctly contrasted to armchair contemplation. Here we can note a bold stand against the superiority of exact science and mathematico-physical reductionism which have dominated the scientific world in the early part of the 20th century. In his view of science, mathematics has no reserved seat at all. First encounter reality through personal experience, next describe it accurately, and then classify it; this procedure may be called Baconian, but what is more significant is the historicism explicitly proclaimed. Ijiri's ultimate aim is to establish a unified methodology of science, taking geology as the model science and encompassing the biological, historical and social sciences. Exact science or physico-chemical reductionism is deliberately excluded from his scheme.

In the actual practice of science the present writer would not concede the excessive merit of 'philosophical' methodology. 'Given an appropriate methodology, then every solution will follow logically' – such a magic recipe with an almighty problem-solving power is hardly conceivable. In some quarters of Japan, especially among young scientists and students right after the Second World War, methodology has been considered all-important. This illusory expectation may be a reflection of an inferiority complex of Japanese scientists, too anxious to catch up with the forefront of world research and to fill overnight the gap between the West and war-devastated Japan.

Some scientists are fond of ornamentation. Just as practitioners in science and technology in premodern Japan decorated their prefaces with *yin-yang* doctrines that had nothing to do with the content of their works, the generation of scientists brought up in the early decades of the present century also seem to have a common affection towards philosophy. Ijiri may not be exceptional. Among the various natural sciences, geology's prestige is low because of its low level of abstraction (measured by distance from tangible daily experience) and its lesser degree of mathematization. To those who suffer from the low prestige of their chosen subject of study, Ijiri's work, providing geology with 'philosophical' profundity, appears to have been received as a long-awaited gospel.

Moreover, unlike conventional philosophy of science, which tries to conform to the established norms of science, Ijiri's philosophy of science is so close to the daily experiences of practising scientists that it exercises a far more positive influence. Influence varies, of course, according to the age and background of the reader. If one is an already established geologist, he may take Ijiri's philosophy rather matter-of-factly; it may arouse in him some sympathy but never influence his established course. If one is at the height of his productive research career, he may find some hints to appropriate in his own work. If one is a student or an immature novice researcher, he may receive it so dogmatically as to be in danger of falling into methodological inflexibility.

METHOD BASED ON PERSONAL EXPERIENCE

Ijiri's philosophy of science starts with a method based on personal experience. In it we find unmistakable priority given to naked sense perception and intuition. By 'experiment' he means commitment to practise, eventually related to an ethic of social action. First, 'One strikes his hammer into the earth, that is, performs some constructive action' and then 'feels and experiences nature instinctively with his whole body and senses in the extreme'. This is the fundamental scientific exercise. Such a methodology appeals most directly to uncomplicated mountain-climbers and amateurs rather than to men in the exact sciences who try to uncover the innermost structure of nature with highly sophisticated mathematical means. In geology, the barrier between a professional and an amateur is relatively low, compared to other scientific disciplines, perhaps because of its lesser degree of conceptual abstraction. A professional scientist tends to raise this barrier in order to authenticate his discipline. But as a professional Ijiri now tries to liquidate the barrier by laying stress on 'scientifically unsophisticated, naked sense perception'. Naturally, then, the frustrated energies of semiprofessionals and amateurs, annoyed by insurmountable professional barriers, can be effectively attracted and mobilized. This possibility was aptly expressed in Chidanken's slogan 'penetrate into and educate the people' (*v-narodism*). I often come across groups of frustrated local amateur astronomers who say, 'If only we could contribute something to the development of astronomy, however insignificant it may be, our efforts would be rewarded. Can you suggest any task that even an amateur can do?' It is much easier for local historians and grass-roots geologists to find appropriate topics to satisfy their zealous desires and to make the necessary organizational arrangements to carry them out. On such a footing has the corporate research method of Chidanken been formulated.

EMPHASIS UPON HYPOTHESIS-VENTURING

Another feature of Ijiri's view of science is its emphasis upon hypothesis-venturing. In his view, 'logical' procedures for verifying a certain hypothesis or law have no significant place. The generation of a brand new idea – or hypothesis – is most important.

A general misconception prevalent among the non-scientific public is that scientists are all engaged in a search for the unknown. Even in the exact sciences, most scientists in reality undertake the search for the 'to-be-known' on the assumption that certain solid data or results are obtainable by following certain rules of inquiry. Except on rare occasions when one is possessed by an extraordinary idea, most scientists, by working along established lines, receive professional recognition and achieve personal satisfaction. In the post-war Japanese scientific community, this more conventional mentality has been predominant, and bold attempts to venture major hypotheses have been very scarce. This situation Ijiri found quite intolerable.

To a much greater extent than in other branches of natural science, the emphasis upon hypothesis-building is characteristic of geology, and especially paleontology, in which Ijiri specializes. In this field, as in archeology, an endless cycle of building and refuting hypotheses takes place without reaching any acknowledged certitude. One gets the impression that scientific progress based on accumulated knowledge, so characteristic of natural science, does not operate well here.

The stress on hypothesis-formation – or more exactly, paradigm-generation – closely parallels the spirit of dissenting from authority, according to Ijiri. Venturing hypotheses demands what he calls a 'denying or dissenting spirit', quite the opposite of a conservative mentality founded on the accumulation of knowledge. He identifies this spirit with 'class-consciousness itself'. Perhaps he identifies the ruling class with the academic establishment of the University of Tokyo and the ruled with those dissenting against it. On the organizational level, the dissenting spirit is embodied in the anti-establishment, anti-ivory tower and anti-bureaucratic attitudes of Chidanken. In the sphere of action, Chidanken has been active along the ideological lines of Minka on the battlefield of the Japan Science Council, and thus has established a unique, democratic system for allocating its research funds. Ijiri's vigorous personality and the way his dissenting spirit stirs even idle minds into action are incarnated in the activities of Chidanken and pervade the organization.

RESEARCH-ISM

Hypotheses are generated in the cerebral membranes of an individual. Therefore Ijiri clearly recognizes, despite his advocacy of corporate research, that research activity in science ultimately depends upon individual activity. Thus, he is not satisfied with the crude kind of Marxist interpretation of the history of science that ties the development of science to that of society at large. In this respect he is more pragmatic than his Marxist colleagues, whose interpretations have nothing to do with the promotion of science, for in them he has found nothing real.

His is the attitude of an individual working scientist. Ijiri asserts, 'The study of science should not be motivated by the utilitarian aim of exploitation; rather, one must maintain a mental readiness to commit double suicide with research.' In this reference to the extreme sacrifice of loyal lovers we see a limitation which Ijiri and his Chidanken have placed on their general three-fold programme for concentrating on research, education and action. That is, the thinking pattern which shapes the organization for concentrating on research – scientism itself – limits the scope of their engagement in scientific campaigns. They cannot become the core of a revolutionary movement by participating in community movements or by cooperating with labour organizations.

HISTORICAL SCIENCE

In Ijiri's outlook on science, geology is closer to the social sciences than to the exact sciences, and in fact he places it at the core of the historical sciences. The goal of geology is description of the history of the earth, and hence an historical approach should be adopted as its research methodology.

Just as a mechanistic view of nature dominated the scientific community in the past, so it seems that at the present time physical reductionism dominates all the natural sciences. Everything in nature, inorganic and organic, must eventually be explainable in terms of the behaviour of molecules, atoms and, ultimately, elementary particles. Hence, there are no independent laws in either the geological or biological sciences.

Against this kind of reductionism, those who resist subordination to physics may raise the following criticism. Physics is a 'low-grade' science, since it deals with the simplest phenomena. Historical sciences, on the other hand, deal with the most complicated entities, and therefore the methodology of low-grade physics cannot be applicable to more complicated levels, which should not be, in any case, considered reducible to physics.

It is merely professional prejudice for physicists to claim that since physics deals with ultimate matter, it is the most fundamental of all sciences. Others may claim equally that the consideration of human nature is most essentially indispensable and fundamental, since it deals with patterns of human behaviour. Or, the humanities may claim to be most important, as they deal with problems of the human soul. By the same token, the earth sciences may claim to be most fundamental, since mankind can survive only on this earth, at least for a conceivably long time to come.

Such debates actually do not constitute discussions of pure research methodology, as they derive largely from value judgements and professional prejudices about some presumed hierarchy among the various branches of science and scholarship. If a smaller element is more fundamental, then personal ethics and family affairs are more fundamental than national administration and international politics; but, in reality, the work of a prime minster of a nation, who presides over the more complex matters of national administration, is usually considered more significant, if not also more prestigious, than that of a village chief or household head responsible for smaller and simpler units.

Despite such comments and criticisms, a physicist can in reality perform his trade without paying any attention to developments in the geological sciences, while geologists and biologists must always watch out for and catch up with developments in physics. Field-oriented or phenomenological scientists may pose new problems and new objects of investigation to the physicist, but they do not provide him with any new methodology. Thus, all fields of natural science eventually become applied physics.

Disturbed by a dominant physical-reductionism of this sort, there have appeared quite a few scholars who lay claim to genuine biological and geological laws free of reductionism, in order to maintain the disciplinary independence of their fields. For instance, among Marxist scientists, Joseph Needham in his earlier writings asserted the existence of a unique approach to the levels of organization, the biologists' stronghold of independence, which the physicist and chemist can neglect. (See my 'Josefu Niidamu-ron', in *Igakushi kenkyū* [Journal of the history of medicine, No. 17, 1965], or the English version, 'Joseph Needham; organic philosopher,' in Nakayama & Sivin, ed., *Chinese science* [M.I.T. Press, 1973], p. 25 ff.)

Though Ijiri explicitly claims that the methodology of the geologists belongs to the historical sciences, the method as defined by him is nothing but concentration on field work. Each historical event has its own individual characteristics in space and time. Hence, unlike timeless universal physical occurrences, historical events cannot be dealt with at one single central research institute, but must be pursued through the accumulation of field investigations at every provincial level and in each locality.

Field work can be carried out with the simplest equipment – a hammer, a clinometer, and a pair of tough feet. As far as a particular locality or particular field is concerned, those whose labours involve endless walking and exhaustive observations, whether amateurs or semi-professional geologists, know more and know it better than an established top-notch geologist settled in Tokyo. In this respect a local historian or a local researcher is able to maintain his identity and pride. Likewise, the corporate research method, by organizing semiprofessionals and students, has great merit, inasmuch as the exploration of mountains and fields requires as many human hands as possible. Thus, the 'historicism' and 'field-concentration' of Chidanken have been brought into a close relationship at its organizational level; and conversely, in order to maintain its organization intact; Chidanken must commit itself to 'historicism' as its official methodology.

Diametrically opposite the extreme of 'field-concentrationism' stands physico-chemical reductionism. The latter claims that, rather than exhaustive investigation by the time-honoured 'hammer and tough feet' method, the most recent findings of experimental physics and chemistry, such as X-ray analysis and high-pressure techniques to reproduce earth history processes, should be widely adopted. In order to carry on such a high-technology approach, it is necessary to acquire sophisticated apparatus so costly as to be beyond the reach of any individual researcher's personal budget. To meet the high costs of such an approach, some sort of centralization is inevitable. A high-pressure laboratory can be installed only at one or a very few central research institutes funded by the government. Geological study at this level turns out to be unattainable for our grass-roots geologist. This issue of high-cost apparatus presents him with a problem far more serious than those faced by his fellow grass-roots historians, anthropologists and sociologists. While research in physical sciences and technology rapidly progresses towards centralization into 'big science' using high-cost research apparatus, the anti-establishment orientation of Chidanken can be understood as a reactionary stand against 'modernization'; it is forced to adhere to its 'historicism' because of the inaccessibility of costly instrumentation.

I fully recognize the significance of research pursued along pluralistic lines. At the same time, we must admit that many precariously popular modernistic approaches can easily sink into oblivion in due time. Meanwhile, the general trend towards reductionism into a physico-chemical approach invades irresistibly into the biological and other sciences alike. Biology and earth sciences can no longer remain aloof from this general trend. However much one emphasizes that geology is one of the historical sciences, the question remains: what is the specific merit of being a historical science? Therein is nothing particularly new, nor any sign of leading towards a new horizon.

It can be legitimately argued that geological science, as well as organic evolution, deals with unrepeatable historical experience, and that it therefore can be neither reproduced nor verified; that is quite unlike physics and chemistry. We have no rigorous proof that Newtonian laws operated in geological ages in exactly the same manner as now. We may also argue that the laws proper to geology are not mechanistic but rather phenomenological and kinematic.

However, as Miyashiro Akio proposes in his 'Chikyū kagaku no rekishi to genjō' (History and present status of earth sciences; in *Shizen*, May 1966); we cannot prove that geological phenomena are beyond the reach of the laws of physics. In the fields of history and biology, annoying factors like free will and organism come into play, but geology is much closer to pure physical or material science, the only difference being its involvement in the kind of time sequence that historical sciences deal with. It may not be too profitable for the geologist to draw his methodological stimulus from the established historical disciplines and thus intentionally reject the intrusion of physico-chemical methods. Unless it is liberated from the limitations of the framework of historical science and is open to enrichment by whatever is useful, Ijiri's philosophy of science will itself turn out to be a historically outmoded fossil.

If one wishes to learn from historical methodology, examination of the historical development of geological explanations and hypotheses may be suggestive. Ijiri's 'history', however, remains natural history. Is it not rather odd that Ijiri and his Chidanken have not displayed as much interest in examining the history of geological sciences as have physicists in the history of physics?

The various elements discussed above may be summed up as follows:

	Philosophy of science	*Philosophy of organization*
1.	Methods based on personal experience; emphasis on practice and action	Emphasis on the role of amateurs
2.	Emphasis on hypothesis venturing	Value of the dissenting spirit and anti-establishment orientation
3.	Research concentration	Scientism
4.	Historical science	Localism, emphasis on field work, and corporate research method.

Ijiri's philosophy of science corresponds beautifully with the organizational principles of Chidanken. For this reason, it appears that both Ijiri's leadership role and the organization of Chidanken manifest a stability unparalleled in similar organizations. Ijiri's unique philosophy of science is primarily the outcome of his own personal experience; his version of dialectical materialism seems to be more or less a philosophical ornament to buttress his own thought. (See his *Chishitsugaku no konpon mondai* [Fundamental problems in geology]; Tokyo, 1949). He was not disturbed so much as other professional leftist thinkers by the criticism of Stalinism; he claims that, even after the incident, we may be able to learn something from Stalin on a case-by-case basis. His self-confidence as a competent natural scientist and as the moving force of Chidanken provides him with a solid ground undisturbed by shifting ideological vicissitudes. After two decades of its existence, though, his organization faces a variety of problems for the future.

TROUBLE SPOTS IN CHIDANKEN

Chidanken embodies two different functional roles, one as a crusading movement, and one as a scientific society. Were its function limited to that of crusading, then it would have faded away along with the decline of Minka. The reason for its surprising longevity may be found in its two-fold nature: crusading activity in time of emergency, and normal research activity during less critical times.

As a scientific society, Chidanken faces the problem of confrontation with physico-chemical reductionism. It can continue to produce at least some meaningful research along normal lines, such as increasing the precision of geological maps, by retaining its present priority on field work. Such work has as much worth as the endless excavation of local historical sources. Yet, however progressive the corporate research method may be, its instrumentation remains old-fashioned and, consequently, at some time in the future Chidanken may no longer sustain its morale as a scientific vanguard.

Its parochialistic approach also has dangerous elements that may lead to narrow isolationism. At the time when the establishment of the University of Tokyo's Institute of Nuclear Physics became a controversial issue, Chidanken proclaimed its slogan of 'science for the (Japanese) people' and stood in opposition to those physicists who were afraid of remaining at an internationally inferior position in the front-line competition of nuclear research.

Ijiri emphasized hypothesis-building and also at one time supported the Michurinian theory. There may be some similarity between geological hypotheses and the Michurinian one; in an area where experimental proof is hardly attainable, hypothesis-venturing is more appealing than verification. The abstraction of a hypothesis is generated out of one's pre-disciplinary life, or ideology, as Ijiri put it. In this area, there are some discernible dangers of indulging in irrationalism.

On the other hand, in spite of the emphasis upon hypothesis-building, corporate research operates in such a way as to curtail individualistic claims to credit. A highly imaginative hypothesis generated in the mind of a single individual may, in the process of mass discussion, become diluted and have its sharp edge moderated into mediocrity.

Chidanken's achievements as a crusade body have been spectacular and dramatic. In the early phase of the post-war democratization movement, it heralded the democratization of the scientific community by making some solid achievements in that direction. As a brake against possible reactionary moves in the future, Chidanken has reason enough for its existence.

Still, democratization of the scientific community does not necessarily require a Marxist methodology; it is more likely a 'bourgeois' revolution. Here we find one limitation of a scientists' movement. Once depived of its morale, it may degenerate into a mere professional society or a pressure group like a Medical Association.

In pursuing the goal of its 'science for the people' movement, the activity of Chidanken has been remarkable. However, scientists have, after all, little chance of becoming the main force of social revolution and, if isolated from the general trend of leftist movements, can hardly achieve anything significant themselves.

Furthermore, it seems that in its scientism, Chidanken imposes a certain limitation on itself in its approach to the people and society. Research-concentration or scientism may be necessary prerequistes for the professional activity of scientists, but are hardly appropriate as goals for popular movements. Would it not be possible for Chidanken to join in tackling such problems as pollution and community development in close cooperation with public health engineers and civil engineers in local government?

Chidanken advocates the dissemination of science. Then, what is to be disseminated? In an earlier age when superstition overshadowed society and no public education was available, enlightenment by science as a rationalistic form of thought was still meaningful. But, in the present age when moon rockets are launched for greater political glory, science is excessively deified among laymen. Thus, Chidanken, the last stronghold of post-war scientists' movements, now appears to have lost sight of its original goal, and its once attractive slogans sound jaded.

Generally speaking, a crusading body should not remain in existence just to perpetuate its own authority; it had better be dissolved. It was only natural that Minka and associated bodies should lose their original glory, but, then, what new vision is emerging to replace the old? With its policy of excluding members over forty years of age from its council, Chidanken metabolizes its executive officers; but how can its policies and activities be metabolized?

It may also be natural that any crusaders' group eventually becomes fossilized; and yet, there may be no compelling reason to declare its dissolution. But the next generation should proceed cautiously by analysing and examining their inheritance, in order to avoid the failings experienced by their predecessors.

Perhaps the scientists' movements are now passing through a winter day, or worse, the hottest time of summer, when everyone is short of breath. Fortunately, Chidanken has so far survived in this suffocating climate. However, a crusading body should, by its very nature, be metabolized all the time; mere survival is not to be simply congratulated. The grass-roots Chidanken must, then, scrutinize the merits and demerits of of the Ijiri legacy, with rejuvenation for the future in mind.

First published by the Science Council of Japan, *Texts of Symposia*, No.1, pp.83-89, XIVth International Congress of the History of Science, 1974

9 Problems of the Professionalization of Science in Late-Nineteenth-Century Japan

Shigeru NAKAYAMA (Japan) and Masao TERASAKI (Japan)

This report is to clarify the characteristics of Japanese scientific profession, which was created through the modernization policy of the Meiji oligarchy government in the late nineteenth century. This event of creating such a new profession may be marked as the first such event in the non-Western world.

1. OLD PROFESSIONS

Under the preceding Tokugawa regime, a number of local fief governments had maintained loosely subordinate feudal relations with the central Shogunate government. At the top of the class hierarchy, the samurai, comprising approximately five per cent of the total population, enjoyed the most prestigious positions and were separated by distinctive differences of social status from other inferior classes – farmers, artisans and merchants. The samurai were primarily administrative bureaucrats employed in either the Shogunate or fief governments.

The dominant intellectual profession was that of Confucian scholars, whose teaching encompassed a wide range from politics to personal ethics. The specialized scientific professionals of the time consisted almost entirely of medical practitioners, traditional mathematicians and a handful of Shogunate astronomers.

Many Confucian scholars maintained private tutorial schools, but gradually these private schools were put under government sponsorship, and many of them in the nineteenth century were turned into governmental schools in which the Confucian scholars became the teaching faculty. At times, they also served as governmental advisers on political affairs.

Some of the top-notch doctors received a sort of professional title, but there was no licence system for practising medicine; that is, in the medical profession, there was no guild autonomy to exercise authority for enforcing a licence system and to exclude the unlicensed. Some doctors were employed by the central or fief governments, the top ones for the Shogun and feudal lords and most of the others for the average samurai class, but their revenue largely depended on private clinics. Even the most famous ones ran private schools and clinics in big cities like

Edo and Kyoto without having any relations with the governments. They were much more free-lance-minded than the Confucian scholars.

Traditional mathematicians found their financial basis in teaching at private school to all classes of people, but the mathematical group was much less vocationally established than other professions. At their schools, they issued academic degree certificates, but these were not necessarily professional licences for practising or teaching. Their fame and prestige was mainly based on leading a special group of mathematical talents whose approach to mathematics was recreational rather than for practical applications. Traditional mathematics was supported by disinterested dilletantism.

Astronomy was purely a governmental enterprise to issue official calendars. The positions of the Shogunate astronomers were hereditary. At the time of big events like calendar reforms, the government recruited talents outside of hereditary astronomers when no talent was found in the latter and usually created new posts within the Shogunate's Bureau of Astronomy for them.

Under the feudalistic regime, there was, as a rule, no social mobility among different classes. An inherited stipend was the most important criterion for judging one's social status. In such a society, even professional activities were very much conditioned by hereditary institutions. All governmental posts, astronomers, Confucian scholars and medical doctors alike, were inherited ones. When they could not find proper talents for carrying on professional duties among their offspring, they usually adopted talented sons-in-law. The mathematical tradition, which required a special talented genius, was most free from the hereditary system, compared to other professions.

2. DEMISE OF THE OLD PROFESSIONS

In 1868 the old Shogunate-fiefs regime was terminated. In replacing it, the new Meiji oligarchy government started with centralizing all administrative functions. The old class hierarchy and particularly the special status given to the samurai class were taken away. Accordingly, the character of the scientific community in Japan underwent a transformation.

All the former governmental posts were categorically abolished. The functions of the Shogunate astronomers were suddenly discontinued. Under the new regime, Confucian study lost its former prestige of being the official learning, and consequently most of the Confucian scholars became teachers in the primary and secondary schools under the new modern educational system. The new government, with its Westernization policy, favoured Western medicine over the traditional Chinese-style medicine. Under the new circumstances, traditional medical practitioners formed a guild in opposition to the adoption of Western medicine at the administrative level, and they waged a desperate political campaign to obtain the favour and recognition of the government up until the 1880s, but traditional medicine was eventually illegalized and forbidden the right to issue licences for practising medicine by government decree.

As Western mathematics gained official sanction in the new educational system, traditional mathematics was excluded from the curriculum of the modern Westernized educational structure. Only abacus calculation remained alive because of its practical importance in daily life. It was often adopted in school curricula but was mainly taught at private schools as an extracurricular activity.

3. THE FORMATION OF NEW SCIENTIFIC PROFESSIONS

Unable to withstand the blast of the 'civilization and enlightenment' policy of the government, all of the old professions gave way to Western science. At the frontiers of newly-forming disciplines in nineteenth-century Europe, the voluntary activities of scientific or professional societies usually preceded their inclusion in the university curriculum. This process was reversed in the case of Meiji Japan, where first the university system was founded in the 1870s and 1880s by the goverment authority and then groups of scientists were nurtured in it. Only then did university graduates in each discipline form their scientific societies.

In the 1870s various vocational schools for scientific training appeared in Tokyo and various local cities, but during the 1880s when centralization of administration and the hierarchical formation of the educational ladder centring on Tokyo University were promoted, these schools were discontinued or rearranged to conform with the new system. From the 1890s, as the Japanese industrial revolution took off, various technical schools were created in order to meet the demands of industry for supplying lower and middle classes of technicians in positions subordinate to the higher-rank university graduates.

The newly created scientific professions had their basis in the nationwide centralized scheme of academic degree and licence system, in contrast to feudalistic Shogunate-fief regime. In cases where professional autonomy was not firmly established and authorization by state power had more significance, academic degrees conferred by the governmental university were automatically considered to be preferable to professional licences.

In principle, the members of new professions were recruited from all classes, independent of birth and former social status. Under the old regime, those professionals who inherited family stipends as the main source of revenue showed unlimited loyalty towards their sponsors, the Shogunate or fiefs, and little to the professional community in which they were members. In the new regime, in contrast, new professionals showed loyalty solely to the single authority of the centralized government. Professional communities were formed under the auspices of state power. Thus, the goals of the government and the professional societies were taken for granted to be in harmonious agreement. Only in the twentieth century did the second generation of scientific professionals become more independent of national identity, assimilating themselves towards the international community of science.

4. CHARACTERISTICS OF THE NEW PROFESSIONALS

According to the statistics compiled by Ikuo Amano — *Kyoiku shakaigaku kenkyu*, vol. 24 (1969), p. 84 — the per centage of samurai birth among graduates of each department of Tokyo University in 1890 is as follows:

Medicine majors:	40.8%
Agriculture:	55.9%
Law:	68.3%
Literature:	75.0%
Science:	80.0%
Technology:	85.7%

We find here a conspicuous difference between medicine/agriculture groups with the lowest rate of samurai origin and science and technology groups with the highest.

In interpreting this difference, we may assume that those graduates from medical and agricultural schools were mostly sons of former medical practitioners and farmers respectively. While doctors employed by the government in former days were counted as belonging to samurai class, others were commoners. All farmers were categorically placed in the commoner's class. Hence, the above statistics should be interpreted as showing the relative continuance of class structures from the old to the new regime among the professional medical and agricultural workers.

The special status of the samurai was completely lost under the new regime, but the sons of former samurai, the proudest class, did not view medicine and agriculture as promising future careers for themselves, as these occupations were already monopolized by the former professionals and landed experts, whom the samurai had undoubtedly looked down upon.

In the early Meiji period, before the Constitution was promulgated in 1889 and the modern bureaucracy was instituted, the legal profession had not yet been established as a respectable career. Hence, many samurai deprived of status and occupation found their most promising career possibilities in the entirely new professions of science teachers and engineers. Science majors as the teachers and promoters of the new Westernization policy, and technology majors as technocrats in building up a modern state, did satisfy the former samurai's aptitude for administrative posts, their accustomed class vocation, as science and technology were, at least in the early Meiji period, purely governmental enterprises of top priority. The respect for Western science was founded on the 'civilization and enlightenment' boom of the early Meiji era, and technology met the needs of the new state ideology of 'industrialization and armaments'. Especially those samurai of fiefs which had opposed the Meiji government, who were refused entry into the new government power structure, found appointments related to science and technology relatively independent of nepotistic government restrictions. Thus, science and technology provided a newly-opened horizon and frontier for such new career hunters.

Thus, Japanese modern science and technology professions were, in the beginning of their formation, very much 'samurai-spirited'. Unlike the British pattern of the formation of technology professions, which came from the lower classes in industrial areas, or unlike the German pattern of scientific professions in the nineteenth century, in which lower middle class persons, such as minor clerks and school teachers, were the main source of future scientists, the modern Japanese scientific and technological professions in the last quarter of the nineteenth century were dominated by the old established samurai class.

Traditional craftsmanship, such as carpentry and saké brewing, remained outside the government enterprise of building modern science and technology. Established scientific and technological professionals worked in complete separation from traditional craftsmanship, and found their appropriate places of work in the government industries, such as mining and military arsenal. Only later in the early twentieth century, as private businesses rose as a consequence of the Japanese industrial revolution, did these professionals gradually shift over to private business.

5. GOVERNMENT FAVOUR OF SCIENCE AND TECHNOLOGY

During the 1870s and 1880s the relative position of science and technology in the whole educational curriculum, from elementary school to university level, was much higher than that of other nations. For instance, mathematics and science occupied about one-third of the school curriculum at the lower grades (first four years) and two-thirds at the upper grades (second four years) of the 8-year elementary education. At the university level too, the emphasis of science and technology appeared in the high per centage of graduates in scientific disciplines of Tokyo University (85% in the 1880s as compared to 40% in the 1920s.)

On the part of the new government, the recognition that science and technology constituted the core of Western superiority was firmly established. On the practical level too, they badly needed qualified teachers and engineers in building up a modern state. Science and technology were not allied with a rising middle class ideology, as often found in the West; it was a matter, rather, of artificial breeding by the government and for the government. The sort of science and technology they tried to import was characteristically that of the late nineteenth-century West; namely, specialized, compartmentalized *Fachwissenschaft* without any ideological implication. Specialization, they thought, was the major merit of Western learning, which was missing in the Confucian tradition.

In 1878, an Educational Act draft admonished:

> As for the discipline in higher education, students should be directed to science, so as not to provoke them into political debate. . . .

In those days, there was a strong anti-governmental civil rights movement pressing for greater political participation. With the deliberate intent of diverting youthful enthusiasm away from subversive ideology, the government favoured non-ideological science.

While suppressing the political opposition of civil rights advocates, the government, now constitutional, planned to recruit new talent into their own camp and to strengthen their power by building up the machinery of a modern bureaucracy. For this purpose they instituted the Civil Service Examination in 1890 open to Tokyo University law graduates. Since then, scientific and technocratic professionals have been placed in a position subordinate to the law-administrative elite.

6. CONCLUSION

Modern scientific and technological professions were the product of artificial creation solely by the hand of the new Western-oriented government. The main constituents of these new professions were former samurai who had received an inherited family stipend in exchange for their loyalty to feudal powers, the Shogunate and fiefs, and thus maintained even after the Meiji Revolution their loyalty to the new state power.

The new government in pursuit of its set aim built universities and factories, trained scientists in these institutions, and sent them off to their posts and businesses. Rather than every scientist following his own individual research interest, priorities were collectively determined to fulfil certain basic tasks, the accomplishment of which was necessary for the government of a modern state to function, matters such as geographical surveys, railways and military installations. Differing from the antecedent Western scientific communities, which were formed by individual scientists drawn together by common interests, the scientific professions in Japan had a 'planned character', planned by and according to state authority. Only one generation later did Japanese scientists and engineers tend to be relatively free from complete loyalty to the state.

▸ First published in R.S. Cohen et al. (eds), *For Dirk Struik, Boston Studies in the Philosophy of Science*, Vol.15, pp.213-224, D.Reidel Publishing Co., 1974

⑩ History of Science: A Subject for the Frustrated

Recent Japanese Experience

---

1. TWO GENERATIONS OF SCIENCE HISTORIANS

Counting active members in the field of history of science in Japan, we are immediately struck by two peaks that stand out in the age-groupings, namely, a generation born in the early 1900s which started their professional careers in the late 1920s and early 1930s (hereafter called the pre-war group), and the other a group born in the late 1920s which entered the field after the Second World War (referred to below as the post-war group).

Unlike a mature academic field where a mechanism for recruitment through the higher education system has been established, a brand new discipline like the history of science has no assured way for continuous production of those professionally committed. Younger scholars have had to commit themselves to the field without guarantees of job prospects, and thus the possibility of being dropped out of the established academic world. Hence, the rise and fall of the production rate of historians of science has necessarily and directly reflected various external (non-institutional) causes as well as an overall *Zeitgeist*. It may not be too far fetched to explain the emergence of these two distinct generations in connection with the two major wars of this century.

Not involved in the First World War to any serious extent, Japan reaped a huge economic harvest in the absence of Western competitors. Just after that War the Japanese government in 1919 issued the 'University Act' with a stated purpose to expand higher education to match the now enlarged national prestige and economic capacity of Japan.

The pre-Second World War generation of historians of science enjoyed the benefits of this Act; numerous students flooded into the expanded system of higher education. When they graduated from the universities in the late 1920s, however, a series of depressions came and a surplus of college graduates suffered from widespread unemployment. This was also a time of a rising Marxist ideological wave. As a matter of course, this generation turned out to be very socially minded and some of them were, no doubt, influenced by the Marxist approach to the history of science, as exemplified by B. Hessen.

This sketch of the typical Japanese historian of science belonging to the pre-war group – albeit an oversimplified one – is superbly confirmed by Tosaka Jun, the leading Marxist philosopher in pre-war Japan, in his article, 'Saikin Nihon no kagakuron' [Recent Japanese Arguments on Science], in *Yuibutsuron kenkyū* [Studies on Materialism], No. 56 (1937) as follows:

> For the last ten years favourable(?) industrial and technological circumstances have led to an increase in the number of young intellectuals who choose careers in science or engineering. This is evident in a comparison of the number of students in higher schools who major in the sciences and humanities. However, even an expanded industrial sector has not in fact been able to absorb sufficiently the rising generation of young engineers and natural scientists. At least this was the situation before the recent passage of the new defence budget, narrowly defined. Hence, young engineers and natural scientists have faced a kind of unemployment problem. Actual unemployment occurs only in extreme cases; in most cases graduates form a quasi-professional class in the natural science field, that is, a reserve supply, or indeed, a congested surplus of those seeking legitimate professional status. This quasi-professional group, unlike established university scientists before them, have been exposed not only to training in post-First World War social thought but also to the various social contradictions directly affecting their personal livelihood and future job prospects. Therefore, they are naturally led to play the role of bringing natural science, heretofore possessed exclusively by bourgeois academies, into the context of social thought.
>
> . . . Thus it is that, despite the limitations described above, the quasi-professionals in natural science are destined to perform an extremely useful social function. Their participation in the 'philosophizing' of, or scientific examination of natural science studies is, whether conscious or not, the (an?) outcome of this function.

The next occasion of encouragement for scientific careers came just before and during the Second World War effort to meet the wartime shortage of scientists and engineers. During the war science majors were regarded as reserve scientific manpower and enjoyed the privileges of exemption from, or postponement of, compulsory military service, while students majoring in the humanities were called into service and nearly eradicated from the campuses.

When the War ended the surplus of scientific manpower found it difficult to obtain regular scientific jobs, as the whole industrial sector was closed down, university laboratories had been destroyed by sustained bombing, and equipment for carrying out scientific research was not available. Again, in the late 1940s, a generation of frustrated young scientists turned out to be very socially conscious. Such were the shared circumstances of historians of science in the post-war generation, to which the present author belongs, though individual motivations to give up a normal career in science and turn instead to its history are complex and varied, and it is not easy to generalize about them.

2. HISTORY OF SCIENCE FOR THE FRUSTRATED

Those who are content with the practising of science, more or less taking for granted the conventional value of the professional community to which he happens to belong, can be a good scientist but not always a good critical historian of science. When such a person turns to writing a history of his profession, the work usually turns out to be a self-congratulatory narrative of the orthogenetic evolution of his discipline.

Only for those who fail to conform or otherwise accommodate themselves to the norms of the existing scientific community does the gap between one's original image of science and the existing mode of research to which he cannot conform become the vital source of a critical attitude towards the practice of the contemporary scientific profession. Some of them may seek possible alternate courses of development of science by returning to historical origins, or by examining the later points of divergence precipitated by particular choices as to developmental courses. In such a search history plays at least the role of liquidating the seemingly fixed authority of the scientific establishment and gives one an advantageous height of perspective.

The sources of discontent among reflective scientists can perhaps be classified into three categories: (1) social, (2) institutional, and (3) internal (intellectual).

Those who cannot find an ideal social milieu in which to carry on their scientific research tend to be critical of their own societies, or of the existing social system as a whole. Such was the case for the pre-war Japanese generation. Their frustration often happened early in their scientific careers, if not at the very time when they chose their future careers, that is, precisely when they had reached their most impressionable years. Consequently, they developed an acute sense of problem relative to the social aspect or the social history of science.

Those who are involved in a particular scientific community yet cannot assimilate themselves to the existing institutional setting tend to develop critical attitudes towards the current institutions, such as academic snobbery, university systems and science policy. In the middle of the nineteenth century, when scientific work was being established as a profession, young scientists actively participated in the advancement of their own professions. But in the twentieth century, now that the professionalization of science has been completed, new recruits are inevitably forced to follow ready-made courses of prescribed curricula and must adjust themselves to standardized behaviour, conventionalized values, and ready-made rules of a given community. Professional behaviour is no longer a matter of personal choice. If one fails to conform, he will be labelled as 'unqualified' and eventually purged from the professional community. The qualification for a historian of science may well be, on the contrary, independence of such conformism. Lacking the necessary critical stance towards current systems, a historian of scientific institutions merely ends up as a dull, bureaucratic archivist.

Those who embark upon inherited courses towards normal scientific careers only to reach a dead end may, in desperation, seek a way out through historical reflection on his trade. This sort of internal crisis can be recognized only after one goes as far as the esoteric depths of a discipline, and usually is not reached until the graduate or post-doctoral levels of study. Thus, a critical attitude towards the internal aspects of (the history of) science often finds support in one's own experience of trouble spots. It is doubtful that future generations of science historians professionally trained in an American graduate school, and lacking experiences of frustration on the research front, can be effectively critical towards the contemporary frontiers of research.

Exceptions are most likely to arise from among the recent generation of American historians of science who, despite assured, if not promising, career possibilities, have 'dropped out' of established scientific professions and entered the history of science, presumably because they have, self-consciously or otherwise, traversed one or more of the above paths leading to professional frustration. In scientifically less developed countries, like Japan in the past, still another element may come into play, namely, dissatisfaction with the pettiness of the research climate in one's environment that prompts one to seek refuge in the historical moments of great discoveries of the past, rather than in present reality.

To illustrate the activities of the frustrated, let us review briefly three important, consecutive periods of the history of science in Japan.

3. BEGINNINGS OF THE MARXIST APPROACH TO THE HISTORY OF SCIENCE

It is commonly recognized among the pre-war group that the man and the paper that shocked them in their youthful days[1] was Ogura Kinnosuke and his article, 'Class Society and Arithmetics', written in 1929, or two years earlier than Hessen and his group of Russian historians of science produced a similar shock to historians of science at the Second International Congress of the History of Science held in London (in 1931). Stimulated by G. V. Plekhanov's 'Art in Class Society', Ogura published successively a series of articles on the relation between mathematics and social classes,[2] marshalling historical evidence to demonstrate that even in mathematics, supposedly the most abstract branch of knowledge, class structures are reflected. This thesis created quite a sensation and was widely acclaimed by Marxist-oriented intellectuals in Japan.

When he wrote his first article, Ogura was not too familiar with Marxist doctrine, as he later confessed.[3] In a private conversation with the present writer, he said that his research just happened to coincide with the Marxist approach. Later, after more conscious study of Marxist literature, he extended his research beyond Japan to China, as he felt a serious disadvantage of inaccessibility to original source materials on Western mathematics. My own view, however, is that he remained throughout his life a liberal democratic thinker and an outspoken critic of authoritarianism, rather than a committed dogmatic communist. This may be one of the sources of his personal attraction and power of persuasion.

His influence was certainly widespread. After the War he was installed as the first president of 'Minshu-shugi kagakusha kyōkai' [Society of Democratic Scientists, a nation-wide movement of Japanese scientists], and also as the second president of the History of Science Society of Japan. His most direct influence was exerted on the pre-war group of historians of mathematics, such as Kondō Yōitsu and Mita Hiroo, who soon after the War published their works on the social history of mathematics, in a more elaborated and systematic fashion.[4]

To this volume particularly I cannot forego the temptation to add a note on the similarities between Ogura and our honourable teacher, Dirk J. Struik. They were both mathematicians, social historians of mathematics distinguished mathematics teachers,[5] superb pathfinders, civil libertarians and brave men. They even shared their original field in mathematics, that is, differential geometry. In his reminiscences, Ogura states that his work during his stay in Paris, 'Sur le champ de gravitation dans l'espace vide' (in *Tohoku Math. J.* **22** (1922)) was later found to coincide partially in its results with the research done independently by a Dutch mathematician – Professor Struik![6] When I first met Struik in 1955 I was immediately impressed by the remarkable resemblance of his manner of speech, his high spirits, and even his physiognomical features to those of Ogura.

When I was requested by Struik in 1965 to provide him with some information concerning a possible translation in the 1870s of a work by the Dutch mathematician A. B. Ghyben, the only source available to me was Ogura's *Sūgakushi kenkyū* [Research on the History of Mathematics; (1935 and 1948)], in which Ogura surveyed early Meiji mathematics and mentioned Ghyben's influence. Alas, Ogura had already passed away in 1962 and I could not personally consult him.

Actually, Ogura was a bit older than Struik's generation, which included J. D. Bernal, J. G. Crowther, and Joseph Needham – men who were heavily influenced by *Science at the Crossroad* (Account of Russian delegates to the Second International Conference on the History of Science). This work was, of course, welcomed in Japan by the pre-war group (being put into two different Japanese translations), but the way towards the development of the social history of science was already paved in Japan by Ogura's earlier works.

Ogura's interpretation of Japanese science, based on his own historical research, can be summarized as follows:

(a) While science in Western Europe played a crucial role in the emancipation of people, Japanese science does not have such a glorious tradition. It was an imported product, and hence naturally imitative and superficial.
(b) Since Japanese science has always been placed under the heavy protection of feudalistic and bureaucratic governments, scientists are cowardly, uncreative, and dependent.
(c) Governmental institutions monopolize all learning, and hence even natural science retains a strong feudalistic and bureaucratic flavour.
(d) The scientific community is controlled by feudalistic academic cliques, minimizing mutual criticism.
(e) Scientists are lacking in social-consciousness. These affirmations may appear hypercritical of his own tradition, while uncritically laudatory towards modern European science (Ogura was a great admirer of the French Revolution and its scientific institutions). But they can be understood as one of the best possible forms of resistance against the background of an authoritarian nationalistic bias in the 1930s and early 1940s in Japan. This way of understanding provided the basis of the post-war leftist science movement, and was adopted as the official interpretation of the Japanese communists.[7]

4. WARTIME ACTIVITIES

From the late 1920s on, the communist party and its sympathizers suffered police persecution. Their open activities were outlawed, so many scholars concentrated their energies on the analysis of the philosophical implications of Marxist doctrine rather than indulge in concrete realities. The 'Yuibutsuron kenkyūkai' [Society for Materialistic Studies, 1932–1938] was most active in such philosophical affairs. Ogura was one of the founding members, and Saigusa Hiroto, who after the war succeeded Ogura as president of the History of Science Society, was an original organizer.

Saigusa, after his arrest for an offence involving 'dangerous thought' in 1933, devoted himself to working on past Japanese thought and to editing a number of Japanese scientific classics. Ostensibly objuring Marxist belief, he maintained and defended rationalistic thought. History was a safe ground – the older, the safer – to explore even amid current repressive measures to control thought. In the late 1930s it became difficult to employ Marxist terms outspokenly, and they were gradually replaced by the vocabulary of 'science'. Inside the cover of 'science', which could not easily be infected even by the wartime demagogues, the history of science provided for leftist liberals a shelter from the eyes of governmental censorship and from the arms of police thought-control.

The History of Science Society in Japan was founded in the same year that the Pearl Harbor Incident occurred. Why so? There was at the time a boom of the history of science in the popular reading world. One of the main causes of this popularity was the effort to enhance and glorify the scientific achievements of the ancestors in Japan's past, by which it was intended to wipe out the inferiority complex of the Japanese towards Western science and to encourage self-confidence in their cultural heritage.

This boom had a parallel in the enhancement of Aryan scientific contributions in Nazi ideology, but it does not appear to have been directly connected to the latter. Japanese scientists, even on the extreme right, were not so irrational as to accept Nazi science at face value, and moreover, the latter was too ethnocentric to be adopted by non-Aryan races. Adolf Hitler in *Mein Kampf* defined Japanese science as uncreative but only culture-supporting. This part was omitted from the pre-war Japanese translation and revived only in the post-war version.[8]

According to the editorial notes of the first issue of *Kagakushi kenkyū* (Official organ of the History of Science Society), the society was founded in order to correct uncritical popular versions of the history of science current at that time and to demonstrate the genuine standards of scholarship in the field. This principle has been well maintained, as testified by the wartime issues of the journal, which was continued up to the end of the war despite various difficulties. It was not affected by any nationalistic bias, as other semi-popular journals were obliged to be.

It is apparent, nonetheless, that founding the society did depend somewhat on taking advantage of popular support. Under the same nationalistic support an extensive series on the history of Japanese science before the Meiji era (*Meiji-zen Nihon kagakushi*) was projected to commemorate the 2600th year in the Japanese chronology in 1940. The editing of this series was seriously interrupted by the war and its publication was postponed to the post-war period, when it finally appeared in 1954–1968 in twenty-six volumes.[9]

5. THE POSTWAR HEYDAY

Postwar Japan has been characterized by the triumph of Marxist doctrines. Primarily because of the pre-war persecutions suffered by Marxist ideology, its post-war prestige has run extremely high among intellectual circles as a counteraction. Many Marxist-oriented authors writing on science who had been suppressed during the war years, now found opportunities for uninhibited expression in post-war literature. Some examples include Kondō's history of mathematics, Taketani Mitsuo's three-stage scheme of scientific development, and Hara Mitsuo's advocacy of Engel's dialectics of nature. As a result, Marxist doctrine became established orthodoxy in the academic world in the period 1945–1950, gaining hegemony especially in the fields of economics and history.[10]

The post-war generation of historians of science received the full impact of the post-war version of Marxism in their youth, but their reaction to it differs somewhat from that of the pre-war generation. While for the earlier generation Marxism meant a new scientific outlook, a new world-view to be advocated and diffused, it was the academic establishment for the second generation. Once established, it tended, because of trying to exterminate unorthodox views, to become stereotyped and to lose its creativity. Hence, the new generation shared with the old a deep sense of problem but is often inclined to be critical towards the 'established Marxist code'.

For instance, Hirosige Tetu criticized Ogura and his followers on the basis of his own analysis that the post-war leftist movement in science depended too heavily on their assessment of the historical conditions of previous Japanese society, and thus failed to face squarely the newly-arisen post-war factors; the earlier group had adhered faithfully to the formula that science is a superstructure determined by its social base, but overlooked the new phase of science established now as a 'social institution'.[11] Nakamura Teiri made a critical appraisal of the Japanese version of the Lysenko controversy,[12] Yamada Keiji tried to find a new perspective on science in New China under the Cultural Revolution as a sign of the bankruptcy of modern Western science.[13]

Generally speaking, though, Japanese historians of science are stronger in the externalist rather than the internalist approach. While a great deal of Marxist and externalist literature in Western languages – such as works of J. D. Bernal, J. G. Crowther, D. Struik, Franz Borkenau, and even Sir Eric Ashby – are available in Japanese version, no work of Alexander Koyré has yet been fully translated. (So far only an article by him in the *Journal of the History of Ideas* has been translated as a part of a collection of essays by many authors.) This externalist characteristic necessarily originates in the historical recollection of the Japanese.

In Western history, historians of science almost unanimously agree that modern science was founded at the time of the Scientific Revolution in the seventeenth century. But this is true only from the viewpoint of intellectual history. The Scientific Revolution was only an intellectual movement of a handful of scientific intellectuals. Institutionally, modern science was founded, on the other hand, in the nineteenth century when scientists tried to advance their social status and eventually succeeded in establishing themselves, when their uninterrupted reproduction through a recruitment mechanism based in institutions of higher education was accomplished, and thus when science was fully professionalized. Some people therefore call the nineteenth century the century of the Second Scientific Revolution. The nineteenth century gave birth to the present-day gigantic, monster-like scientific establishment, and hence it is worthwhile to examine the historical origins of the various complex problems existing between contemporary science and society.

When we try to locate the point of departure for modern science in Japanese history, our attention is riveted to a big revolutionary break at the Meiji Restoration in the latter half of the nineteenth century when Japan embarked on its wholesale importation of Western science and related institutions. Prior to that time there had been, of course, Japanese scientific endeavours; but it is totally anachronistic to think of Japanese participation in the seventeenth-century Scientific Revolution. Premodern Japanese had nothing whatsoever to do with the Scientific Revolution in the internal sense. The Japanese encountered modern science from the very beginning as the developed Western institution of the nineteenth century. Thus, this century is doubly important to the Japanese since both the internal and external Scientific Revolutions coincide in the same period in their history. Hence, the Japanese may have a keener sense of the problems of external history of science in recent history than the average Western historians of science who flock at the seventeenth century or even before.

In the historical recollection of the Japanese, the term 'modern' may summon up the image of American gunboats anchored in Tokyo Bay in 1853; or it may be synonymous with the economic capacity and technological innovation – products of the Industrial Revolution – that supported the military superiority of the West. To the 'now-concerned' Japanese, technology has almost equal status with science and therefore the Japanese History of Science Society has a stronger complement of historians of technology than is found in its Western counterparts.

This sense of problem was materialized in the recently completed 25-volume *Nihon kagaku-gijutsushi taikei* [Compendium of the History of Science and Technology in Japan (1964–1970)], which deals with Japanese scientific development since the middle of the nineteenth century. It is, at least in bulk, an astonishing achievement, though judgements as to its quality are in the hands of future historians of science.

6. *NOUVELLE VAGUE* IN THE HISTORY OF SCIENCE?

Which way the future generation of Japanese historians of science will proceed is not easily predicted. We have seen few fruits from the labours of the 'mid-war' generation who came between the pre-war and post-war groups, because the devastation of war sapped their youthful energies. We find still less harvest from the new recruits in more recent days, who are, thanks to the science and technology boom, unwilling to follow such a precarious career as that of a historian of science. If new recruits do in fact appear, we can expect that they may be more 'normal and academic' and commit themselves to the field only when conditions of normal and academic success are assured to them.

Only very recently, however, can there be seen symptoms of a new generation coming forth – due perhaps to recent university struggles and pollution problems. They will face science with entirely new perceptions.

For the pre-war generation science was the last gleam of rational thinking to be defended from stormy wartime demagogism. For the post-war group science still stood at the hub of the democratization choir. For the new generation science may appear as a monolithic establishment not to be easily undermined: it is no paradise to be discovered, but only a harsh and inescapable reality. What new picture of science, then, may come out of this 'paradise-lost' generation?

In the cold war atmosphere two sciences have prevailed: Japanese science has in actuality consistently and definitely accommodated itself to the American model, despite adherence in some circles to the socialist version of science. Now that the American model has faltered seriously, are there any frustrated youths searching for a third model of science, a new paradise?

University of Tokyo

NOTES

1. Kondō Yōitsu, one of his pupils, testifies to this shock in his article 'Ogura Kinnosuke sensei no sūgakushi kenkyū' [Ogura's Research in the History of Mathematics], *Kagadushi kenkyū*, 1963, p. 17.
2. They are: 'Kaikyūshakai no sanjutsu' [Arithmetics in Class Society], *Shisō*, No. 1 (August, 1929) – a treatise on arithmetics in the Renaissance period; No. 2 (December, 1929) – a treatise on Colonial American mathematics.
   'Sanjutsu no shakaisei' [Social Character of Arithmetics], *Kizō* (September, 1929) – British society and economy in the sixteenth century as observed through books of arithmetics.
   'Kaikyū shakai no sūgaku' [Mathematics in Class Society], *Shisō* (March, 1930) – a treatise on the history of mathematics in France.
3. Ogura Kinnosuke, *Ichisūgakusha no kaisō* [Reminiscences of a Mathematician], p. 102.
4. For instance, Kondō Yōitsu, *Kikagaku shisōshi josetsu* [A history of Geometrical Thought; a Preliminary Study, 1946, Itō Shoten].
5. Ogura's *Sūgaku kyōikushi* [History of Mathematical Education, 1932, Iwanami, Tokyo] is reviewed in *Isis*.
6. *Ichisūgakusha no kaisō*, p. 89.
7. Hirosige Tetu, 'Ogura Kinnosuke to Nihon kagakushi' [Ogura Kinnosuke and the History of Japanese Science], *Kagakushi kenkyū* (1963) 9–16.
8. *Nihon kagaku-gijutsushi taikei, kokusaihen* [Compendium of the History of Japanese Science and Technology, International Relations], (ed. by S. Nakayama et al.), 1968, p.333.
9. See Suketoshi Yajima's review in *Japanese Studies in the History of Science*, No. 7 (1968) p. 159.
10. *Nihon kagaku-gijutsushi taikei, Shisōhen* [Compendium of the Japanese History of Science and Technology, Ideology] (ed. by T. Tsuji), p. 396.
11. Hirosige, 'Ogura Kinnosuke to Nihon kagakushi',
12. Nakamura Teiri, *Ruisenko ronsō* [Lycenko Controversy], Misuzu, Tokyo, 1967.
13. Yamada Keiji, *Mirai e no toi* [Question for the Future], Chikuma, Tokyo, 1968.

First published in William Beranek Jr. and Gustav Ranis (eds), *Science, Technology and Economic Development: A Historical and Comparative Study*, pp.202–232, Praeger, 1978

11 Science and Technology in Modern Japanese Development

In those countries that have most recently attained the modern stage of economic development, science and technology have usually been introduced by the public sector and transferred to the private sector when they have become productive. The Japanese experience in the late nineteenth century was typical and one of the earliest examples of this process. In the process of transferring a technology to the private sector, there are critical points at which it must be determined whether imported science and technology will become established in the private sector and flourish in an indigenous, self-perpetuating form, or whether their continued importation into the public sector will be necessary.

The next question is who was instrumental in initiating and transplanting science and technology and who supported this effort. I will discuss this in terms of the following tentative model: First, there must have been a group of native people who could become professional scientists and engineers, and who could act as leaders in the mass education of the people and in material construction. Next, the level of basic education must have been substantially raised from that of a nondeveloped economy for a successful transfer to occur; a modern state needs scientific manpower at all levels of society. Finally, in this process social mobility begins and class differences tend to diminish. Modern science and technology is then diffused and rooted among the populace, and its traditions become perpetuated.

The achievement of these three steps may be the major factor differentiating the countries that are to become technological from those that must experience an ever-increasing technological gap.

We must also consider that the degree of success in the transfer depends to a great extent on the kind of technology chosen. To find the conditions for successful transplantation, we shall examine several selected technologies of entirely different types, some imported and some autochthonous.

EARLY EFFORTS, 1868–85

The Utilitarian Image of Science

It is a common belief among historians of Western science that in premodern times science and technology were distinct activities with different social origins. In spite of the effort made by the Encyclopaedists to liquidate this social interface, the dual structure of science and technology was still maintained, even in the nineteenth century, by socially separated groups. This was exemplified by such institutional separations as that between the German university and the polytechnical college. However, there was no particular reason for the mid-nineteenth-century Japanese to distinguish between science and technology when facing the impact of the modern West. To the Japanese it appeared that modern science and modern technology grew in a single Western tradition. It was not the science-versus-technology dichotomy but, rather, the traditional-versus-Western dichotomy with which the Japanese were seriously concerned.*

While science in nineteenth-century Europe was still in the main a cultural activity rather than a practical means of achieving economic growth, as is well illustrated by the issue of the theory of evolution, the Japanese image of science in the late nineteenth century was quite modern. It was exclusively utilitarian and pragmatic, planned for national interests if not purely for profit-making, specialized and compartmentalized. Emphasis was on physical and applied science rather than on biological, and hence the style was closest, for that period, to contemporary scientific technology.

The Institutionalization of Science

After the Meiji Restoration of 1868, the Japanese response to Western science was dramatically transformed. In the preceding Tokugawa period, Western science was initiated and advocated mainly by scholars in the private sector. Only in the last period of Tokugawa rule were official training institutes for Western naval technology and related sciences opened at Nagasaki and elsewhere. Being hard-pressed by the urgent defence needs of the country, the administration lacked the foresight to build institutions providing modern, systematic education in science and technology. During the first two or three years (1868–70) of the new Meiji government, there was still an influential group that wanted a faithful restoration of the ancient imperial system of government; but the modernist leadership soon followed the policy of Westernization. Scientific education was completely institutionalized. It became firmly programme d in such a way that an institution was first created, and then European and American scientific and technical specialists were invited to meet selected Japanese students within the institution. Outstanding graduates were sent abroad for further study.[1]

Guidelines for Westernization

In the draft rules for sending students to study abroad, prepared in 1870, the following subjects and preferred countries of study are listed:[2]

Britain: machinery, geology and mining, steelmaking, architecture, shipbuilding, cattle farming, commerce, poor-relief

France: zoology and botany, astronomy, mathematics, physics, chemistry, architecture, law, international relations, promotion of public welfare

* Until the 1880s the Japanese language did not distinguish clearly between 'science' and 'technology'. The separation between the concepts became real only towards the close of the century, when autonomous scientific communities were formed at the university-faculty level.

Germany: physics, astronomy, geology and mineralogy, chemistry, zoology and botany, medicine, pharmacology, educational system, political science, economics
Netherlands: irrigation, architecture, shipbuilding, political science, economics, poor-relief
United States: industrial laws, agriculture, cattle farming, mining, communications, commercial law.

In view of the history of nineteenth-century science, the above assessment was for the most part correct and objective. Presumably the policies for science and technology, including the above recommendation, were drafted mainly on the suggestion of G. F. Verbeck and other foreign advisers to the government.

Employment of Foreign Scientists and Engineeers
About 1870, a policy was adopted for the employment of Westerners to guide teaching in schools and governmental enterprises. French engineers were employed in the army, in mines, and in dockyards; Germans taught medicine and basic sciences in the schools; and a number of Americans were agricultural specialists for the Commission for the Colonization of Hokkaido (the island on which American large-scale cultivation was attempted).

The British were numerous in many fields. Generally, in engineering they made the greatest contributions. Nearly 100, for example, were employed in the construction of railways. At the Imperial College of Engineering, which existed from 1873 to 1886, the teaching staff was mostly British and was directed by a Scottish engineer, Henry Dyer.

The number of government employees reached its peak in 1874. The total of the employees' salaries reached its peak in the same year, its ratio to the total national expenditure being about 2 per cent. Apparently the government wanted to reduce the number and cost of foreign employees, who were paid salaries ten times greater than those of the Japanese, and to replace them with natives who had received scientific training in the West. The Ministry of Education allocated funds for this purpose, but the amount was still far less than the foreigners' salaries.

Japanese Students Abroad
The initial programme of sending students abroad at government expense was not well organized, and in 1873 it was abolished. In 1875, in its place, overseas scholarships were instituted by the Ministry of Education. The Ministry of Technology sponsored a similar programme beginning in 1880. Since they had already received basic training from foreign teachers, science students abroad concentrated on research, and technology students used their time to observe the actual practice in the cities.[3] When they returned, professorial (or equivalent) posts awaited them. This first generation of Japanese university professors displaced the foreign instructors.

Until about 1877 education in Japan put more emphasis on language study than on one's chosen specialty. There was, therefore, a tendency among the first students who went abroad to choose the country on the basis of the language they had studied rather than on the basis of a particular subject. After 1881, however, the Japanese government began to copy the German administrative system and political institutions; accordingly, many students turned their attention to Germany. It was the heyday of German science – students from England, France, and the United States were going there for advanced study – and the Japanese students must have been impressed by the German scientific leadership. Out of 26 Japanese doctorates of science conferred by 1891, 13 went to men with study and research experience in Germany (five had studied in the United States and three in Britain).[4]

Emphasis on Physical Science and Specialization

Japanese studying Western science during the late Tokugawa period were impressed by the Western process of inquiry into physical laws, rather than by the aggregate of facts and objects of nature. They came to feel keenly that although there was no great gap between East and West as far as the classificatory knowledge of natural history was concerned, Chinese and Japanese cultures seriously lacked the belief in the underlying regularity of nature and the 'investigation of its principles': natural philosophy.* Along with this view of Western science, the first primary school curriculum prescribed by the Ministry of Education was not oriented to natural history and biology, as was the case in American primary education, but to physics.[5]

During the 1870s and 1880s the relative position of science and technology in the Japanese educational curriculum, from elementary school to university level, was much higher than in any other nation. For instance, mathematics and science occupied about one-third of the school curriculum in the lower grades (first four years) and two-thirds in the upper grades of the eight-year elementary education. However, the shortage of qualified teachers makes it somewhat questionable to what extent these ideal plans were put into practice. At the university level the emphasis on science and technology was evident in the high per centage of graduates in scientific disciplines from Tokyo University (85 per cent in the 1880s, compared with 40 per cent in the 1920s.

The Official and Planned Character of Science

At the frontiers of newly forming disciplines in nineteenth-century Europe, the scientist was free to select for his research any problem that interested him; and the scientific communities were formed by individual scientists drawn together by a common interest. The voluntary research activities of scientific or professional societies usually preceded their inclusion in the university curriculum. In Meiji Japan (and generally in the case of the artificial transplantation of a foreign discipline under state sponsorship) this process was reversed. The government first created institutions for training scientific personnel; only then did

* This intellectual tradition started with B. Miura and T. Shizuki and extended until the time of Y. Fukuzawa. Its exponents appropriated the neo-Confucian concept of *kyuri* to translate 'natural philosophy' (in Dutch, *natuurkunde*), but later, especially in the Meiji period, this term was specified to mean physics.

university graduates in each discipline form scientific societies, not purely for academic purposes but mainly with common interests in their new and still very weak scientific careers. Thus the scientific community in Japan had a 'planned character'; it was planned for the purpose of catching up with the Western standard of science as quickly as possible.

In the nineteenth century, however, it was uncommon, even among advanced nations, to find established precedents or formulas for a national science policy. Thus, the Meiji government had to find its own way by trial and error.

On the practical level, it urgently needed qualified teachers and engineers to build a modern state. The new government built schools and factories, trained scientists and engineers in these institutions, and sent them off to their posts. Rather than having every scientist follow his own research interest, priority was given to certain basic tasks, the accomplishment of which was necessary for the operation of a modern state: geographical and geological surveys, weights and measures, meteorological observation, sanitation, printing, installation of telegraph and telephone lines, military works, railways, and surveys of natural resources. All these enterprises were carried out by the Ministry of Technology, the Ministry of Interior, the Ministry of Finance, the Commission for the Colonization of Hokkaido, the army, the navy, and other governmental agencies under the supervision of such engineers. To conduct such nationwide projects, these agencies had to have their own short training programmes to provide field assistants for foreign employees. These agencies did not depend exclusively on the Ministry of Education, which was responsible for regular long-term educational programmes.

We may label these activities as 'public science' initiated by the government. This step in science for public service was the indispensable prerequisite for the industrialization carried out by the next generation. In addition, the government entered into private entrepreneurship, constructing and managing pilot plants, guiding and subsidizing new kinds of industries. The Ministry of Technology and the Commission for the Colonization of Hokkaido were two major institutional innovations. They carried out experimental programmes and were centres of Westernization and modernization.

Their enterprises were from the outset, however, exposed to financial risk. Many of the Ministry of Technology's projects eventually proved to be too far ahead of their times, for they tended to introduce the technology of an industrialized society into a preindustrial environment. For instance, the railway construction enterprise was economically unsuccessful at the time, and paid off commercially only after 1885, in the next phase of industrialization.

Thus, Y. Fukuzawa, Japan's foremost exponent of industrial revolution, concluded that 'we should not blame them too much for their financial failure. After all, it was a costly tuition fee for the Japanese to learn civilization.'[6]

Lack of Research

This planned character of Japanese science has one notable defect. The institutionalization of science and technology at the governmental initiative was certainly an efficient tool for transplanting and introducing foreign science and technology, but it was less effective for fostering original creative activity. Institutions were divided into specialized disciplines, each scientist and engineer assuming his specialized role, a situation permitting little cross-fertilization of ideas.

In fact, in comparison with traditional Confucian learning, the Japanese in the early Meiji period found the strongest point of Western science to be its specialization. This was a particular nineteenth-century aspect of science, not found earlier. The term coined by the Japanese in the 1880s as a translation of the word 'science' was *kagaku*, which originally meant 'study of one of the hundred departments' and was the equivalent of the German term *Fachwissenschaft*.

Thus, each scientist or engineer was concerned with absorbing the foremost achievements of the Western world, within a narrowly prescribed disciplinary barrier, rather than with cooperating with his Japanese colleagues in different fields.

The other side of science policy, research policy, was tactfully avoided. Research policy in any systematic form was the product of the mobilization of science during the First World War. The Meiji government gave practically no attention to scientific research. One point should be noted, however. Unlike the European monarchies, Japan had no great interest in founding a national academy of science as a status symbol. The practically oriented Japanese totally subordinated pure research to manpower policy; thus Tokyo University enjoyed the position of the top educational institution as well as the highest academic prestige in the country.*

For the first generation of scientists in the early Meiji period, it was more important to establish institutions than to pursue piecemeal individual research topics. There were some internationally notable contributions made by early Japanese scientists, such as S. Kitasato's works on microbiology, but these usually originated during research apprenticeship in the West. In Japan research workers could not find colleagues for discussion, unless they trained their own students in the same research tradition. Thus they became exclusively involved in administration and education, eventually leaving their research activities completely aside.

The first generation of scientists concentrated on local science, the application of modern scientific methods to the analysis of local flora and fauna and earthquake and geological observations. Such disciplines as zoology, botany and geology remained local sciences at least until the Second World War. Even in such method-oriented disciplines as physics and chemistry, the early scientists' concern was with geophysical observations and the chemical analysis of local products. Until 1885 more than half of the research articles published by the faculty members of the College of Science at Tokyo University were on local science.[7]

The Samurai Spirit in Science and Technology

Modern scientific and technological professions were the artificial creation of the new, Western-oriented government. The main practitioners of these new professions were former samurai, who were warriors by tradition but who, during the Tokugawa period, became primarily administrative bureaucrats. In the past they had received hereditary family stipends in exchange for their loyalty to the

* The Japan Academy, founded in 1879, was a gathering of obsolete scholars of general Western learning and was often ridiculed as being a home for the retired.

shogunate or local feudatories. They were the class long accustomed to thinking in terms of public affairs and to holding public office.

In the 1870s efforts were made by the Meiji government to curtail the inherited family stipends of the samurai class. While farmers, artisans, and merchants could continue their inherited vocations, the samurai lost their traditional source of revenue. The modern government needed less than 10 per cent of the samurai population to meet its personnel needs, so the rest were forced to find an entirely new means of living. Since samurai could not compete with other classes in traditional business, they were invited to engage in such new business projects as the agricultural exploitation of Hokkaido, silk manufacturing and textile industries. This was especially true after 1876 when the former privileges of the samurai were completely abolished. Science and technology was one of the new fields into which jobless samurai were attracted. It is reported that almost all of the early graduates of the engineering college were samurai.[8]

As late as in 1890, the per centage of samurai-born graduates from each school of the Imperial University was as follows: engineering majors, 85.7 per cent; science majors, 80.0 per cent; literature majors, 75.0 per cent; law majors, 68.3 per cent; agriculture majors, 55.9 per cent; medicine majors, 40.8 per cent.[9] The science and technology group had the highest rate of samurai-born graduates and the medical group the lowest. We may assume that graduates of the medical and agricultural schools were, respectively, mainly sons of medical practitioners and farmers, whereas many samurai, deprived of status and occupation, found their most promising career possibilities in the new professions of science and engineering. As the science teachers and promoters of the new Westernization policy and as technocrats building a modern state, both purely governmental enterprises of top priority, the former samurai were able to use their aptitude for administration to achieve a class vocation.

Apart from ideological considerations, and perhaps more important, the sons of the impoverished samurai class were attracted to government-sponsored careers in science and technology because tuition was free, grants were provided, and government service was obligatory on graduation. (For graduates of the Imperial College of Technology, seven years' service in government factories was required.) Thus, science and technology provided new careers for a substantial stratum of society.

Hence, modern Japanese scientific and technological professions were, at the beginning of their development, very 'samurai-spirited'. Unlike the European pattern, in which science and technology attracted the rising middle class, Japanese scientific and technological professions in the last quarter of the nineteenth century were dominated by the samurai class, the top 5 per cent of the total population.

REORGANIZATION AND TAKE-OFF: 1886–1914

Reorganization of Institutions Around 1886

After two decades of intensive development programmes, including the institutionalization of science and technology, Japan was prepared to become a modern industrial state. Around 1885 the Meiji government reassessed, reformulated and reorganized many of its early policies. The most significant event was the dissolution of the Ministry of Technology in 1885. Many government enterprises were transferred to the private sector, mainly because both the government and its critics realized the inefficiency of government factories in profit-making industrial activities. Public-minded samurai engineers accordingly moved into the private sector.

In 1886, when Tokyo University was reorganized into the Imperial University by uniting the Imperial College of Engineering with the Ministry of Technology, most foreign teachers were replaced by native scientists who had just returned from studying abroad.

Until 1885 the scientific public services, such as surveying, had been conducted by transitional manpower who had received short, intensive instruction and had trained mainly on the job. After such services were reformulated, many transitional schools and offices were closed and, around the time of the founding of the Imperial University in 1886, the vocational engineering schools* were rearranged more systematically to provide engineers for Japanese industry, which at that time badly needed lower- and middle-level engineers.

The days of enthusiasm for Western science were over, and the government became more interested in having good-quality administrative bureaucrats, rather than entrepreneurial technologists, for the maintenance of regime. Law graduates of the Imperial University received governmental favour, and engineers were placed in positions subordinate to them.

The Language Problem

In the early days of the Meiji period, college teaching was still conducted, even by native Japanese teachers, in such foreign languages as English, German and French, according to the country in which they were trained. By 1900 the Japanese language was dominant, though technical terms remained untranslated.

It was once seriously proposed by A. Mori, an exponent of modernization and Westernization, that the Japanese language should be wholly replaced by English in order to follow the international standard of knowledge as closely as possible, but his plan was never put into practice. If it had been enforced, it might have created a dual culture with an English speaking elite. It would have blocked or considerably delayed the diffusion of Western scientific culture among the populace, who would continue to speak Japanese.

In the nineteenth-century scientific community, English was not so predominant a language as it is now. At the Imperial University, while English prevailed in the School of Engineering, the Medical school adopted German as its academic language as early as 1869 and continued to use it until after the Second World War. Further, German was second only to English as the second language taught in schools that prepared students for the university. Japanese students were required to learn it as a scientific language.

* Major examples are Ashikaga Textile Training Centre (founded in 1885); Kyoritsu Women's Vocational Training School, Tokyo Apprenticeship School of Commerce and Technology, and Kyoto Dyeing Training Centre (founded in 1886); and Hachioji Textile and Dyeing Centre and Kanazawa Engineering School (founded in 1887).

The Dual Structure of Japanese Technology

Modern engineering and technology may be said to have two entirely different origins. One is the 'community-centred' or 'public-centred' engineering service practised in the public sector, best exemplified by the military engineering taught and practised at the French École Polytechnique. The other is 'self-centred,' profit-making, capitalistic engineering practised in the private sector, as exemplified by the Watt-Boulton type of enterprises, power engineering, pharmaceutical technology, and so on.

The traditional Western dichotomy created in the Japanese technological world can be translated in socioeconomic terms into the private/public dichotomy. Private technology is not only 'self-centred' and profit-making but also traditional and domestic, not based on science, and transmitted through apprenticeship in the non-samurai sector, such as in saké brewing and in ceramics. Public science and technology are, on the other hand, not only 'community-centred' but of Western origin, university-based (and hence science-based), and practised mainly by the samurai elite.

Traditional craftsmanship, such as carpentry and fishery, remained outside the government system of modern science and technology. Established scientific and technological professionals were completely separated from traditional craftsmen and found employment in such government agencies as the office of the geological survey and in military arsenals.

I will now examine the actual state of this dual structure by using the list of patent applications. In 1885 the Patent Law was issued, and 100 patents were granted in the first year of its existence. Its purpose was to protect the honour and profits of native Japanese inventors against other Japanese, and had no international implications.

In the early Meiji period, Western industrial designs and inventions were not legally protected in Japan, and the Japanese were free to appropriate them. Westerners claimed that Japan should subscribe to the international patent regulations, since otherwise they could not give their technological advice freely, without the fear of unlicensed Japanese copying. But those who welcomed the enforcement of international regulations were, of course, the Westerners rather than the Japanese. Around 1885 a major political and diplomatic issue in negotiation with major Western powers was the adjustment of unequal treaties. The government intended to give patents to foreigners in exchange for Western denunciation of unequal treaty items,[10] and made a preliminary draft of a domestic patent system. The treaty amendment was postponed to 1899, when Japan subscribed to the International Industrial Property Regulations. Not until then did foreigners obtain Japanese patents.

Out of the first 100 patents issued in 1885, applications by former samurai accounted for only 17 per cent.[11] This is in clear contrast with the fact that samurai made up a high per centage of the College of Engineering graduates. Many of the samurai applicants signed only as investors, rather than as inventors.

This contrast may well be explained by supposing that while elite graduate engineers were busy introducing and translating the technology imported from the West, inventive skill in the private sector was being demonstrated in local technological adaptation. (A collection of lives of Japanese inventors to 1935 shows that only 7 per cent were college graduates.) College graduates were engaged in public works, and thus it was perhaps not proper for them to apply for profit-making patent rights, whereas private-sector inventors were eager for the honour of official recognition, even if the invention was not put into profit-making practice. Only in the twentieth century, after Japan had subscribed to the International Industrial Property Regulations in 1899, were high-level engineers concerned with patent rights.

Since 1899 Japan has received a number of foreign applications for patent protection. (See Table 11.1.) The United States was particularly strong in electrical equipment and Germany was strong in weapon manufacturing and chemicals. British interests were diverse but particularly strong in shipbuilding.

Table 11.1 Foreign Holdings of Japanese Patents, by Country, March 1910

Country	*Number of Patents*	*Per cent*
United States	1,373	38.4
Britain	948	26.5
Germany	647	18.1
France	198	5.5
Italy	68	1.9
Sweden	56	1.6
Austria	47	1.3
Denmark	36	1.0
Hungary	36	1.0
Netherlands	33	0.9
Belgium	33	0.9
Switzerland	31	0.9
Russia	23	0.6
Norway	22	0.6
Other	29	0.8
Total	3,580	100.0

Source: *Daigoji tokkyokyoku nenpo* (Fifth annual report of the Patent Office, Tokyo: Tokkyokyoku, 1911), Table 12.

During the first seven years under the International Regulations, Japan experienced a gradual invasion by foreign patent rights. In 1899, foreigners owned about 17 per cent of the Japanese patents. In 1905, 28 per cent of the patents granted were to foreigners.

The Patent Office had its own classification system, dividing items into 136 kinds. Table 11.2 presents the major items and classifies them into three categories, according to the degree of native contribution.

In category C the native contribution is more than two-thirds. The items are mostly small, useful gadgets for local use that were developed in the practice of traditional craftsmanship, mostly in the private sector, devised by those without college and science backgrounds. Agricultural implements, such as a rice-polishing machine, belong to this category. Some technologies, such as that for making matches, had been completely transferred by the turn of the century.

Table 11.2 Degree of Native Japanese Patent Holdings, 1899–1905

Category	*Number of Native*	*Patents Total*	*Per cent Native*
Category A			
Electric lamps	6	46	13.0
Electrochemical industry	5	34	14.7
Miscellaneous electrical applications	10	61	16.4
Steam motors	14	73	19.2
Metallurgy	6	31	19.4
Guns, bows	28	128	21.9
Telegraphy, telephony, electric signals	30	118	25.4
Regulating, distributing electricity	22	73	30.1
Category B			
Electric batteries	7	20	35.0
Steam generators	47	121	38.8
Chemicals and chemical products	36	86	41.9
Mechanisms and mill gearing	55	130	42.3
Mining machinery	14	33	42.4
Ships and boats	32	75	42.7
Signals and communicating apparatus	18	40	45.0
Furnaces and kilns	22	44	50.0
Rope tramways and the like	18	35	51.4
Carriages	73	138	52.9
Civil engineering	26	48	54.2
Printing machines and appliances	18	50	64.0
Miscellaneous chemical works	36	56	64.3
Spinning	27	42	64.3
Match manufacturing machines	24	37	64.9
Category C			
Miscellaneous manufacturing machines	199	295	67.5
Buildings and structures	34	49	69.4
Ceramic industry	10	14	71.4
Pumps, other means of raising liquids	48	67	71.6
Tools	71	94	75.5
Measuring, indicating, registering	84	111	75.7
Food, beverage, confectionary machines	48	63	76.2
Kitchen appliances	73	93	78.5
Matches	9	11	81.8
Medical instruments	34	38	89.5
Safes and locks	77	86	89.5
Illuminating apparatus	190	212	89.6
Dyestuffs	50	55	90.9
Paints, varnishes, lacquers	40	43	93.0
Toilet, hairdressing articles	56	60	93.3
Corn husking implements	58	62	93.5
Knitting and netting	89	95	93.7
Dyeing machines and appliances	60	64	93.8
Sanitary appliances	64	68	94.1
Furniture	99	105	94.3
Tobacco manufacture	39	41	95.1

Category	*Number of Native*	*Patents Total*	*Per cent Native*
Sugar manufacture	41	43	95.3
Looms and weaving	273	285	95.5
Tea processing machines	69	72	95.8
Winding and twisting machines	85	88	96.6
Traps	61	63	96.8
Umbrellas and walking sticks	34	35	97.1
Bedroom furniture	44	45	97.8
Stationery	158	161	98.1
Boots, shoes, wooden clogs	116	118	98.3
Agricultural appliances	238	242	98.3
Boxes, cases, trunks, bags	123	125	98.4
Grain polishing machines	69	71	98.6
Sericultural appliances	56	56	100.0
Weaving and knitting	59	59	100.0

Source: *Daiichiji tokkyokyoku nenpo* (First annual report of the Patent Office, Tokyo: Tokkyokyoku, 1907), Table 17.

In category B the native contribution is reckoned at between one-third and two-thirds. It includes Western technologies that are nearly assimilated but still being transferred, such as cotton-spinning machinery, shipbuilding technology, and printing machines. In these fields the native contribution is near the level of international technology.

In category A the native contribution is less than one-third. It includes new Western inventions, such as electrical equipment, and such large-scale technologies as metallurgy and weapon manufacturing. It is science-based and occurs mainly in the public sector, which has qualified scientific and technological personnel.

By selecting cases of agricultural technology, spinning and textile engineering, and military engineering and its associated mechanical engineering from categories A, B and C, respectively, I shall try to show how the public/private and Western/traditional dichotomies were removed.

Agriculture

The most important traditional technology was, of course, agricultural technology. Most of the government revenue in the early Meiji period before industrialization came from a land tax on the agricultural sector. It was a prerequisite for modern economic growth that agricultural technology be improved and farm production increased, in order to use the agricultural surplus to defray the cost of modernization and capital investment for industrialization. Most agricultural technology was practised in the private sector through the private initiative of experienced farmers, and the Meiji government tried to introduce its new agricultural experts trained in Western technology.

The new agriculturalists were educated at two agricultural colleges, one in Komaba (Tokyo) and the other in Sapporo (Hokkaido). Students were mostly sons of impoverished samurai who rushed to apply to the colleges, attracted by government scholarships that were better than in other schools.

They had no previous background in agriculture and were trained by foreign teachers.

In the earliest period of the Komaba Agricultural College, which was founded in 1877, British and German agricultural scientists tried to introduce the European style of animal husbandry and to import Western plants, livestock and farm implements. They had nothing to offer for the improvement of paddy-rice production, even though rice was the major crop raised in Japan. The Komaba College proved to be unsuccessful in the 1880s when farmers realized that Western techniques were difficult to adapt to traditional, small-scale, land-saving paddy-rice agriculture. They turned to the veteran farmers for instruction.

There was a struggle between college graduates and veteran farmers for leadership of the agrarian reform movements.[12] The policy of local and central governments to experiment with newly imported methods and techniques was for the most part abandoned in the 1880s. The application of the newly acquired technology of the college graduates now focused on paddy-rice production, but it was limited to such academic studies as chemical analysis and biological studies.[13] The initiative in technological improvement of production was seized by veteran farmers, who took advantage of some of the conclusions drawn by the new agricultural scientists and applied them on the basis of their own largely empirical knowledge, acquired through long experience. The new research findings were taken into consideration insofar as they supported the farmers' knowledge. An example is the chemical analysis of traditional fertilizer by O. Kellner, a German chemist hired by the Komaba Agricultural College.[14] Except for research on plant diseases, basic college agricultural research in the early Meiji period did not contribute to the reform of agricultural technology.

The exploitation of Hokkaido, on the other hand, led to successful transfer of foreign methods. The government policy of establishing farms in the newly settled areas, designed to aid jobless samurai, encouraged the experimental introduction of farm appliances and new varieties, which were generally adopted by those unexperienced in conventional farming.[15]

In the early Meiji period rather superficial efforts were made to import the visible results of Western achievements, but without much practical success; however, since the 1890s scientific research has made great contributions to Japanese agricultural production in two important fields: the promotion of soil fertilizer and the breeding of improved varieties. It was only in the twentieth century that academically originated research in soil chemistry and the application of empirical breeding knowledge and Mendelian genetics[16] yielded practical results. Japanese agricultural science and technology concentrated on the land-saving intensive growth of crops without radical land reform or a change in economic framework.

Indigenous Industry

In textile enterprises, college-trained engineers mainly served as official supervisors assessing the work of entrepreneurs. In 1903 a government official criticized their attitude in an address entitled 'The Encouragement of Indigenous Industries':

> People and industrialists do not know yet how to utilize indigenous inventions. Most inventors are mechanics or industrialists, without adequate scientific background, or schoolteachers without technical experience. Those who had been trained in specialized disciplines are concerned only to import and imitate new industry. They have the mental habit of feeling ashamed to devote themselves to the reform of indigenous industry and stay out of it.[17]

He noted the manufacture of matches as a good example of an indigenous industry.

Though incomparably smaller than major industries like spinning, match manufacturing provides an exceptionally interesting case of technology transfer. It had no roots in traditional technology but was entirely imported from Europe in the early Meiji period. Since it did not require much capital stock, it achieved business success by acquiring technical know-how quickly, reaching export capability by the turn of the century.[18] As shown in Table 11.2, most of the patents on match manufacturing were native ones, and those foreigners who applied for Japanese patents were taking advantage of the Japanese business boom in the industry.

Textiles

The major industry of Meiji Japan was textiles (spinning, silk-reeling and weaving), which, according to the statistics of 1900, employed 233,000 people and had machines of 39,000 horsepower – more than ten times the size of the government armament industry.[19] The government encouraged business by establishing pilot plants with imported spinning machines and then turning these into private businesses, which obtained government loans for further importation of machinery. At the turn of the century, while the quantity of imported machinery was rapidly increasing in government heavy industry, the importation of spinning machines decreased considerably as capital investment increased. Textile products were turned from imports to exports.[20] As indicated in Table 11.2, at this time native patents on spinning machines considerably exceeded foreign patents. This may be one indication of the 'domestication' of spinning-machine technology, though wholly domestic production of modern spinning machines started only in the 1910s.

The textile industry is basically a profit-making activity of the private sector, and hence its technological advancement has been accomplished in that sphere. The technology was basically completed before modern engineering science was formulated in the middle of the nineteenth century. No course was given in the Imperial College of Engineering.[21] In the early Meiji period, spinning technology was looked down upon by college graduates, who considered it to be craftsmen's work and not the business of an intellectual class.[22]

The invention of the *garaboki* (a simple spinning machine) in 1876 was accomplished in the context of traditional home spinning, without any foreign influence. However low its manufacturing cost, its inability to produce quality-controlled fine yarn prevented its nationwide diffusion. In 1890 *garaboki* spinning reached its peak, 11.8 per cent of total production. It declined soon after the importation of a ring spinning frame, which made possible both high speed and continuous production.[23]

In contrast with the spinning industry, which was still completely dependent on imported machinery, the hand loom was produced domestically and played a great part in promoting the textile industry. Compared with the spinning industry, a weaving business is too small to afford costly imported machinery. Also, it is more dependent on domestic technology to satisfy the local demand for varieties of fabric. Looms made mainly by local carpenters were basically of wood with metal fittings. Local ironsmiths and machine shops for farm implements could repair such simple machines. S. Toyota, a son of a carpenter, devised a power loom in 1890 that proved to be suitable for making cheap cloth of narrow breadth for export to Korea and Manchuria. In 1914 he again designed an automatic loom; but, mainly because of cheap and abundant labour available at the time, it was not widely employed until 1929, when a factory act to prohibit night labour was enforced.[24]

Military and Mechanical Engineering

It is a cliché among historians of Japanese technology to characterize pre-war Japanese science and technology as 'armament-centred and unbalanced.' Yet we do not know what is balanced and what not. In the absence of an adequate international criterion, any account tends inevitably to be impressionistic. In addition, the Japanese military sector did not permit examination from outside. Hence, reliable data are still scarce.

Prevailing theory says that since private enterprises in the advanced capitalist countries are governed by the pursuit of profit, their interest in the armaments industry is indirect. In contrast with this, the Japanese industry was led by a government preoccupied with the goal of directly increasing the wealth and military strength of the nation through the efforts of government-owned industries. Japanese industry and technology were from the beginning strongly coloured by the militaristic policy. For example, the exhibits at the National Industrial Promotion Fair held in 1877 were composed mainly of the best-quality facilities and machinery belonging to the army and navy arsenals, as well as of other government-operated plants. Exhibits from the privately owned plants were few in number and poor in quality.

Priority for military engineering was reflected in the departmental organization of the Imperial College of Engineering. The core of modern scientific technology was mechanical engineering, from which engineering science was formulated. Japan was not too backward to catch up with this rising science, which was systematized only in the nineteenth century. Mechanical engineering was also fundamental to naval architecture and military engineering. In the early Meiji period, however, the government shipbuilding industry dominated the job market for college graduates in mechanical engineering. It was wholly and directly connected with the armaments programme of the government. Though it was not included in the original programme of the college, drafted by Henry Dyer on the model of the Zurich Technische Hochschule, the new department of, shipbuilding was added in 1882, at the request of the navy.

The Meiji government's interest in strengthening national power can be seen in the creation of special departments for the study of armaments at the university level, an act without precedent in advanced countries. In 1887, within the Imperial University, there were founded a department of arms technology, at the suggestion of the army, and a department of explosives, at the request of the navy. These were opened in the College of Engineering in order to secure engineering personnel for military arsenals. Since these departments did not involve existing university disciplines, no adequate teaching staffs were available at first. Only with the coming of the twentieth century did they have full-fledged professorships and regular students. In the meantime, the arms technology department developed a discipline of precision engineering, while the explosives department was led to chemical engineering.[25]

In 1887–96 many state-owned plants were transferred to private ownership, so that private capital increased steadily. However, there was no appreciable decrease in the total horsepower or the number of employees of governmental plants.[26]

On the other hand, the manufacture of machine tools, which formed the foundation of the machinery industry, was left in the hands of small and medium-size private enterprises, where no college graduate engineers were available until the Second World War.[27] This neglect revealed the basic weakness of Japanese technology during wartime, when high-quality machine tools ceased to be imported. Simple lathes and other standard machine tools had been homemade, but machine tools for automation and precision had been imported from Germany and the United States.

RISE OF TECHNOCRATS, FIRST WORLD WAR AND AFTER

Start of Financing Scientific Research

Until the 1880s, the scientific institutions created by the government were mostly for geophysical survey work: the Navy Hydrographic Office (founded in 1871), the Tokyo Meteorological Observatory (1875), the Geological Survey (1882), the Army Ordnance Survey (1884) and the Tokyo Astronomical Observatory (1888). Their work was essential for a nonindustrial modern state. After the 1890s, however, many national research institutions were established for fostering industrial developments, such as the Electrical Laboratory (created in 1891), the Central Inspection Institution for Weights and Measures (1903), the Fermentation Laboratory (1903), and the Railway Research Institute (1907).

Before the First World War, Japanese industrial laboratories existed only in the public sector, in order to give guidance in the technology needed by private enterprises. The mobilization of research in Europe and the United States during the First World War provided great stimulus and opportunity for government and scientists in Japan to think about the financing of scientific research. The Japanese, who had stayed out of major First World War hostilities, took advantage of the wartime economic boom. Scientists changed from academic bureaucrats to technocratically oriented bureaucrats who advocated planning scientific research for national goals.[28] The latter were backed by the rising industrialist class. The creation of the *Riken*[29] (Institute for Physical and Chemical Research) in 1917 was a landmark in this change, since the major source of funds

(85 per cent) was the industrial sector rather than the government. During and after the First World War private enterprises, notably in the chemical industry, established their own full-scale industrial laboratories. This signified the maturity of capital formation as well as the rise of interest in R&D in the private sector.

It was also during the First World War that two other arrangements to promote scientific research were formed in the public sector: university-affiliated research institutes and governmental research funds.

Owing to increasing specialization and the advancement of research in science and technology, Humboldt's ideal of the unity of education and research at the university level was virtually destroyed in the early part of the twentieth century. The American solution of the problem was to create graduate schools, so that the unity was delayed from the undergraduate to the graduate level, where students could be educated by formal curriculum as well as by apprenticeship in research. The other solution was the German method of creating a governmental research institute, called Kaiser-Wilhelm-Gesellschaft, to liberate scientists from their educational burden. This model was followed by the Soviet Academy of Science.[30]

The Japanese tried to circumvent this difficulty by creating the Riken, modelled after the Kaiser-Wilhelm-Gesellschaft. Still more significant was the affiliation of research institutes to a government university, where professors were full-time researchers without formal obligations to teach while enjoying academic freedom and higher standing than those working in the government and industrial laboratories.

The Ministry of Agriculture and Commerce created an Invention Fund in 1917 to encourage inventive researchers and, later, an Industrial Research Fund. The Ministry of Education's Science Research Grant programme was started in 1918. Other small philanthropic foundations were also created during the war to promote scientific research. This heyday of scientific research funding was rather short-lived; and when the economic recession started in 1920, the amount of government funds either remained the same or decreased.

The worldwide depression of 1929 had relatively little influence on Japanese business, however, because Japan's involvement in the Manchurian Incident of 1931 caused a munitions boom. A study of the industrial production indices of Japan and the United States shows that while the United States was unable to recover its 1920 level of prosperity until the start of the Second World War, Japan soon regained normality. Heavy industry was mainly responsible for the economic recovery. By about 1935, heavy and chemical industry production exceeded that of light industry.

With the aim of avoiding economic depression by means of scientific research and of promoting power of its industry to compete internationally, the Japan Foundation for the Promotion of Science was created in 1931. It arose from a proposal by scientists and started functioning in 1933 with government funds dramatically bigger (about ten times) than any ever contemplated before. Nationalization of science was a worldwide trend at this time, but Japanese science took particular advantage of this trend to reach international standards.

In spite of the heavy government intervention in scientific research, we should not overlook the trend in the machine industry, constant since the early Meiji period, towards moving industrial technology from the public sector to the private sector. As a 1931 survey indicates, nearly half of the researchers in engineering fields worked for private businesses.

Historians are concerned with finding an indicator of the military inclination of pre-war Japanese scientific research; but in the absence of adequate data, we must be content with Hiroshige Tetu's rough estimate that in 1931 nearly half of the national research investment in engineering went to military research establishments.[31]

Thus, we may characterize the interwar period as one of involvement in the worldwide trend to nationalization and militarization of scientific research, superimposed on the background trend of the transfer of industry from the public to the private sector.

THE SECOND WORLD WAR AND AFTER

After 1939 the wartime recognition of the need for science mobilization and armaments led to the expansion of education programmes for top-level scientists and engineers (according to one estimate, three times the usual number of scientific manpower was produced). As the war intensified, students of the humanities were called to active military service, while students of science and technology were exempt.

Demands from the military and from business made engineering research particularly favoured during the war. According to a questionnaire sent to engineering scientists after the war, their happiest time was that of the war mobilization, when they were given preferential treatment, in the form of abundant funds and materials. Furthermore, the complete isolation from Western scientific circles inevitably provided an opportunity for public recognition of those who had formerly been psychologically dependent on Western authority in their fields.

Immediately after the war, many Japanese spokesmen of science stated that the lack of recognition in pre-war Japanese society of the importance of scientific research had contributed to the Japanese defeat. The experience of military supremacy, thought control and the virtual failure of war mobilization must have given many scientists bitter memories and led to a negative assessment of the war effort. Wartime isolation especially resulted in a hiatus that presented the post-war generation with tremendous difficulties. Only recently have more critical attempts to reevaluate wartime scientific efforts as the source of the post-war boom in Japanese technological advancement and economic growth appeared.[32]

During the occupation years some strategic research, such as nuclear and airplane research, was forbidden by the occupation forces; the two Japanese cyclotrons were thrown into the sea. Military research was almost totally absent. Demilitarization and the demobilization cut most of the government support of science, which may have been partially responsible for the post-war transfer of scientific and technological leadership to the private sector.

After the war, governments in other advanced countries supplied the largest share of funds for scientific research. In contrast, in post-war Japan private enterprise, which formerly did not supply a large proportion of research funds, underwrote a significant portion of scientific research.

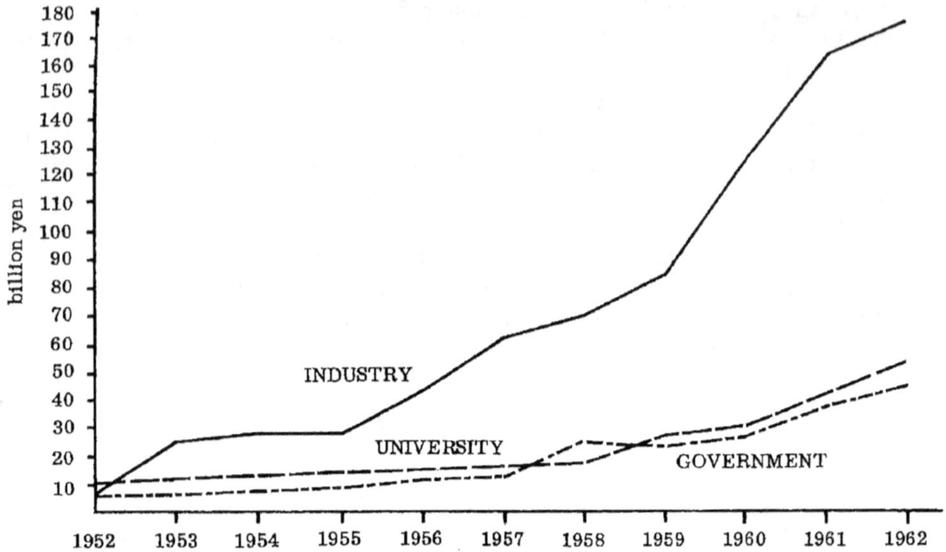

Figure 11.1 Japanese Research Expenditure, by Type of Institution, 1952–62
Source: *Nihon no kagaku gi jutsu katsudo* (Activities of Japanese science and technology) (Tokyo: Maruzen, 1965), p. 97.

Japanese private enterprise has grown to the point where it must, and can, sustain its own research. Against a universal trend in the post-war period to move towards 'government science' or 'nationally sponsored science,' such as the major space and atomic research programmes, the centre of gravity of Japanese science moved from the public to the private sector, as is shown in Figure 11.1.

Table 11.3 reveals that the pre-war public spending for defence and industry has now moved to social welfare, development and education. This trend is reflected in the structure of scientific and technological research in such a way that post-war Japan is said to have stayed out of big science in the public sector and to have concentrated on profit-making, economical development on the basis of

Table 11.3 Economic and Functional Classification of Japanese Public Expenditure, 1900 and 1960 (per cent)

	1900	*1960*
Directly productive activities	20.78	3.63
Agriculture, fishery	1.32	2.24
Commerce, mining, manufacturing	2.54	0.80
Transport, communication	16.92	0.60
Education	2.30	11.33
Defence	45.29	8.50
Social welfare	1.93	21.82
Other (general and local administration)	29.70	54.72

Source: Choki keizai tokei, zaisei shishutsu (Long-term economic statistics, Public expenditure, Tokyo: Toyokeizai, 1966).

imported know-how. However, Japan is now facing the problem of a laissez-faire science policy, which may be partly responsible for heavy pollution and other public nuisances. At the same time, when the source of technological borrowing is nearly exhausted, Japan may have to find new, costly ways of breaking through by itself. The Japanese government is beginning to take note of this reality, and the national budget for scientific research is being increased gradually and steadily.

CONCLUSION

Considering the past of Japan's science and technology, which began under government sponsorship in the early Meiji years, the direction of the scientific information flow has been always one-way, from abroad to the public sector, and in turn from public sector to private sector. The public sector has always turned Japanese science and technology in the directions that are its most serious concern: enriching and strengthening the country.

The course of this endeavour was first determined by the samurai intellectual class, which was seeking a new source of income after being deprived of family stipends. They took the initiative and were responsible for the systematic introduction of technology into public institutions. The process of modernization could, however, never be successfully completed unless the efforts in the public sector reached the private sector. Early enthusiasm for Western science led to the diffusion of the ideology of modernization to the social grass roots. At the same time this educational effort resulted in the liquidation of samurai/commoner class difference, to the extent that by the 1920s commoners had a higher representation among the students at the Imperial University.

As a device for explaining the transfer process, I introduce the following matrix:

	Western-origin	*Tradition-bound*
Public	I (military)	II (mining)
Private	III (matches/textiles)	IV (agriculture/ brewing)

Technology of Western origin was usually practised in the public sector, and tradition-bound technology in the private sector. In the process of the transfer from section I some adaptive technologies emerge in sections II and III. We have examined the success or failure of technology transfer from the imported section I to other sections.

In the early Meiji period, college graduates, mostly of samurai origin, were busy reading and translating Western scientific works, visiting and studying in the West, learning Western science, copying high technology, assisting Western advisers, teachers and engineers in surveying and construction jobs, and planning for Japan to become an industrial nation, catching up with the most advanced high technology (section I). On the other hand, more practical inventions and improvements at the grass-roots level were made by non-samurai, artisans and experienced farmers often under the influence of the Westernization policy (section IV).

Government enterprises focused their efforts on areas where no existing system was available, such as railway construction and telecommunications (I to III) or where large-scale capital and mechanization were needed, such as government-owned mines (I to II). In such fields all essential aspects of enterprises were introduced *in toto*; both chief engineers and less skilled workers were recruited abroad. If production technology and the production system were radically different from the domestic ones, even where traditional domestic engineering and industry existed, as in shipbuilding (I to II) and the spinning industry (I to III), the mechanization system and the managerial form were imported as a single, inseparable unit.

In the Japanese experience, a special adaptive technology was selected in order to compete with more capital-intensive advanced countries by using Japan's cheap labour. This aspect was not consciously emphasized by the Meiji government, whose foreign advisers, with very few exceptions, lacked much imagination in adapting Western technology to local situations. Their samurai followers, without much experience in domestic technology, tried to copy and demonstrate the Western technology, often with disastrous results. A notable labour policy of the Meiji government was to provide jobless samurai with new means of livelihood, but in doing so the government did not seem to have conceived of labour-intensive intermediate technology. Rather, it directed them into more labour-saving new technologies, such as large-scale farming in the colonization of Hokkaido (I to III) and/or mechanized mining (I to II).

On the other hand, in purely domestic private-sector industries where no similar Western counterpart existed, such as paddy-rice agriculture, saké brewing, and lacquerware and ceramic production (IV), the innovation process was naturally very slow and gradual. Only the generation trained in science and technology under the modern Western-style school curriculum began to utilize the knowledge acquired at school and to apply it, if practicable, in their domestic businesses.

In between these two extremes were many intermediate technologies. There were cases in which the production scale was too small for the enterprise to be in the public sector, yet local demands made the private sector quite receptive to the imported knowledge. Typical of this was the match-manufacturing industry, where at first cheap labour and matchsticks were the only exploitable factors; later potassium chlorate and phosphorus became available domestically, thus making it possible to complete all processes in Japan. Silk reeling and textile manufacture were also such that imported machinery in government pilot plants served to demonstrate the way to technological progress, although for a considerable period production was carried out by means of homemade wooden machines and small-scale production systems. The local synthesis of the two extremes was often found in this area of the transfer section (I to III).

The most important technology – arms technology – should not be forgotten. It has always remained within the public sector (I). The major areas of mechanical engineering and naval architecture were monopolized by military arsenals and large government industries without much effect on the private sector. Machine-tool engineering, despite its basic character, was relatively neglected, and continued to be a small-scale private industry.

If we characterize 100 years of Japanese industrial development by one phrase, it was a constant process of transfer from the public to the private sector. This policy was publicly announced and was followed during the 1880s. It was basically, if not explicitly, continuous, although often interrupted by the effects of the government arms industry.

This effect totally disappeared after the Second World War. With economically motivated, private industry based on profit-making, a laissez-faire kind of science and technology became dominant in the budgeting structure of post-war Japanese R&D. This contrasted with other advanced nations, such as the United States and Britain, where public expenditure for science and technology became the major and predominant source of research budgets.

Postwar Japan also presents an example that conflicts with the prevailing belief that science and technology input contributes to economic growth in proportion to its size. Among advanced nations, Japan paid less for science and technology and gained more in economic growth. Unless we specify the quality of science and technology and the quality of economic growth, we may not be able to say anything definite; but the post-war private leadership in Japanese industry and technology may give a clue to this question.

On the level of catching up and profit-making, post-war Japan proved that borrowing is less expensive than spending on costly original research. In so doing, it needed more intermediate scientific manpower than highly talented Nobel laureates. This could be accomplished only through a social revolution in which vertical and horizontal mobilities should be accelerated for the recruitment of scientific and technological manpower.

Let me conclude with a metaphor. Science and technology have usually been received in the upper social strata in the developing countries, and then have penetrated the society as a whole. These two steps were unusually orderly in Japan. If a class system is inflexible, the new knowledge fails to reach the lower levels. If we could complete the above two steps at once, the growth of science and technology might be more efficient and far-reaching.

The introduction of science and technology into Japan has taken place with unprecedented speed, but our studies suggest that the transformation can take place even more rapidly and more thoroughly if the conditions are right.

NOTES

1. S. Nakayama, 'Kokuei kagaku' (Nationalized science), in I. Sugimoto, ed., *Kagakushi* (History of sciences) (Tokyo: Yamakawa, 1967), pp. 351 ff.
2. The original is reprinted in S. Nakayama et al., eds., *Nihon kagaku gijutsushi taikei, kokusai* (Source books of the history of science and technology in Japan: International relations) (Tokyo: Daiichihoki, 1968), pp. 35–36. Hereafter cited as *NKGT*.
3. *NKGT: Kyoiku I* (Education no. 1) (Tokyo: Daiichi-hoki, 1964), p. 353.
4. Ibid., pp. 392–393.
5. Ibid., p. 202.
6. *NKGT: Tsushi I* (Outline history, I, Tokyo: Daiichi hoki, 1964), pp. 180–86.
7. *Tokyo Teikoku Daigaku gakujutsu taiken, Rigakubu* (Survey of research activities in the Tokyo Imperial University, Faculty of Science, Tokyo: Tokyo Daigaku, 1942). In the chemistry department, the change from local science to internationally recognized subjects became noticeable in 1885.
8. T. Muramatsu, 'Nihon no kogaku soseiki no jakkan no mondaiten' (Some problems of the early Japanese engineers), *Kagakushi kenkyu* (Journal of the history of science, no. 32, 1954): 8–14.

9. I. Amano, 'Kindai nihon ni okeru koto kyoiku to shakai ido' (Higher education and social mobility in modern Japan'), *Kyoiku shakaigaku kenkyu* (Journal of educational sociology, no. 24 (1969): 84.
10. *Takahashi Korekiyo jiden* (Autobiography of Takahashi Korekiyo [founding president of the Japanese patent office]). The part relevant in this context is reprinted in *NKGI: Tsushi I*, pp. 339–40.
11. 'Tokkyo dai-100 go no meisho oyobi shutsugannin ichiran' (Titles and applicants' names of patents no. 1 to no. 100), in ibid., pp. 334–36.
12. J. Iinuma, *Nihon nogyo no saihakken* (Rediscovery of Japanese agriculture) (Tokyo: NHY Books, 1975), ch. 3.
13. *NKGT: Nogaku I* (Agricultural sciences) (Tokyo: Daiichi hoki, 1967), p. 26.
14. Ibid., p. 14.
15. Ibid., pp. 122 ff.
16. *NKGT: Seibutsu kagaku* (Biological sciences) (Tokyo: Daiichi hoki, 1965), ch. 5, sec. 1.
17. *NKGT: Tsushi II* (Tokyo: Daiichi hoki, 1967), p. 240.
18. T. Furushima, *Sangyoshi II* (History of industry) (Tokyo: Yamakawa, 1966), pp. 474 ff.
19. T. Tsuchiya, *Zoku nihon keizaishi gaiyo* (Outline of economic history of Japan), II (Tokyo: Iwanami, 1939), pp. 180–81.
20. Ibid., pp. 178–79.
21. *Tokyo Teikoku Daigaku gakujutsu taikan, Kogakubu* (Survey of research activities in the Tokyo Imperial University, College of Engineering, Tokyo: Tokyo Daigaku, 1942), p. 91. The courses of instruction are given in English in *The Engineer*, December 3, 1897, p. 544.
22. *NKGT: Tsushi I*, p. 466.
23. *NKGT: Kikai gijutsu* (Mechanical engineering) (Tokyo: Daiichi hoki, 1966), pp. 119–20, 124–26.
24. Ibid., pp. 136–37.
25. *Tokyo Teikoku Daigaku gakujutsu taikan, Kogakubu*, chs. 6, 10; see also *The Engineer*, loc. cit.
26. *Nihon keizai tokei soran* (Survey of Japanese economic statistics) (Tokyo: Asahi Shinbun, 1930), p. 723.
27. Muramatsu, op. cit.
28. Hiroshige Tetu, *Kagaku no shakaishi* (Social history of science) (Tokyo: Chuokoron, 1973), ch. 3.
29. K. Itakura and E. Yagi, 'The Japanese Research System and the Establishment of the Institute of Physical and Chemical Research', in S. Nakayama et al., eds., *Science and Society in Modern Japan* (Cambridge, Mass.: MIT Press; Tokyo: Tokyo University Press, 1974), pp. 158 ff.
30. Loren Graham, 'The Formation of Soviet Research Institutes: A Combination of Revolutionary Innovation and International Borrowing', in *XIVth International Congress of the History of Science, Proceedings*, I (Tokyo: Science Council of Japan, 1974).
31. Hiroshige, op. cit., p. 116.
32. Ibid., pp. 216–20.

▸ First published in K.D.Sharma and M.A.Qureshi (eds), *Science, Technology and Development*, pp.435–448, Sterling, 1978

⓬ Public Science in the Modernization of Japan

With the Meiji Restoration in 1868, the aspect of the scientific community in Japan undergoes a transformation. Until the last period of Tokugawa rule, Western science had been entering Japan in a stealthy way, subject to political restrictions and taking its place by the side of Chinese traditional science and technology. In contrast, during the Meiji period, modern science surged in unimpeded. The scientific community was exposed to the storm of 'civilization and enlightenment'. Not able to stand this blast, *wasan* (Japanese traditional mathematics), *rakigaku* (calendrical science), *honzo* (materia medica), *kanpo* (Chinese medicine) and other traditional sciences wilted and gave place to Western science. This does not mean, however, that the traditional sciences had exhausted all their potentialities and therefore declined rapidly. On the contrary the essential course of the decline lay elsewhere; it lay in the revolutionary changes that had come about in the institution of science.

Western science had penetrated and spread in Japan even in the preceding Tokugawa period. It has its beginning in the private interests and activities of the Western-oriented scholars. By the last period of Tokugawa rule, it had come to be studied more or less methodically at such schools as the ones in Nagasaki founded by a Dutch physician, Pompe van Meerdervoort. But being hard pressed to attend to the urgent defence needs of the country, they could not, however, develop into institutions giving fully modern and systematic education in science and technology. With the Meiji period, the picture changes, however. Under the new government scientific education was institutionalized and established firmly. Whereas formerly study of Western science had had its origins in the private interest of the scholars or public practical necessity, now first the institutions were created and then modern scientific and technical specialists were brought up within the institutions.

Owing to the Tokugawa 'closed country' policy and the geographical and other limitations stemming from that policy, even Western learning developed in a limited, shrunken fashion, acquiring in the process a peculiarly Japanese character. It may therefore be called 'Japanese science'. After the beginning of the Meiji period, however, all the restraining conditions were removed and the patch was cleared for free growth. Meiji science was the same as the Western science; the methods and objects of research came to be the same as those of the science of the rest of the world. Seen in this light, it can no longer be called Japanese science; it should be called 'Science in Japan'. The specifically Japanese characteristics now are noticeable only in institutional aspects of this science. Therefore the history of Japanese science after the beginning of the Meiji period is viewed as being in the main, history of the development of the methods of cultivating scientific institutions. On this understanding, it is possible to observe science in Japan from an international point of view and grasp its characteristics.

THE PERIOD OF 'THE SCHOLARS OF WESTERN LEARNING' (1868–1873)

The policy changed after the 'august renovation'. Concommitantly a contemporary vacuum was formed in the power structure as well as in thought. There was misunderstanding and confusion among the people concerning the general policy of the new government and the direction of the new times. The people wanted for some insight into the nature and direction of the fast changing times. The men who appeared on the stage in answer to the popular need were the 'enlighteners' of the Western learning, beginning with Yanagawa Shunsan, Fukuzawa Yukichi and others. It was the intellectual activities of these men that gave rise to the 'civilization and enlightenment' movement.

During the last period of the Tokugawa rule Western learning (Western science) had still been the possession of a section of the intellectuals. In the early Meiji years, however, the scholars of Western learning wrote freely and copiously, introducing the West to the Japanese people. There were to be sure some hacks among them who, without any understanding of Western science, followed the times and made a living, dashing off stuff based on a knowledge no deeper than eating beef and wonderment at telegraph. On the other hand, Fukuzawa and others were really filled with zeal for enlightenment and devoted all their energies and passions to the spread of Western learning.

As to the new government, during the first two or three years of the Meiji period (1868–1870) there were still influential men in it who wanted a faithful restoration of the ancient imperial system of government. But the reformist leadership of the new government firmly recognized that to build up the nation in the contemporary harsh international environment, adoption of 'Westernism' was necessary. Preoccupied as the new government was with problems of its own organization and other urgent matters, it could not give much attention to the education and enlightenment of the people. But it welcomed the 'civilization and enlightenment' movement, though in this connection it had to depend upon the scholars who belonged to the tradition of Tokugawa Western learning.

Certainly the policy of 'rich country, strong army' was the primary concern of the government; but this ideology was not directly reflected in the enlightenment activities of the Western learning scholars. For one thing the enlightenment books were aimed at the individual intellectuals or the general mass of the people; and for another the authors, the Western learning scholars, were not disposed to servilely receive and follow the policies of the government. They looked upon post-industrial-revolution material civilization of the Western countries with rather naive admiration; and further tried to convey the essence of the natural science behind it. There is some academic controversy about the role of the

Western learning scholar under the Tokugawa system: whether he was the critic of the feudal system or an intellectual technician useful in the defence of the system. The true shape of the Western learning scholar became manifest, however, once the Tokugawa feudal system was dismantled. He was revealed to be a healthy enlightener. It is not that they did not accommodate themselves and truckled to the political establishment and powers. It is that their real height and *raison d'être* lay elsewhere. The people knew little of the world situation and had no share in the benefits of civilization and science. The Western learning scholar's essential function was to be a step ahead of the people and, independently of the politics of the day, to guide and teach them.

The Western learning scholars did not expect to be read by Westerners; from beginning to end they addressed their writings to the general educated Japanese. For this reason, far from having, as the later scientists could have, any feelings of inferiority towards the Westerners, they had the elation born of the conviction that they were the pioneers and leaders of their age. Their thought was filled with the belief that science was the standard of human progress. They were not interested in a faithful, academic exposition of the results of Western science. Representative books like Fukuzawa Yukichi's *Kunmo Kyuri zukai* (which appeared in the first year of Meiji) and Obata Tokujiro's *Tempen chïi* were not simple translations. What such books sought to do was to explain to the people in easy language, to make them understand the difference between the traditional ways of thinking on the one hand and the Western view of nature on the other, and the essential modern thought at the bottom of science. In other words, they tried to grasp and present science as thought, as enlightening thought. In the approach to scholarship and learning the West tended to inquire into physical laws rather than the aggregate of facts and objects of nature, into essence rather than phenomenon; the tendency of the East was the opposite. The Western way came to be called 'investigation of natural principles', and following upon the publication of Fukuzawa's *Kunmo kydri sukai* there was an 'investigation of natural principles rage'. Perhaps this was because the Western learning scholar came to feel keenly that, although there was no great gap between the East and West as far as natural historical kind of knowledge was concerned Japanese culture seriously lacked the element of 'investigation of principles', namely natural philosophy. When the Western learning scholars themselves compiled – not translated – their *shogaku kyosoku* (primary school curriculum – prescribed by the Education Ministry in 1872), the book was not natural history oriented in the American fashion, but investigation of principles (physics) oriented. One can say that this was so because they clearly perceived that no simple accumulation of positivistic knowledge but the structure basing, in Fukuzawa's words, 'physics as its foundation' was what characterized modern science.

Thus in the world of education and enlightenment, a world that differed from telegraph and railway enterprises in that it did not require much state funds, the movement of the Western learning scholars from below was freely and directly reflected. This was certainly the case during the time when the textbooks were not yet being published officially. At that time the government had neither time nor resources to operate a firm educational policy.

As time passed, however, the new government abolished the feudal domains and established in 1872 the prefectural system, having in general made its foundation secure, came to interest itself in education as well. It adopted a policy of centralization in education. After the establishment of the new educational system in 1872, it was possible to make the textbooks reflect the policies of the government. In 1872 normal education was started. Foreign teachers were hired and placed in charge of this; and it was to conform completely to the methods of the Western primary schools. As the Meiji government's educational organization was established, the Western learning scholars came to count for much in the government circles, not least because of the dearth of teachers with any knowledge of Western learning. In consequence the Western scholar changed aspect from one who hired by his pen to a salaried man, and for the second time found his place in the system. After 1873 the enlightenment both as thought and movement lost in great measure its liveliness and shrivelled in scale and spirit.

The period when the Western learning enlighteners were most active was a transitional period. They were first of all translators and popularizers, never technicians doing specific work of scientists in the laboratory engaged in actual research. Their knowledge came in the main from books, though a few of them received instruction from foreigners. It never met much beyond elementary level; many of them may have been too old to turn into specialist professional scientists. They were in their element in the fields of enlightenment and education, but were not filled to give basic and systematic instruction in the natural science, which alone could produce specialist scientists. For this reason, as the school system gradually opened out and began to turn out scientific specialists, who would have gone through the regular route from the elementary education up, the Western learning scholars progressively yielded place to the products of the new school system. Men who conceive of and treat science as thought may at certain stages in the development of science play leading parts, but soon must draw back from the frontier of research.

SCIENCE AND STATE ENTERPRISE

Japan reacted most sensitively to the advance of the Western nations in the East in the nineteenth century. In China too after the Opium War the military supremacy of the West gave rise to the feeling of crisis. There was a difference between China and Japan, however. The Ce'ing China was ruled and administered by civilian officials, whereas in Japan the rulers were the warrior class. The latter, being professional military men, recognized quickly that expansion and modernization of armaments were needed for national defence. This realization led later on to the 'rich country, strong army' policy of the new Meiji government.

Futher, it was very quickly understood in Japan (i) that in military terms the West was unchallengeably superior, (ii) that behind this superiority lay the industrial and technological superiority (a result of the Western industrial revolution), (iii) that again at the bottom of it all was the modern scientific thought, and lastly (iv) that therefore what was needed above all was the fostering and training of native scientific and technological talent. In nineteenth century China some advanced bureaucrats did make plans for the introduction of Western military technology and even for modernization of industry, but, being inhibited by the 'agriculture first' principle and the civil service examination system basic to the Chinese order, could not in earnest set about the task of genuine modernization, which would have required them to grasp above points (i), (ii) and (iv). Meiji Japan offered a sharp contrast to nineteenth century China in all respects. Here the Meiji government itself took the lead and, employing all the powers it gathered into its own hands in the course of its centralization of power policy, set to the tasks of promoting industry, reforming the various institutions, and cultivating human talent. At a time when China was stuck at the stage of making 'things', Japan went ahead to make 'men who make things'.

WESTERNIZATION POLICY

Science in Japan in and after the Meiji period was not something that grew out from inside the society, but something forcibly transplanted from outside at the initiative of the government. In view of this characteristic, from the standpoint of the history of science the organization and institutions of the transplantation assume more importance than the content of the science itself.

It was not the case, however, that the new Government took in the organizational work concerning science from its very first year. In 1868 and 1869 the direct power of the new government did not reach much beyond the territory which had been the immediate domain of the Tokugawa Shogunal house. The antiquity-minded imperial loyalists established in Kyoto institutes of Chinese learning and Imperial learning. In Edo (which was renamed Tokyo in 1869) *Shohei gakko* and *Daigakko* (both of which succeeded to the old Tokugawa university of Confucian learning, *Shoheiko*) became the leading educational institution, with *Kaisel gakko* and *Igakusho* (which descended from old Tokugawa translation bureau and the school of Western medicine) occupying a subordinate place. Rather than any of these, it was perhaps *Numazu heigakko* (a military academy, established with the old Tokugawa personnel) that maintained the highest standards of science and technology at that time. Its staff included Nishi Amane, Akamatsu Noriyoshi and other noted scholars drawn from the old Tokugawa schools such as *Nagasaki denshujo* (naval academy) and *Bansho shirabesho* (translation bureau); and it taught even differential and integral calculus. And then there were private academies such as the Keio Academy, and the schools belonging to the various *han* or feudal domains. This co-existence of several kinds of schools, many of them not under the control of the central government, was a feature of that time.

Since the new government did not have in its service many Western learning experts, it sought to effect its policy of centralization even in education and knowledge by such means as drafting into its service experts like Nishi Amane of the Tokugawa Western learning tradition and by reorganizing existing institutions (for example it incorporated the *Numazu heigakko*, a Tokugawa institution, into its own Naval Training Institute in Tokyo).

Among the scholars so drafted and employed by the government, there were both men belonging to the *Bansho shirabesho* line mostly translators, and those who had studied with foreigners at *Nagasaki denshusho* and other places. Perhaps there were more of the latter than of the former. These technical bureaucrats had now, for the first time, the opportunity to exercise their abilities to the full. In this sense there was a continuity in the world of these technicians. For them the end of Tokugawa rule and the Meiji restoration meant no more than change of political masters.

HIRED FOREIGNERS

The new government adopted several policies towards its objective of nation-building. Its science and technology policy was one of them; and probably it owed much to the suggestions of foreigners, chiefly H.S. Parkes, the English Minister to Japan and G.H.F. Verbeck, adviser to the Meiji government.

The decision to establish educational organs under the guidance of Americans and Europeans and on the model of European and American institutions was taken about 1870 when the 'restoration of antiquity' party in the government had begun to decline. Since the Western learning scholars, with their existing knowledge, could hardly manage anything above elementary education, foreign instructors had to be hired for the higher levels of education. At the same time it was felt that Japanese students should be sent abroad, to Europe and America so that on their return they might replace the foreigners and take charge of the higher education. From this time on the government pursued an educational policy, the two chief principles of which were as described above: the employment of foreign instructors in the establishment of the Japanese educational system and the sending abroad for higher education and training of Japanese students.

In China, especially after the Opium War, missionary societies such as the London missionary association of Alexander Wylie and others played a large role in higher education and the medical field. Differing from China, in Japan, even though the ban on Christianity had been lifted, the missionaries could not play a significant part. In Japan the government itself took the initiative and hired the foreign teachers it needed. As a result the government and Tokyo University, a government university, came to take the leadership in science. Little room was left to the missionaries and on the whole they were forced to concentrate their activities on women's education, a field held in low esteem and neglected by the government.

Generally speaking, most of the white men who came to the colonies were men of a rather low grade. There were some exceptional people, however, whose ambition was to seize the opportunity offered by the underdeveloped area and build up new kinds of educational institutions, which it was impossible to do in their own countries in the face of opposition from old established institutions. An Englishman, T.S. Raffles, who was active in Asia in the early years of the nineteenth century, was a good example.

Some of the foreign teachers were really men of original ideas. Henry Dyer, who established the Engineering College was one such. There were not many such institutions even in Europe at that time, combining science and technology and employing new methods of education. These men were not, however, given free scope. All real power for building up new institutions remained in the hands of the government; and the foreigners were never more than temporary advisers.

Even before the Meiji restoration, there had been foreigners in Japan invited and employed by the Tokugawa government as well as by the other feudal authorities. After the restoration, their number rapidly increased. In the first years of the Meiji period the employment of the foreigners was not yet on an organized basis. There were even cases of rough and ready hiring of foreigners who had found their way somehow into Japan. After about 1872, the help of Japanese missions abroad was enlisted and the selection of foreigners became systematic and careful. The foreigners so selected were used in the establishment of the Japanese higher educational system, which, starting with *Kaisei gakko*, culminated in Tokyo University.

The foreigners employed by the new government came from several different countries. Frenchmen were employed in the army, mining, and shipbuilding; Germans in medicine and pure science; and Americans in the development of Hokkaido, of which agricultural technology was the most important part. But all in all Englishmen were most numerous. Nearly one hundred Englishmen were hired in railway construction. They also contributed more than any other group in the technical work connected with the Ministry of Works. The number of hired foreigners reached the peak in 1874. (In the years following 1877, the government adopted a deflationary financial policy and sold its enterprises to private owners; and the foreign technical advisers also increasingly moved to non-governmental employment. The number of privately hired foreigners reached the maximum in the decade after 1877.)

Unlike the colonies, and semi-colonies, Japan kept all rights and all powers of decision with regard to hiring foreigners in its own hands. It was not bound to take its experts from any one particular country, but retained its absolute freedom of choice; and tried to make use of the special strong points of the various countries. For example England was drawn upon for technology and Germany for medicine. At the same time, because of the huge cost of hiring foreigners (a foreigner was paid five to ten times as much as a Japanese in a corresponding position), the government kept sending Japanese students for studies abroad with the idea of making them, on their return, replace the foreigners as soon as possible.

The foreign scientists who were interested in Japan as professional scientists, were primarily interested in phenomena peculiar to Japan, such as Japanese flora and fauna, and earthquakes. From the point of view of universal sciences like physics and chemistry, Japan naturally could not call forth any special interest. An American biologist, W.S. Morse carried on his researches in biology with Japan as his field, and an English physicist, J.A. Ewing, invented the seismograph and became the founder of modern seismology. Here, there is an analogy with the colonial educational policy of Japan in Taiwan at a later time. When Japan established Taihoku University in Taiwan, its science faculty had departments of zoology and botany, but no provision was made for physics and related sciences. This may have had something to do with the broad considerations of colonial policy; but the explanation given was that there was nothing in Taiwan and its geography to attract physical scientists from Japan.

SCIENTISTS SENT ABROAD FOR STUDIES

The rules for study abroad, drafted in 1870, make the following prescriptions as to what subjects ought to be studied in which country.

England:	engineering, geology and mineralogy, iron manufacture, architecture, shipbuilding, cattle-breeding, commerce, poor relief.
France:	botany and zoology, astronomy, mathematics, physics, chemistry, architecture, law, international relations, public welfare.
Germany:	physics, astronomy, geology and mineralogy, chemistry, botany and zoology, medicine, pharmaceutics, educational system, political science, economics.
Holland:	hydraulics, architecture, shipbuilding, political science, economics, poor relief.
America:	industrial law, agriculture, cattle-breeding, mining, communications, commercial law.

The list indicates a preference for Germany and France in science and for England in technology. It shows an appreciation of the highest standards obtained in any given field in particular countries. From this time on the introduction of the various sciences and techniques would follow the lines evident in the list. This appreciation of the international scientific standards perhaps was based on the advice of the foreign advisers.

In the matter of sending students abroad, at first the new government and the various *han* vied with each other, and the sending was not well-organized. After the abolition of the *han* in 1872, the expenditure came to be borne by the central government. This in itself introduced a measure of order; and the government soon planned an organized programme . In 1873 the system of sending students abroad with all expenses borne by the government was abolished; and in its place, in 1875, the Ministry of Education overseas scholarships were instituted. Only a limited number of students were chosen for these scholarships after a very severe examination (the first year, eleven and the second, ten). In the early stages of this practice, as the higher education in Japan was not yet firmly established, the students sent abroad had to study for as much as four to six years abroad to master their special subjects. When they returned professorial or corresponding posts were awaiting them in Japan. They constituted the first generation of Japanese professors at Tokyo University, displacing the foreign instructors. From about 1887 the period of study abroad got shortened to two or three years, and the returned scholars had to be satisfied with less prestigious posts – as the posts in the highest educational institution had already been filled and many of them could only become teachers at professional schools.

The students could freely choose the country in which to study. Until about 1877, however, education in Japan put more stress on language study than on the study of one's special subject. There was therefore a tendency among the students to choose the country, the language of which they had studied. Terao Hisashi, who graduated in 'French physics' course and went to France for further study, was an example. A number of students moved to Germany from the country they had initially chosen. Fujisawa Rikitarō (mathematics) and Tanakadate Aikitsu (physics) who transferred to Germany from England were examples of this. The explanation was of course that it was the heyday of German science – students from England, France and America were going to Germany for advanced studies – and the Japanese students abroad must have been impressed by German scientific leadership.

OFFICIAL CHARACTER

It is said that capitalism in its developed stage turns monopolistic, and imperialistic and develops military technology. However that may be, when a country, where capitalism is not yet fully developed, wants to catch up rapidly with the advanced capitalist countries a policy of forcible transplantation under the sponsorship of those in power is the only way. In such cases the absolute government adopts a policy of 'rich country, strong army' and its will to resist against and catch up with the advanced countries will be expressed as a direct and urgent interest in armaments, to be translated into a reality through the government-owned industries. This contrasts with the position in the advanced capitalist countries where the interest of the private enterprises in arms production is indirect, mediated by the pursuit of profit, the primary demand of capital. It seems to be an historically inevitable course that Japan, following the precedents of Germany and Russia, should have started with government ownership of industry and that its industry and technology should have from the beginning a strong military colour.

But military technology alone cannot make a country also a modern state. The government, engaged in building up a modern state, becomes early and greatly interested in communications, transportation, mensuration, and other aspects of the science and technology of communication necessary for internal administration and centralisation of power. Therefore differences are bound to develop between the industrial structures of *laissez-faire*, profit-pursuing, liberal-capitalistic countries like Britain and America on one hand and backward countries like Japan, where the initiative belongs to the government, on the other. Moreover, corresponding to the difference in industrial structure, differences appear in the structures of technological and further, scientific communities.

In the advanced countries there was an element of spontaneity to the scientific communities and the study and research of the individual scientists. The scientist was free to take up for his research any problem that interested him; and the scientific communities were formed by individual scientists drawn together by a common interest. Differing from this the scientific community in Japan had a 'planned character'. Here the new government in pursuit of its set aim built universities and factories, trained scientists in these institutions, and sent them off to their places of business. Rather than every scientist following his own research interest, in a collective way priority was given to certain basic tasks, the accomplishment of which was necessary for the government of a modern state to operate: matters such as geographical and geological surveys, weights and measures, meteorological observation, sanitation, printing, telegraph and telephone, military works, railways, survey of natural resources, etc. For example in 1868 work on telegraph service was begun; in 1870 the Ministry of Public Works and the Hydrographic Bureau were established; in 1872 the new educational system was promulgated, and the Julian calendar was introduced; in 1874 the drug control agency was established, as also examination of imported drugs and medicines; in 1875 the Tokyo Meteorological Observatory was inaugurated; in 1876 B.S. Lyman, a hired foreign geologist began the survey for oil; in 1880 the rules for the prevention of infectious diseases were formulated and also rules were drawn up for the licensing of medical doctors. All these enterprises were carried out by the Ministry of Works, Ministry of Interior, Ministry of Popular Affairs, the Commission of Hokkaido Colonization, the Army, the Navy and other governmental agencies functioning actually and daily. For these kinds of practical work the Western learning translator-scholars were not of much use and the universities were too unripe. These urgent tasks could not wait for specialists who had gone through a regular university education; there was not that much time.

The men who filled the need of this transitional period for trained talent were those *Yanagi Narayoshi* who had been trained at the Nagasaki *denshusho* (naval school) before the end of the Tokugawa period, and the foreign instructors. These men entered the government agencies and ably tackled the basic enterprises. The technical leadership in all these works of course was assumed by the foreigners. Among the helpers who worked at the middle level and below, many were graduates of the Commissioner of Hokkaido Colonization School and the Ministry of Works Engineering School.

While the elite were away studying and acquiring a high-level education abroad, these people were consolidating the ground work of the modern state. Speaking in terms of the symbols of the structural character of the science and technology of the various periods, if one called the period up to 1872 as the age of Western learning scholars, and the years after 1886 the age of the Imperial University, then the time between the above two periods can be called the age of foreigners and working government agencies.

Therefore free research could appear not with the scientists and technicians of the initial pioneering time but with the specialist-scientists who began to graduate and come out of the universities since about 1886. Around the same time, the appearance of capitalistic technicians, who aim at the invention and improvement of consumer goods, also took place, when the government sold its industrial establishments to private capitalists and when private capital began to accumulate and ripen. Until 1882, when the requirement was removed, the graduates of the Engineering School had to serve in the Ministry for seven years on graduation. These men, carefully trained and sponsored by the government were active in the state enterprises during the last quarter of the nineteenth century (chiefly the heavy industries). The graduates of the Tokyo Trade School and others, middle and lower level technicians, were in comparison late in starting to enter private industrial enterprises.

In general, when one regards the process of the impact of the advanced countries on the less advanced countries from the point of view of science, the following three stages are discernible:

(1) Collection of materials meant for reports to the Western learned societies.

(2) Development work in the country concerned, and preparatory survey, investigation, etc. towards that.
(3) Education of the people of the country concerned.

In relation to Japan, the first stage began when, during the Tokugawa 'closed country' period, Japan was introduced to the West by such men as E. Kaempfer and P. Siebold and came to be looked upon as the terminal point of the world-ranging Western expeditions. After M. Perry's visit in 1853 it stimulated the curiosity of many Western scholars and intellectuals. During the early Meiji years, several reports were published on the manners, customs, flora, fauna, nature, industrial arts and crafts, etc. of Japan in journals like *Nature*. These belonged not so much to modern natural science, which quests about 'universal truth', but rather to Oriental studies both in method and object.

As the next step, generally from about the beginning of the nineteenth century, survey, investigation of minerals, flora and fauna and other natural resources and other such activities were carried on to provide materials for colonial policy. These activities did not spring from the personal curiosity of the individual scientists, but were organized activities, with the government at the back of the enterprise.

Further, in order to get capitalistic production going, both general education and technical education of natives in crafts and engineering would be necessary.

Japan checked the 'impact' of the West at the first stage taking the second and third stages into its own hands; science was let into Japan in the early Meiji period on Japan's own initiative and without long-lasting Western 'influence'. On the other hand, in countries under complete colonial rule, like Indonesia, Philippines and the former colonies of Japan, the ruling country was in control at every stage, from stage one to three.

In the case of China, an independent country, an attempt was made, in some quarters during the late nineteenth century, to carry out the work of the third stage, the educational work, by its own hands, and to make the scientists so trained and produced at home take in hand the work of the second stage. In Japan the execution of the tasks of stages two and three ran parallel with each other: while the technicians who had been taken into the various government agencies were carrying on the work of stage two, they were at the same time engaged in consolidating and providing for stage three. Such was the situation in Japan during the 1870s and 1880s. In the case of China, however, stage two did not begin until the abolition of the civil service examination system in the early twentieth century; and it was only during or after the First World War that the scientists who had studied in Europe or America (a few in Japan) came home to take up the work of stage three.

▶ First published in *Dictionary of Scientific Biography*, Vol.XV, pp.728-58, Charles Scribner's, 1978

⓭ Japanese Scientific Thought

PROFESSIONAL VIEWS OF NATURE

In seventeenth-century Europe the goals and approaches of modern science were established by the scientific revolution. The professionalization of science in the nineteenth century, sometimes called the second scientific revolution, was no less important. When science became a full-time vocational activity, even the perception of nature was reorganized.

The second scientific revolution has not attracted the interest of many intellectual historians because it has seemed to be a revolution only of the social system of science. How intellectual its motivations were is a moot point; it has important intellectual implications. As the disciplines separated out of the ancient unity of science, each professional learned to view nature from the standpoint of his own field. The physicist's nature overlapped little with that of the botanist; the same could be said of the mechanical engineer and the professional philosopher (who in the nineteenth century often, especially in Germany, thought of himself as a *Wissenschaftler*). Although the assumption survived that one nature underlay the work of all scientists, there was no longer a consensus among professionals as to what it was and how it was articulated. The unity of nature was abandoned to the layman, but the technical perspectives of scientists moved so quickly and divergently that there remained no standpoint from which an overview was possible. The fragmented conceptions of and assumptions about nature, centred in academic specialties and heavily coloured by their prejudices, will be referred to here as professional views of nature.[1] People within the specialities tend to see them as universal. The metallurgist or biochemist, when he sets out his opinion on broad questions of natural philosophy, often is unable to transcend the ideology of his own scientific community.

The views of nature imported into Japan in the Meiji modernization period (late nineteenth century) were of the kind described above. When the theory of organic evolution was introduced, the fact that Japan was not a Christian country was largely beside the point in determining its reception. By and large the Japanese sedulously assimilated views that had become international in character. The confrontations of values that one might expect did not occur. Furthermore, since science was professionalized by fiat of the central government rather than by the piecemeal effort of scientists, there was no natural continuity between traditional Japanese outlooks on nature and the new disciplinary views. The new

Japanese scientist could not draw directly upon his native heritage in establishing a career; all that mattered was to conform to the conventional view from abroad.

In order to find views of nature that are in some sense genuinely Japanese, it will be necessary to examine the state of affairs before the Meiji reforms cut off the confrontation of Eastern and Western science. This is not to say that there was a single characteristic Japanese view of nature that can be reconstructed by reading the works of individual thinkers and examining anecdotes about them. The view of nature varied from one period to the next and depended greatly upon social background. It is, in fact, worth pursuing the hypothesis that even in traditional Japan, views of nature were rooted in and bound to professions – although the definition of a profession and the character of membership were very different from those of modern times.

During the Tokugawa shogunate (1603–1867) class hierarchy was tightly maintained, with the hereditary warrior class, the samurai (about 5 per cent of the population), at the top of a rigid structure of farmers, artisans and merchants. The major professions, independent of the four classes, were those of the Confucian scholar, the physician and the Buddhist priest. Vocation was hereditary in feudal Japan, and even professionals were bound by their inherited callings to a partial view of nature. Given the lack of social mobility, collective, static views of reality are more prominent than the individualistic activity that certainly existed at the same time.

In seeking to identify views of nature, we shall pay particular attention to the prefaces and postscripts of scientific writings. The texts generally were concerned with stylized technical subjects, and there was no place for direct and outspoken expression of the author's assumptions. In the front matter and end matter, on the other hand, basic philosophical matters often were argued. Frequently, these discussions of fundamentals were merely ornamental, adapted from Chinese arguments using the yin-yang principle and the Five Phases doctrine. Nevertheless, a comparison of Chinese and Japanese prefaces reveals differences in views of nature.

Also useful are prefaces contributed by writers other than the authors. Such prefaces, which tend to be complimentary rather than critical, provide excellent sources for the criteria of evaluation in each profession. Even today these vary; good philosophers are profound, scholars are erudite and mathematicians bring elegance to their proofs. In old Japan, astronomers, mathematicians, medical doctors, natural historians, linguists and Confucian scholars differed in their excellences. The differences, I argue, reflect different views of nature.

ASTRONOMY

Pseudoscience and Science

Technical professions began in Japan with the immigration of Korean and Chinese experts in the sixth to eighth centuries. As their Chinese view of nature began to take root in Japan, it was institutionalized on continental models in the College of Confucian Studies and the Board of Yin-yang (divination) Art. The Yin-yang Board had three departments: observational astrology, calendar making (mathematical astronomy), and yin-yang divination. In principle it was a miniature reproduction of the Astronomical Bureau at the Chinese court. A closer look discloses significant remodelling to meet local requirements. In China the Divination Bureau was administratively separate from, and much inferior in status to, the organizations that computed the ephemerides and observed celestial phenomena for astrological purposes. In Japan all of these activities were subsumed under the single Yin-yang Board, the name of which indicates a clear priority for divination.[2]

The Japanese yin-yang art was a complex of magical divination techniques. These techniques had little in common with the portent astrology practised in the Chinese and Japanese courts, which was based on a belief that the natural and political orders influenced each other in such a way that changes in the former could be taken as warnings about inadequacies in the latter. Throughout the sphere of Chinese culture, calendar making was the paradigmatic exact science, used for computing solar, lunar and planetary positions, forecasting eclipses and composing a complete lunisolar ephemerides.[3]

Although both the yin-yang art and portent astrology were ways of forecasting changes in human affairs, the latter depended upon unpredictable omens, such as comets and irregular eclipses, as indexes of mundane crises. The yin-yang art, because it is less passive, was more important in the everyday life of the court – it determined the dates of court ceremonies, fixed propitious directions in which to begin journeys, and so on.

Much of this divination (in particular, the kind called hemerology) was based on cyclic notations of the year, month and day, and therefore was an outgrowth of calendar making. In general the goals of mathematical astronomy are universal; local differences in the motions of the sun and moon are trivial. Given a Chinese manual from which to determine basic parameters and computational procedures, there was little that local talent or preference could add, at least to the routine work of making the yearly ephemerides. On the other hand, when unforeseen and ominous celestial phenomena were observed, they had to be interpreted without delay. No Chinese book could cover every contingency, and there was no time to consult with the astrologers of the Chinese court. In astrology the Japanese were thrown upon their own resources. In Japan, as elsewhere, the practical applications of imported knowledge were valued over basic theory.

Theoretical elements from Chinese natural philosophy played an important part in the interpretations of the yin-yang art and of astrology. There is an old belief, for instance, that the north-east (*Kimon*, the gate of the demon) was a channel of unlucky influences. The yin-yang art explained it as 'the direction from which the god of Yang enters and the *ch'i* [energy] of Yin goes out.' But this notion did not imply a strictly deterministic causal principle. It was merely a warning, so that countermeasures might be devised; otherwise the art would have no practical use. For instance, people avoided building houses at places where the configuration of the land opened towards the north-east. Nor was astrology thoroughly deterministic. Before a predicted solar eclipse appeared to cast doubt upon the emperor's virtue, he could defend himself by calling in Buddhist monks to exorcise it.

Although astrology and alchemy often are called pseudosciences, they are neither misconceived sciences nor forerunners of modern science. Their goals are in no sense those of science, which may be defined as a pursuit of regularities that underlie natural processes. Astrology assumes that the future may be predicted and that defensive measures may avert undesirable futures; alchemy of the Chinese kind assumes that eternal life is possible. It is true that they assume certain regularities in nature, but these are means rather than goals.

Although the astronomers who computed the calendar were considered to be mere functionaries, the master of astrology and the practitioner of alchemy were regarded by laymen as possessing more than ordinary wisdom. As in early Europe, astrology was the higher art. Only in the Tokugawa period, when Confucian rationalism became intellectually predominant, could the official astronomers of the shogunate (the military government) attain high status by monopolizing the scientific aspects of calendar reform. Even so, traditional prerogatives kept them formally subordinate to the Abe family, hereditary masters of the yin-yang art.

Western Orientation Towards the Regular and Japanese Orientation Towards the Extraordinary

As we have just seen, it was not the regularities or eternal truths of mathematical astronomy, but the unforeseeable omens of the astrologers, that attracted attention in Japan. Only after exposure to Western influence did academic disciplines (*gakumon*; Chinese, *hsueh-wen*) come to be considered predominantly as parts of a converging search for eternal laws and for enduring realities.

We may juxtapose these two tendencies as orientation towards the regular and orientation towards the extraordinary. Exaggerating the difference for heuristic reasons, we may say that the former assumes that there are eternal and universal truths, and seeks to formulate underlying laws. The latter denies that such truths are attainable and therefore is not disposed to debate their existence. Those who relentlessly pursue regularity overlook the individual and the accidental. Those who value the extraordinary pay little attention to persons or events that conform without deviation to stereo-typed patterns. If the former are unresponsive to change because of their preoccupation with order and system, the latter reject change reflexively because they lack set principles against which to measure it. Philosophers, especially natural philosophers, strive to discover underlying laws; while historians, including students of natural history are attracted to discontinuities.

These two tendencies were rivals in the formative period of philosophy in both Europe and the Far East. Over time the emphases became divergent in the two cultural spheres. The main current of the Western academic tradition remained centred upon philosophical and logical inquiries in the Platonic and Aristotelian traditions. Eastern scholarship definitely inclined to history, with the *Shih-chi* (ca. 100 B.C.) of Ssu-ma Ch'ien as the prototype.[4]

Various conjectures have attempted to explain this bifurcation. Joseph Needham argues that the absence of an anthropomorphic lawgiver in their religion left the Chinese with little motivation to conceive laws of nature.[5] While intellectual centres shifted from state to state in the West, in China there was one polity and one elite culture; historical records were accumulated in a single language. The record of the past was conceived as that of a single people. The early appearance of a true bureaucracy encouraged the development of chronology and the systematic compilation and classification of administrative precedents. The early availability of paper and the currency of printing by A.D. 1000 made historical literature more subject to ideological control and thus more central to political and social concerns.[6]

Historical analogy rather than tightly constructed chains of logical reasoning became predominant in Eastern learning. This was true even in natural science, so that although there were considerable overlaps of subject matter in East and West, there were enormous differences of style. Abstraction and involved theoretical argument were by no means rare in Chinese science but, as already noted, they were vastly less important in Japan.

It is well known that in ancient China historical scholarship grew out of the recording of astrological portents to provide an empirical foundation for future prognostications. From the time of Ssu-ma Ch'ien, whose duties included astrology and history, such omens were an important component of the imperial chronicles. The positivistic view of history predicts that the horror of celestial omens, such as eclipses, should evaporate with the development of rationality. This was not the case in China, because such foreboding is a social rather than a psychological phenomenon. The astrologer-historians also were mathematical astronomers and strove to remove phenomena from the realm of the ominous by making them predictable. Once that happened, such events lost their significance in astrology. What could be predicted no longer had news value and no longer needed to be individually recorded in the annals.

The Platonic conviction that eternal patterns underlie the flux of nature is so central to the Western tradition that it might seem no science is possible without it. Nevertheless, although Chinese science assumed that regularities were there for the finding, they believed that the ultimate texture of reality was too subtle to be fully measured or comprehended by empirical investigation. The Japanese paid less attention to the general but showed a keener curiosity about the particular and the evanescent. In the early West, in keeping with the orientation towards regularity, phenomena that could not be explained by contemporary theory, such as comets and novae, were classified as anomalous and received scant attention. In the history-oriented East, extraordinary phenomena were keenly observed and carefully recorded. The incomparable mass of carefully dated astrological records that thus accumulated has proved invaluable to astronomers today.

In the classical Western tradition there is an urge to fit every phenomenon into a single box; those unassimilable to the pattern thus formed are rejected. In the Eastern tradition, in addition to the box in which all the regular pieces are assembled, there are a great number of others in which irregularities can be classified. Sorting exceptional phenomena into proper boxes was as satisfying for the Japanese as fitting together the puzzle in his single box was for the Platonist (the Chinese preference was intermediate). If science is defined, as Europeans conventionally do, as the pursuit of natural regularities, the Far Eastern tradition is bound to appear weak because it lacked analytical rigour.

Judged less ethnocentrically, there is some merit in its relatively catholic and unprejudiced interest in natural phenomena.[7]

When a court astronomer in Peking or Kyoto found that the position of the moon was radically different from what he computed, one would expect him to consider his theory to be compromised. Such crises often occurred in China, but there was an alternative that can be seen with some frequency both there and, quite often, in Japan. The phenomenon could simply be labelled 'irregular.' It was not the astronomer's fault that the moon moved erratically.

This attitude may be seen in the career of Shibukawa Harumi, the first official astronomer to the Japanese shogun, in a form more distinct than that which existed among his Chinese contemporaries. In the preface to his early treatise *Shunju jutsureki* ('Discussions on the Calendar Reflected in the *Spring and Autumn Annals*', the oldest Chinese chronicle), he stated:

> Astronomers have rigidly maintained that when Confucius dated the events in his *Annals of the Spring and Autumn Era* he made conventional use of the current calendar with little care for its astronomical meaning, so that the dates are not very reliable. This error is due to their commitment to mathematical astronomy, so that they do not admit that extraordinary events happened in the sky. . . . Extraordinary phenomena do in fact take place in the heavens. We should therefore not doubt the authenticity of [Confucius'] sacred writing-brush.

In his own work on mathematical astronomy, Shibukawa remained thoroughly positivistic; but he also left a somewhat problematic astrological treatise, *Tenmon keito* ('Treasury of Astrology', 1698). Careful examination of these eight volumes of astrological formulas and interpretations of recorded portents discloses that a large portion was inattentively copied from a famous Chinese handbook, Huang Ting's *T'ien-wen ta-ch'eng kuan-k'uei chi-yao* ('Essentials of Astrology', 1653).[8] In this work he repeatedly expressed the scepticism towards astrological interpretations that might be expected of a practical astronomer, and would often write 'We do not know the basis [of this interpretation]. . . . Is this unreliable?'

Shibukawa believed that a professional astronomer must be thoroughly competent in both major branches of celestial studies: portent astrology and calendrical science. His Jokyo calendrical reform (ca. 1684) provided a 'box' for regularities. It was no less important to furnish the means by which astrological portents might be classified. He was convinced that the heavens could not be fully comprehended through mathematical regularity. The sky was a unity of such depth that the tools of no single discipline could plumb it. Although he found astrological interpretations to be often equivocal, the vast historical accumulation of omen records suggested that portent astrology had to be taken seriously. There must have been, he thought, justified passion and reason behind that tireless activity of the ancients.

Once admitting, as Shibukawa did, that regular motion was too limited an assumption, one could easily conceive such notions as that astronomical parameters could vary from century to century. In the official Chinese calendar of the thirteenth century and earlier, the discrepancy between ancient records and recent observations was explained by a secular variation in the length of the tropical year.[9] Shibukawa revived this variation in the Japanese calendar, and Asada Gōryū extended it[10] to other basic parameters to account for both Western and Eastern observations then available to him.

The variation terms used in Chinese and Japanese astronomy were too large to survive empirical testing and eventually were discarded. Wherever the Aristotelian notion of an unalterable universe was followed rigorously, irregular motions in the sky were inconceivable. Even the mathematically justified variation in the precession of the equinoxes, which had a brief acceptance in Europe, came from Islam. After Newton, variations in parameters were acceptable to the extent that they could be given mechanical explanations. In the West, the first systematic study of variations in basic parameters was delayed until Laplace, in the late eighteenth century. It is significant for the history of ideas that in China and Japan there was no reason to resist such variations.

In the Far East, not only were irregular motions of the celestial bodies admissible, but the algebraic approach to mathematical astronomy made it unnecessary to take a stand on the spatial relations of the sun, moon and planets. The earliest astronomical schemes in China (first century B.C.) depended heavily upon a cyclical view of nature. These numerical models explained all of the calendrical phenomena by a vast construction of interlocking constant periods. The cycles of the sun and moon, the synodic periods of the planets, and cycles of recurrence for such phenomena as eclipses were tied together by larger cycles determined by their least common multiples. By the end of the Han period, however, improved observational precision and recording accuracy made it clear that the heavenly courses were too complex to fit such simple assumptions. Eventually the metaphysical commitment to cyclical recurrence was abandoned.[11] Periods of recurrence became no more than algebraic constants to be used alongside a great variety of other numerical devices. Neither celestial morphology nor cosmic ontology was of further professional interest to astronomers.

The Chinese tradition of astronomy, including its offshoots in Japan, Korea and Vietnam, thus did not depend for its direction of development upon a dialectical relation between metaphysics and observation. Computational schemes neither challenged nor strengthened philosophers' conceptions of cosmic design.

Differences in Chinese and Japanese Views of Science

Ogiu Sorai (1666–1728), the most influential Japanese Confucian philosopher, had some interest in astronomy. He commented on the variation of astronomical parameters (*Gakusoku furoku*) 'Sky and earth, sun and moon are living bodies. According to the Chinese calendrical technique the length of the tropical year was greater in the past and will decrease in the future. As for me, I cannot comprehend events a million years ahead.' Since the heavens were imbued with vital force the length of the year could change freely, and constancy was not to be expected in the sky. Indeed, only a dead universe could be governed by law and regularity. The study of such a world would be of no interest to the natural philosopher. Since it was precisely the vital aspects of nature that interested Ogiu, he remained an agnostic in physical cosmology.

Indifference towards the search for regularities in nature prevailed in the School of Ancient Learning (*kogaku*), which emerged in the late seventeenth century led by Ogiu. Its anarchistic and dynamic cosmology was bathed in historicism. 'All scholarship should finally converge in historical studies,' said Ogiu.[12] Because he was a Confucian philosopher, 'history' meant humanistic history. Whenever the philosophers of the Ancient Learning School looked at nature, they saw it in the light of social and ethical concerns.

This moralistic, anthropocentric, and often anthropomorphic view of nature was common among Confucian thinkers throughout the Far East. Many of them were unable to imagine that mathematical astronomy could make any greater contribution than to provide an accurate calendar.[13] Nevertheless, there were some notable differences between the views of Chinese and Japanese Confucians on the search for regularities in nature, especially with reference to calendrical science. Although these views were not imposed upon astronomers as corresponding ideas were in the West, the importance of philosophy in education makes them worth examining.

In China, computational astronomy was an integral part of the imperial bureaucracy. The head of the Astronomical Bureau, unlike his subordinates, was not a technical expert but a civil service generalist on his way up the career ladder. Many Confucian scholars wrote competently on astronomy, and books on the subject often were ornamented with prefaces and colophons by high officials.

In feudal Japan occupations were hereditary. The post of official astronomer to the shogun was created to recognize the personal achievement of Shibukawa Harumi and was passed down to his descendants. It had no significance beyond the technical acumen and thus was of no interest to the generalist. Technical posts of this kind were from their origin separated from the general samurai bureaucracy. When the official astronomer and his subordinates were compiling the ephemerides, Confucian scholars were not consulted. Even the Tsuchimikado family, for many centuries astrologers to the imperial court in Kyoto, was accorded the courtesy of an invitation to contribute a preface to the ephemerides.

A popular Chinese book of negligible depth, the *T'ien-ching huo-wen* ('Queries on the Astronomical Classics', 1675) by Yu I, exerted considerable influence on Japanese cosmological notions. Among the many Japanese editions, the only preface by a Confucian scholar was that of Irie Shukei. Irie states in his preface that he was motivated to write a commentary on the simple work because most astronomical writings were so full of mathematics and technical terms that, although they might be 'useful for the narrow calculations of small men engaged in the divinatory and computational arts, they are of no use for the greater mathematical concerns of gentlemen and scholars'. It was no doubt commonly believed in China as well that calendrical astronomy, which Irie looked down upon, had lost its ideological implications and had become nothing but a collection of techniques. Nevertheless, the Chinese, particularly from the mid-seventeenth century on, continued to think of astronomy as part of the Confucian system of learning. As Juan Yüan (1764–1849), a high official and patron of learning, put it, mathematics and astronomy were 'a proper study for those scholars who search out the facts to get at the truth, and not a tool for technicians scraping up a living'. In China many Confucian scholars contributed prefaces to

T'ien-ching huo-wen, not merely for ornament but often to discuss fundamental technical matters.

What accounts for this difference? Almost without exception the computational schemes and theories used in eastern Asia were discovered by the Chinese. To the Chinese they were integral parts of native culture; to the Japanese they were importations. In China the lingering excitement of discovery clung to knowledge of regularities in nature. In Japan these foreign regularities constituted one more routine skill prerequisite to established occupations.

This was true not only of science but also of Confucianism itself. In China, Confucianism was more than a philosophy; it was the basis of political legitimacy. The government's use of it as a political ideology demanded that great care be given to defining what orthodox Confucianism should be – just as the imperial monopoly of the calendar made it necessary to have one official system of astronomical computation. Official philosophy and official astronomy were exported to maintain China's cultural suzerainty over her satellites and neighbours. Confucian philosophy in its contemporary interpretations endorsed and justified these concerns. The commitment of the Chinese elite to civil service channelled a great deal of intellectual energy in this direction. What interpretations should be orthodox and what sorts of learning should be propagated were central subjects of philosophic inquiry. Not all thinkers shared the official view at any given time; but because it determined the content of the civil service examinations, about which much of early education was organized, the official view was enormously influential.

In Japan there was no social or political reason for philosophic orthodoxy to be an important issue. Although nominally based on the centralized Chinese model, Japanese government until the mid-nineteenth century was imposed upon a feudal society and thus remained multifocal. Although dynastic legitimacy could not be taken from the imperial court in Kyoto, real political power lay entirely in the hands of the military dictator, the shogun, in what is now Tokyo. He was able to retain that power only by leaving local authority in the fiefs (*han*). Certain prerogatives in astronomy belonged by tradition to the Tsuchimikado family, the imperial court astrologers; and others were divided between the shogunate astronomers and those of the fiefs. Satsuma, one of the larger fiefs, issued its own calendar. There was no occasion to establish a single orthodoxy, either political or intellectual.

Just as political and astronomical orthodoxy were related in China, so their absence was related in Japan. This contrast is apparent even in the art of divination. The great Chinese treatise *Wu-hsing ta-i* ('Fundamental Principles of the Five Phases', *ca.* 600) set out a coherent synthesis of contemporary knowledge and belief. The early Japanese treatise *Hoki naiden* ('Ritual Implement', undated), equally influential upon later practice, was an undigested juxtaposition of hemerological practices from Shinto, Buddhism and perhaps Taoism. In Japan freedom to choose between several paradigms seems to have been as desirable as the search for a unitary principle was in China.

When the Japanese did originate something, there was no expectation that it would be universally accepted or that it would have influence outside Japan. Although the Chinese did occasionally acknowledge Japanese originality in connection with one development or another, the Japanese did not believe, prior to the twentieth century, that they could contribute to universal systems of knowledge.

From the seventeenth century on, when Western knowledge began to exert claims of its own in the background of Chinese learning, Japanese thinkers were critically attentive. Once convinced that European technical knowledge was superior, they promptly switched to the new paradigm. This was not the first time that the Japanese had modified their attitudes smoothly and quickly to conform to desirable goals presented from abroad. For the Chinese the encounter with European ideas was traumatic; to accept them was to reject traditional values, and to reject them would leave no defence against dismemberment by the Western powers.

The Academicism of Shibukawa Kagesuke

When Shibukawa Harumi founded the shogunate astronomical office, he was in the rare position of initiating a tradition. The older astronomical institutions of Japan were devoted to inherited responsibilities from which the incumbent could not freely deviate. Shibukawa defined his responsibilities in a way that involved considerable political activity. He enlisted the support of powerful figures for a calendar reform not based upon a borrowing from China. He had to overcome the opposition of Confucian scholars, who saw no merit in native independence. The problem that Shibukawa set and the solution that he envisioned constituted a paradigm (in Thomas Kuhn's sense of the word) – a paradigm of purely local applicability.

As an exact science, calendrical science could be given the solid foundations that enabled normal science to proceed. Shibukawa left to his descendants and disciples the responsibility to work towards greater precision and to improve agreement between observation and theory. For some time there was no need to question the validity of his conception of astronomy. Historical questions that also interested Shibukawa, such as the gradual variation in astronomical parameters, did not retain interest for later generations. The historical orientation of such problems made it impossible to adapt them to the concern of 'normal science' with the improvement of precision.

The success of Shibukawa's programme depended on the quality of his successors. Since the family stipends of samurai, even those in professional posts, were inherited only by eldest sons, there was a conflict between the rigid ideal of social structure in feudal Japan and the need for technical talent. There was more than one astronomical institution because new ones had been created at various times to make room for gifted scientists. Certain established families took advantage of another means to maintain their intellectual standards: adopting as successor to the head an intelligent second or third son of a samurai family. The Shibukawa family maintained its tradition in this way. Adopted sons probably contributed more to the cumulative achievement of the family than did eldest natural sons.

Shibukawa Kagesuke (1787–1856), an adopted heir, was perhaps the greatest professional, as well as the last important figure, of traditional astronomy. In his youth he suffered bitterly when his talented brother, Takahashi Kagesuke, was executed after being involved in an attempt by a foreigner, P.F. von Siebold, to smuggle forbidden materials out of Japan. Shibukawa never deviated from the behaviour expected of a professional astronomer and maintained unblemished authority in his discipline.

Shibukawa's passion for rectitude made him particularly apprehensive about criticisms of the official calendar. In order to safeguard the prestige of his office, it was essential that he have first access to newly imported astronomical literature. This was a time when foreign threat, social change and natural calamity jeopardized the authority of the shogun, who attempted to minimize unrest and maintain a useful monopoly by prohibiting the diffusion of Western knowledge. The translation of Western literature for official use was confined to the Astronomical Bureau. Shibukawa took full advantage of his privileged position. He left voluminous notes on his wide reading in books forbidden to others – Chinese and European, ancient and contemporary.

Shibukawa's motivations emerge clearly from the story of his conflict with Koide Shuki (1797–1865), a scholar of astronomy who did not hold a position in any of the government astronomical institutions.[14]

The Kansei calendar, then in official use, was almost completely based on *Li-hsiang k'ao-ch'eng* ('Compendium of Observational and Computational Astronomy', draft completed 1722, printed 1724), in which the outdated European astronomy introduced to China by the Jesuits was adapted to the needs of a traditional calendar reform. In addition, the compilers of the Kansei computational system had incorporated Asada Gōryū's cyclical variation in length of the tropical year. The official calendrical treatise was not published, for laymen had no business using it. On the other hand, nonreligious writings of the Jesuits in Chinese had recently been exempted from the ban on importation, so the *Li-hsiang k'ao-ch'eng* was available for study by private scholars. If the Kansei calendar had been based entirely upon it, anyone could have checked the validity of the official calendar. In the late 1820s, Koide had an opportunity to calculate an ephemerides on the basis of the *Li-hsiang k'aoch'eng* and compare it with the official calendar. He found considerable disagreement. He suspected that Shibukawa Kagesuke's father, a disciple of Asada, had introduced Asada's variation in the last calendar reform, in order to conceal the system of computation under a cloak of security and thus avoid criticism from amateurs. He was unable to prove this suspicion, since the value of Asada's variation was unavailable to him; it belonged to the private teaching of Asada's successors.

Determined to obtain the formula, Koide enrolled in the academy of the Tsuchimikado clan, the hereditary imperial astrologers, where one of Asada's four major disciples had taught. In 1834, after eight years of discipleship, he was permitted access to the details of Asada's variation. His heart's desire having been fulfilled, as he wrote, he immediately calculated the current value for the length of the tropical year. Koide next made precise observations from which to determine the winter solstice, and found a discrepancy of a quarter of an hour between his computations and those given by the official calendrical system. When he ignored Asada's variation and calculated on the basis of the *Li-hsiang k'ao-ch'eng* alone, his results coincided closely with observation. Thus he determined that Asada's formula was a 'fake supported by blind belief'.[15] In 1835 he visited Edo (now Tokyo), became a disciple of Shibukawa Kagesuke, and questioned him about his view of Asada's variation. Shibukawa would not give him a clear answer.

Shibukawa had already read Lalande's *Astronomie* and had a good understanding of the European astronomy of the previous century. He knew perfectly well that Asada had misled his successors. He did not, however, dare to voice his doubts publicly. In 1835, in fact, Shibukawa drafted a manuscript entitled 'Saishu shochō kō' ('On the Variation of the Length of the Tropical Year'),[16] in which he analysed the origin of, and tried to give a rationale for, Asada's variation.

Why did he write this essay in that particular year? He must have been influenced by Koide's criticisms. Although Koide had no official standing, his connection with the Tsuchimikado (an older institution and, to some extent, a rival for power) gave him some authority.

Koide was not prepared to do more than point out numerical discrepancies. Shibukawa had access to vastly greater information and technical skills, and was able to form an analytical overview of the variation question. Since his bureau had come to be responsible for the actual calculation of ephemerides, he felt that a frank answer to Koide's inquiries might compromise his own official responsibilities. But he was now aware that Koide had found the weak point in the official calendar and would be driven by his remarkable determination to press an attack that was bound to be successful. Although Shibukawa Kagesuke was not responsible for the failure of the Kansei calendar, his office would suffer for it.

Shibukawa had two lines of defence. The only permanent defence was to carry out a calendar reform as soon as possible, discarding Asada's variation in the interest of accuracy. But calendar reforms in Japan were so infrequent that they could not be proposed and carried out overnight. If a crisis arose too soon, Shibukawa wanted to be prepared with an official interpretation of Asada's variation to interpose against attack. That was apparently the point of his treatise.

Shibukawa's fears were soon realized. Koide's prediction of a solar eclipse in 1839 (based entirely on the *Li-hsiang k'ao-ch'eng*) proved to be more accurate than that given in the official ephemerides. Koide submitted to the prime minister (*rōjū*) a proposal for a calendar reform based on the Chinese source.

The official astronomers had no choice but to hurry their own reform. The new Tenpo system of computation became official in 1842. Shibukawa, unlike Koide, did not have to depend upon a century-old treatise. He had the authority to mobilize the best translators in Japan so that his system could be based upon contemporary European data.

Through the years of mounting conflict between Shibukawa and Koide, the goal of both was complete agreement between computation and observation. Nevertheless, they differed in their conceptions of the astronomer's work and thus of astronomy. From Shibukawa's viewpoint, Koide's painstaking efforts had no significance whatever for the advancement of astronomy. Koide was simply duplicating outmoded results. The only positive service he could perform was to prevent the corruption of the astronomical functionaries who monopolized research facilities and tools under government protection. Koide began with a simple puzzle: why the official calendar failed to agree with observation. Since he had only limited access to scientific and political information, all he could do was deal with each step as it unfolded from the last. What began as a technical exercise in the testing of computational theory was reduced to the unravelling of bureaucratic prerogatives.

Here the conflict between Koide and Shibukawa finally lay. Although both accepted the commitment of the astronomer to accurate prediction, Shibukawa's dedication to his profession gave the administrative order precedence over the celestial order. This Koide could not accept; but he was bound to be drawn into intrigue, for Shibukawa's professional standing reserved to him alone the initiative to define the rules of the game.

Western Cosmology and the Traditional Calendrical Science

Elaboration and precision were the criteria by which calendrical treatises were evaluated in prefaces written by astronomers. The holders of astronomical sinecures needed to fear no crisis so long as these criteria were satisfied. In view of the moderate level of precision needed for the traditional ephemerides, why should Shibukawa have wanted to involve the Astronomical Bureau in the active dissemination of European natural philosophy? Such a major departure from his inherited duties would seem to carry as much risk as gain. Nevertheless, Shibukawa wrote an account of Copernican theory and Newtonian mechanics for government use (not for publication) in his 'Shinpo rekisho zokuhen' ('Sequel to the Astronomical Treatises According to the New Methods' [to the series of works compiled by the Jesuits in China before 1635]).[17] Again his motive seems to have been bureaucratic caution. It was essential to formulate an official view of European cosmology as awareness of it gradually became more general in Japan; otherwise a query by a high official might result in grievous embarrassment.

Heliocentric theory had been previously introduced by Motoki Ryoei (1735–1794), an official interpreter; Shizuki Tadao (1760–1806), an independent intellectual; and Shiba Kokan (1739–1819), a free-lance popularizer. Partly because cosmology was not traditionally an important topic in Japan, and partly because they understood Copernicanism only superficially, these writers and those who read their works were not shaken by heliocentricism as Europe had been. Motoki considered it merely an exotic European curiosity. Shiba adhered to it for the sake of sensation. Shizuki treated it incidentally to his main interest, the philosophical implications of Newtonianism, which he tried to assimilate to Chinese natural philosophy.

Shibukawa disdained these amateur writings. For his account of Copernicanism he depended heavily upon an explanation written in Chinese by the French Jesuit missionary Michel Benoist (1760). It was little known in Japan and did not so much as mention Newton. This work allowed Shibukawa to deal with the subject while completely ignoring the Japanese literature (based upon much later secular sources). Although not enthusiastic about the sun-centred system, Shibukawa did accept it with critical reservations. He agreed with Benoist that the difference between heliocentricism and geocentricism was merely a matter of transforming coordinate systems. He contributed the notion that theories of a moving earth were not original in the modern West but had existed long before in China and India – a theory first advanced by Shizuki Tadao.[18] Shibukawa's discussion of the technical aspects of Newtonianism was superior to that of Shizuki, but the latter's philosophic depth was totally missing. For Shibukawa, Newton's work was not the foundation of a world view but a peripheral issue on the margins of calendrical science. Shibukawa was successfully guided by his motto, 'Let us melt down the mathematical principles of the West and recast them in the mould of our tradition,' a cliché in earlier Chinese writing. Newtonianism was so well assimilated in Shibukawa's presentation that it did not perceptibly challenge the traditional definition of mathematical astronomy. Like his ancestors who had headed the Astronomical Bureau before him, Shibukawa devoted his erudition and energy to the perfection of routine, not to the development of new fields of investigation.

In the scientific revolution of seventeenth-century Europe, astronomy played a great role because of its implications for cosmology and world view. Japanese astronomy had no such implications, and its social matrix gave it no scope for free inquiry into nature. During the seventeenth and eighteenth centuries it remained the vanguard of disciplines oriented upon mathematical regularity. Its paradigm was so well insulated from challenge by professionalism that even the impact of Western exact science did not throw it into doubt. The most important function of European astronomy was to furnish numerical data and computational techniques, by use of which traditional calendrical goals could be better met.

The Tenpo calendar reform brought the accuracy of prediction to as high a level as the traditional calendrical art envisioned (solar eclipses to the nearest quarter-hour, for instance). It was no coincidence that the esteem of intellectuals for astronomy was practically lost in the process. At a time when the revolutionary implications of foreign science were gradually coming to be understood, the astronomical profession was seen as too routinized and unimaginative to make any important contribution to change. It was finally the physicians, who lacked both the advantages and the disadvantages of a tightly organized profession, and whose proficiency in the exact sciences was inferior to that of the astronomers, who were able to introduce the true core of Western scientific thought. The structure of the scientific revolution they brought about in Japan was bound to be different from the one led by astronomy in seventeenth-century Europe.

MEDICINE

The Chest as the Seat of the Mind

Even as late as the middle of the nineteenth century, the Japanese did not believe that thought takes place in the head. As Shibukawa Kagesuke put it, 'Mathematical principles all originate in the breast of the mathematical astronomer. . . . [Thoughts] are stored in the chest.'[19] He was typical in situating both memory and arithmetical thinking in the chest.

To the Japanese such expressions as 'a dull head' or 'a clear head' have a modern and exotically occidental flavour; they were never used until the Tokugawa period. The cognitive and imaginative functions of the brain were unknown, and their anatomical substrate undemonstrated in Japan, until the beginning of Westernization.

Traditional Chinese medicine, upon which the learned tradition of Japan depended, was concerned primarily with function and only secondarily with tissues and organs. The sites of function to which most attention was paid were two groups in the thorax, a set of six *fu* that fermented food, separated energy from waste, and excreted the latter, and five *tsang* that stored the refined energy. These spheres of function were identified with the familiar viscera; but the physiological nature of the latter was of such minor importance that little was known about it, and it played only a small abiding role in medical discourse. Occasional drawings of the body ignore both the interior of the head and the nerve tissues, neither of which was assigned specific functions, at least in the Chinese medical writing that was influential in Japan.

To my knowledge, the brain's function was first discussed in China in *Wu-li hsiao-chih* ('Notes on the Principles of the Phenomena', 1664), by the idiosyncratic Fang I-chih, who was acquainted with the writings of the Jesuit missionaries. His knowledge originated in Western medicine.

In Japan thought was first located in the brain in an Amakusa edition (1593) of Aesop's *Fables*, in which it was said, 'if we have intelligence in our heads. . . .' Again it is clear that this idea was imported into Japan along with Catholicism. The notion did not spread until the study of the Dutch language (and consequently of secular sources) became widespread in Japanese society; and in the early period there was not the slightest influence upon developing knowledge of anatomy, physiology or pathology.

It is well to remember that the interrelation between the brain and mental processes could not be proved before the development of cerebral physiology. The idea has a long history in Europe, but it is the history of a belief rather than of a fact.[20] Plato and Aristotle held quite different views on the location of mental processes. Plato believed that the immortal and holy rational soul is located in the brain. Aristotle placed the centre of sensation and perception in the heart, and did not believe that it was related to the brain or spinal cord in any way.

These were not isolated opinions but were integral with coherent views of the body and its functions. Plato and others who placed mental functions in the head have tended to think of them as quite separate from the physical body; schemata that consider the heart and mind as identical have tended to think of mind and body as integral. Traditional Chinese and Japanese views must be classed with the organismic and naturalistic group to which Aristotle belongs rather than with Plato's idealists. The experimental work of Galen settled the matter in the West, providing an authoritative basis for the doctrine that the brain is the centre of perception and of all other mental processes. The introduction of this idea into the Far East had implications at least as revolutionary as those of Copernican astronomy. It challenged the doctrine of bodily functions and the rather negligible notions about internal organs related to them, and posed a range of questions about the physical basis of sensation that had not been previously considered.

Ogawa Teizo has located[21] the first appearance in Japanese of terms that correspond to 'nerve' and to 'consciousness' as associated with the brain in the *Kaitai shinsho* ('New Book of Anatomy', 1774), by Sugita Genpaku and others. Otsuki Gentaku, in his *Chotei kaitai shinsho* ('New Book of Anatomy, Revised' [compiled 1798, published 1826]), enthusiastically described the significance of the new study of the cerebral and nervous system in this way:

> We have not come across anyone in the long Chinese and Japanese traditions who discussed the active functions of these organs of sentience. They were taken up superficially and in an elementary way only at the close of the Ming dynasty [early seventeenth century]. Most regrettably, in the two hundred years since that time no one has taken up the problem in closer detail. It is a great pleasure that now we are able to explore it more deeply. This is not particularly due to my personal endeavour, since we are all influenced by the trends of the times.

The learned treatises of the Chinese and Japanese medical traditions lacked terminology not only for brain functions but also for mental processes. Conventional Chinese discourse was not much concerned with what we would consider epistemology. Some late Confucian schools were somewhat concerned with how knowledge becomes certain but tended to connect this problem with that of attaining enlightenment. The vocabulary for mental operations remained rudimentary and to a considerable extent was borrowed from Indian Buddhism.

Although the spheres of function within the body were thought to process nutriment and to store the energy refined from it – and Japanese terms that predate Chinese influence, such as *kusowatafukuro* and *yuharifukuro*, are literally types of containers – knowledge was never thought of as localized and stored. There was no reason to investigate the physiological basis of cognition.

In short, the need to explore the relations between mind and brain did not exist in China because the Chinese assumed neither the mind-matter dualism nor the dualism between the self and the outer world. They saw all of nature as united in a single pattern of function in which the patterns of function of individual things (*li*) participate. The dualistic terminology used today in Japan, except for a few terms borrowed from fundamentally religious Buddhist dualism, was for the most part devised by Nishi Amane and others at the beginning of the Meiji period (1870–1890).

Anatomy and Energetics

In the East, the apparently rudimentary association of physical functions with internal organs does not indicate a low state of medical theory. Although the Chinese lacked the sophistication of Galen's anatomy, attempts to study rigorously the Chinese language of function in its own terms, a very recent development, suggest an artfulness that was obscured by the imposition of modern viewpoints. From the historical point of view, the fundamental question is not whether, before modern times, the Chinese or the European tradition incorporated the greater number of correct facts, but how their theoretical paradigms, and the views of nature on which they depended, differed.

In Western medicine the rivalry before modern times between solidists and humouralists is well known. The aim of the former was to locate the seat of a disorder in a solid part of the body, such as the stomach or brain. The motivation to pursue anatomical research is obvious. The humouralists, on the other hand, thought of health primarily as a balance of the various humours that circulate in the body. Anatomy had a great deal less to contribute to their holistic diagnoses.

Traditional Chinese theories of bodily function and of pathology are closer to the humouralist tradition than to that of the solidists. Health was related to the balance of *ch'i*, which is the basis of material organization and function not only in the human body but throughout the physical universe.

Ch'i was not a ponderable fluid, as the humours were. It originated very early as the word for air – not as an inorganic gas but was an enveloping substance that maintains vital functions. Its closest analogue in the West was the Stoic pneuma. In medical theory its vital or energetic aspect – in a purely qualitative sense – became preponderant in discussions of etiology. *Ch'i* was not only inspired air but also the energy refined from food that circulated throughout the body and was responsible for all vital functions. The concept of *ch'i* as a material pneuma was to some extent reconciled with this energetic approach and was never abandoned; for instance, certain tumours and internal swellings were thought to be stagnated or congealed *ch'i*. Indeed, the seventeenth-century Japanese physician Goto Gonzan attempted to explain the cause of all medical disorders by stagnation of this kind. Since *ch'i* was involved in processes in physical nature and in the body, it assumed different qualities or characteristics in different phases of such processes. If the whole process was analysed into two phases, the two different types of *ch'i* were characterized as yin and yang; if a fivefold analysis was used, the five types of *ch'i* were described by the language of the Five Phases theory. A dynamic balance between the two or five types of *ch'i* defined health; ethical disorders were always identified with an imbalance.

The language of yin-yang and the Five Phases theory was used to establish sets of correspondences that governed bodily function. For instance, the Five Phases corresponded to the five spheres of function (loosely identified with the heart, lungs, spleen, liver and kidneys). But discourse about health and pathology was never anatomical. The system identified with the spleen amounted to the ensemble of functions that would be ascribed today to the urogenital system and was thought of in functional terms.

Internal disorders were never local in Chinese medicine. Although they might be concentrated in a particular sphere of function; the connection of the spheres by the energetic circulation system meant that the whole body was affected and that as the pathological process developed, its seat would move. There was no value in local treatment, for the site of treatment often was far removed from the centre where the disorder was concentrated for the moment. Abstract correspondences often were used in discussions of pathology and therapy – for instance, Five Phases correspondences between the heart and the ears and between the liver and the eyes. These were not so much statements about physical connections (although such connections were claimed to make the model plausible pneumatically) as about similarities and analogies of function.

Early Far Eastern anatomical charts were extremely simple and crude. As Lu Gwei-djen and Joseph Needham have said,[22] they incorporate a much more rudimentary level of knowledge than the texts that they accompany. Why were Chinese physicians satisfied with them? Their purpose obviously was different from that of Western anatomical diagrams. They were simply meant to depict the broad outlines of the general system of physical function. One might think of them as half anatomical diagram and half flow chart.

According to this view of the theoretical entities of Chinese medicine, reconstructed largely through the painstaking work of Manfred Porkert,[23] it is possible to conclude that it was closer to the European humouralist point of view, although it was pneumatic in a sense that does not fit the European theory. It had no use for exact anatomy. For the latter to be accepted in the Far East, its utility would have to be proved; and it could be proved only by an appeal to a different conception of nature and of the human body.

Anatomy in Japan

When anatomical inquiry began in mid-eighteenth-century Japan, did its demand for an analytical approach to the human body have revolutionary consequences?

Even before the serious study of anatomy began in Japan, Yamawaki Toyo questioned traditional Chinese anatomical charts on the basis of his own anatomical findings (1759), and his criticism entitles him to be considered the forerunner of anatomical students. Yamawaki's interest in anatomy must have been stimulated by access to a Western anatomical chart. Although he could not read the legends, his experience must have convinced him that the Western schema was a great deal more accurate than the Chinese. What led him to evaluate both schemas as anatomical rather than as functional?

First of all, when Western knowledge began filtering into Japan during the Tokugawa period, it was natural that it should have been compared with the official Chinese academic knowledge, since the latter had become firmly entrenched not very long before. It was also natural to ask which set of ideas was better – the situation was different from the case in China, where traditional ideas were so strongly rooted that such a question could only be radical. In mathematical astronomy, criteria of predictive accuracy were so obvious that Western superiority was quickly recognized. This was equally true in China, since the criteria for that recognition could be traditional ones. In medicine, however, there is good reason to doubt that there was any difference in therapeutic efficacy between the two systems before the late nineteenth century. It is above all in the comparison of anatomical charts that the strength of Western medicine would be apparent. But if the foregoing argument is correct, the difference between Chinese and European ideas about the interior of the body would be anatomically significant only after acceptance of the idea of anatomy and of the more general medical and philosophical ideas on which it was based.

Yamawaki Toyo was one of the leaders of a new group called the Koihō ('Back to Ancient Medicine School'), who rejected the theoretical entities of Chinese medicine and undertook an empirical approach to clinical treatment. Their utilitarian goals made the very elaborate conceptual superstructure of the Chinese tradition seem an impediment. Because they wished to confront as directly as possible the ills of the body, its role as a microcosm of physical nature could be rejected. As Yoshimasu Todo (1701–1773), the foremost figure of this school, declared, 'Yin and yang are the *ch'i* of the universe, and thus have nothing to do with medicine.'[24] This group was prepared, then, to take a position much closer to that of the solidists than had previously been possible in Japan. Functional analysis lost its importance, and the physical organs could be studied for their own sake. From this point of view the traditional anatomical charts were recognized as crude and inaccurate representations of material organs. This was nothing less than a change of gestalt.

Nevertheless, this conceptual radicalism was circumscribed. Yamawaki's accomplishment was not to do away with the old scheme of six processing spheres and five storage spheres but, rather, to alter them so that they made sense in anatomical terms. He had no reason to be curious about the contents of the skull. It was only later figures with considerable knowledge of Western anatomy, such as Sugita Genpaku, who could abandon the Chinese tradition entirely and display as much anatomical interest in the brain as in the viscera. At that point the confrontation between champions of the two systems becomes interesting.

Lately there has been a tendency to emphasize the value of organismic and synthetic thought, of the sort that Joseph Needham has found predominant in Chinese science, to the detriment of the early modern habits of physical reductionism and remorseless analysis. Although the pneumatic *ch'i* doctrine, and the yin-yang and Five Phases theories that qualified it, are not precisely reductionist, they are not modern either. All of these concepts, although originally taken from everyday phenomena, were too abstract to have fixed empirical significances. They remained satisfactory only because, as N. Sivin has shown,[25] the goal of Chinese science was not complete understanding of the natural world but limited knowledge for practical purposes. The body was clearly not the cosmos, but the correspondence between the two set limits upon what could be asserted about the body.

Because of the special character of Chinese medical thought as it was received in Japan, we have examined in some detail the reasons that traditional doctors would find anatomy, and thus dissection, irrelevant to the improvement of medical therapy. There were other objections as well. The idea was deeply ingrained in Confucian ethics that keeping intact the body one has received from one's parents is a major obligation of filial piety. This prohibition against mutilating the body effectively ruled out dissection in China. In Japan, however, it had practically no effect on medical specialists.

A second objection to dissection originated in traditional physiology. In his *Hi zoshi*[26] ('A Refutation of the Anatomical Charts', 1760) Sano Antei said, 'What the *tsang* [the word for the spheres of function and their associated viscera] truly signify is not a matter of morphology; they are containers in which vital energy with various functions is stored. Lacking that energy, the *tsang* become no more than emptied containers.' In other words, the internal organs were characterized not by their morphology but by the differences in their functions, which were defined by the energy they stored. Nothing could be learned by dissecting a cadaver, since it lacked this vital energy. The anatomical charts that captured the imagination of Yamawaki, since they were based on dissection, could cast no light on the dynamic functions of the body. The same point emerges in another criticism that Sano made. He noted that Yamawaki's anatomical charts did not distinguish the large and small intestines. He did not believe, in fact, that those organs were morphologically or physiologically dissimilar. What made them utterly different was that the large intestine was responsible for absorbing and excreting solid wastes, while the small intestine performed those functions for fluid wastes. He emphasized that this crucial difference would be undetectable in a dead body. Figure and appearance could be significant only to the extent that they were related to function. Sano, unlike the Koihō radicals, had no use for pure empiricism. 'The observation of two obvious facts is of much less value than groping speculation . . . even a child is as good an observer as an adult.' A scholar who refrained from speculatively tracing the connections between form and function was no better than a child.

Beginnings of a Solidist Approach

After accurate European anatomical charts were introduced into China, even the traditional schools of medicine that admitted anatomy as the basis of surgery (an undeveloped art in the Chinese tradition) still adhered to an energetic and functional view as the basis for internal medicine. But a shift to a solidist approach had at least begun.

It would be a mistake to see this shift purely in terms of the increasing accuracy of anatomical description. The Koihō school, like Western empiricism, could not dispense with metaphysical entities, but depended upon them without acknowledging them. Its physiological and pathological ideas were not only less explicit than those of earlier speculative Chinese medicine but also a great deal less sophisticated. The move towards solidism was not a rejection of models, but the construction of a new model.

Yoshimasu Todo, for instance, rejected the elaborate Chinese theories but was unable to translate his solidist thinking into diagnosis without the aid of a theory that Chinese doctors would have considered primitive: he saw all disease as the action of one fundamental poison on the various organs and tissues of the body. This was not really a pharmacological theory about the effect of poison, but merely a rationale for locating the part of the body on which treatment should be concentrated. He also rejected the traditional pulse diagnosis, which had served as a way of reading functional characteristics of the *ch'i* circulation. Thus faced with the problem of how to determine the condition of the internal organs without dissection, he did not so much eliminate pulse reading as substitute abdominal palpation for it. This technique had been used to a very limited extent in traditional medicine, chiefly to determine whether existing abdominal pain increased or decreased when the belly was pressed. Yoshimasu enormously increased its importance as the most direct way of learning about the conditions of the internal organs and thus founded a Japanese diagnostic tradition that still flourishes among traditional doctors.

The solidist tradition begun by the Koihō school eased the way for Western anatomy. In the second half of the eighteenth century, Sugita Genpaku took up the study of anatomy because it seemed the most tangible, and therefore the most comprehensible, part of Dutch medicine. Following the solidist breakthrough, the successors of Sugita in medicine studied physics and chemistry, thus opening up the world of modern science. The Copernican influence was minor by comparison, because the Japanese cosmos had not been defined by religious authority. The impact of anatomy challenged the energetic and functional commitments not only of medicine but also of natural philosophy. Its effect was bound to be revolutionary.

Medicine and Science After 'Kaitai Shinsho'

Publication of *Kaitai shinsho* (1774), the first Japanese anatomical treatise based directly on Western materials, not only led to recognition that Western knowledge of the interior of the body was superior to that of China but also provided a new paradigm for Japanese science. Once the Japanese were prepared to compare accounts of the interior of the body from a purely morphological point of view, the superiority of the West became obvious. Chinese-style conservatives could dismiss European anatomical charts as superficial, but they could not convince others.

The first Japanese to realize the power of Western anatomical knowledge naturally assumed that the European system also was therapeutically more effective, although there is no reason to believe that this proved to be the case. Indeed, on therapeutic grounds there is very little to choose between the systems of internal medicine evolved in the various high civilizations before the end of the nineteenth century. Moreover, it is unlikely that the relatively frequent resort to surgery in the European tradition led to consistently higher recovery rates before the introduction of anesthesia and asepsis. Some scholars actually give the edge to Chinese internal medicine because it tended to use milder and less drastic drugs than were prevalent in Europe. It is ironic that one of Yoshimasu Todo's innovations was the frequent use of poisonous drugs to 'fight poison with poison'.[27]

Among the great diversity of schools in Japan were eclectic groups that prescribed both Chinese and Western drugs for a single symptom, but that was about as far as eclecticism could go. The views that underlay Chinese and Western medicine, or even Koihō medicine and practice of a more traditional kind, were irreconcilable.

It was quite possible to introduce European data into traditional calendrical astronomy without challenging the paradigm on which the latter was based. An analogous accommodation was impossible in internal medicine, for there was little overlap of the conceptions of relevance. Acceptance of the European view of the body came only with the publication of *Kaitai shinsho*. The Koihō school can be considered a vanguard in this scientific revolution. Such a transition did not take place in China because the Chinese maintained their traditional medical world view much more rigidly than did the Japanese.

An important characteristic of early modern science was mechanical reductionism, in which every phenomenon was believed to be ultimately explainable in terms of matter and motion. This reductionism gave birth to the positivists' hierarchical arrangement of the sciences. Comte ranked the abstract sciences in the order in which he believed they would be entirely quantified, beginning with mathematics, then astronomy, then physics, with sociology at the end of the list. At about the same time Japanese physicians were constructing an analogous but very different schema.

After the publication of *Kaitai shinsho*, some medical practitioners, exploring the newly available writings on European physical science, recognized and responded to its reductionism. In the prefaces to Aochi Rinso's *Kikai kanran* ('Contemplating the Waves in the Ocean of *Ch'i*', 1825) and Kawamoto Komin's sequel, *Kikai kanran kogi* (1851), the authors claimed that physics must be the basis of medicine and the other practical sciences. Kawamoto described a hierarchical order from physics to physiology to pathology, eventually encompassing practical therapy. It is not clear how seriously his fellow practitioners took him. It is likely that they saw no clear role for physics in medicine, except perhaps for embellishing prefaces, as disquisitions on yin-yang had done in traditional books of therapy.[28] Physics and chemistry were introduced into Japan by European medical men only for their limited direct value to clinical medicine, just as anatomy, physiology and pathology were subordinated to the same use. Hoashi Banri, a natural philosopher whose background was in medicine, was disappointed and disillusioned with Western science when he examined books on microscopy and chemistry and found them of no help in the understanding of drug therapy. The disciplines such books represented underlay techniques of measurement in materia medica and of extracting the active essences from herbs. Those applications were the basis for their initial study by physicians. Their value for a new philosophy unfolded only gradually.

Social Status of Medical Practitioners

In Japan mathematical astronomers were minor bureaucrats, responsible for preparing the national ephemerides. Although they were the earliest to recognize superior aspects of Western science, they overlooked its basic paradigms and remained within the traditional mould. Their academic style, as we have seen, tended to be greatly shaped by their proximity to sources of power.

Medical practitioners, who first took up the challenge of the Western sciences, constituted the largest scientific profession during the Tokugawa period. Medicine, unlike astronomy, was a private concern and thus free of one kind of constraint upon the response to new ideas. Because there was no public health programme at the time, medical practice was essentially a relationship with individual patients. There usually was a private physician for each community. The samurai class had its government doctors and fief doctors, and townsmen and peasants had their local practitioners; but the profession was not tied together or controlled by the central government. Although Edo, as the seat of the shogunate, was a centre of professional activity, the important schools of medicine were scattered as far as Nagasaki. This decentralization made medicine one of the few geographically mobile professions in Japan.

Towards the close of the Tokugawa period, in the first half of the nineteenth century, it became conventional for medical students to visit the various centres of instruction and to be initiated into the different schools of clinical medicine. Moreover, practitioners who distinguished themselves often were called to serve the fief governments or the shogunate. Although their stipends were small, the prestige they gained raised their fees in subsequent private practice. The competitive market for medical practices in Japan was most untypical of the society as a whole. Western-style physicians took advantage of it as they moved into spheres of health care that previously had been monopolized by traditional practitioners. There is a loose analogy between this situation and that of the nineteenth-century German academic market described by Joseph Ben-David.[29] In Japan, furthermore, there was no guild organization of physicians to limit or control competition.

Unlike the medical profession in Europe, which was well integrated into society and could reproduce itself in the universities, Japanese doctors were socially marginal. Their mobility was anomalous in a society where status was supposed to be hereditary and where the only elite was supposed to be the hereditary warrior class, the samurai.

The Taki family, hereditary physicians to the shogunate, once tried to centralize medical standards by founding an official medical school that all sons of doctors were to attend and at which they were to be examined for a licence to practise medicine. This attempt failed, in contrast with the ease with which central authority was established in other fields. The main impediment to uncontrolled competition in medicine was not guild or government organization but the hereditary system on which the Tokugawa social order was based.

The samurai, the military elite, inherited ranks and stipends that depended upon the contributions of their ancestors to the foundation of the Tokugawa shogunate in the early seventeenth century. Merchants, artisans and others did not depend upon fixed stipends as the samurai did, but their social class was fixed through inheritance. This system could not easily find a place for intellectual professions, which could flourish only in situations where advancement was based on talent rather than birth. In the Tokugawa period there were three such professions: Confucian philosopher, medical doctor and mathematical astronomer. In all of them people of outstanding ability often remained subordinate to incompetent samurai and, if they worked for the government, received lower stipends.

Attempts were continually made to subordinate these professions on the hereditary principle. It was expected, for instance, that the son of a doctor would eventually be registered as a doctor, regardless of how little intelligence or motivation he might have. At the same time, the shogunate and the fief governments needed talented professionals. The conflict often was resolved by the governmental authorities, who would advise a professional family to adopt a gifted youngster.

Government employment was only one possible source of income for a physician. Osaka, for instance, was famous for a medical centre patronized mainly by merchants. The clinical experience of the therapist mattered a great deal more than his formal education. The son of a village doctor would begin by working with his father, then spend many years as an apprentice to more distinguished doctors, and finally return to his native village to take over his father's practice. From generation to generation the number of patients would gradually increase until such a medical family was expected to provide doctors for the whole village. Because such hereditary traditions were quite independent of the government hierarchy, doctors were among those most responsive to liberal thought in the period shortly before the modernization of Japan. Mathematical astronomers were not independent in the same way.

Although Confucian scholars formed a professional group, they lacked the social mobility and economic security of physicians. Their official social status was a good deal higher, but the revenue that they might earn by private teaching could not compare with the fees of the doctors. In essays of the Tokugawa period the social commitments of the two often were compared, to the detriment of physicians. Confucian scholars were concerned with society as a whole, and physicians only with individuals; Confucian scholars dealt with the mind, and physicians only with the body; Confucian scholars were generally poor, while physicians gouged fees and lived in luxury. It is clear from such remarks that establishment values favoured the scholars, and found the practical skills of the doctor a little too close to those of the artisan.

This difference should not be overstressed, since Chinese-style medicine emphasized that practice must be based upon Confucian ethics. Young men who chafed under this devaluation of medicine as a pursuit in its own right were especially attracted to Western medicine, which seemed free of philosophic and moral constraints.

Those attracted to intellectual pursuits found them most attainable if they left clinical medicine and became teachers or public figures. Western-style medical doctors gradually distinguished themselves into two groups, one that concentrated upon medical practice and one that mainly taught foreign languages and Western science. In the difficult international situation following the Opium War (1839–1842), it was from medical schools of the latter group, such as that of Ogata Koan, that there appeared political activists such as Hashimoto Sanai who renounced their inherited professions to pursue political careers.

Of the three intellectual professions in the Tokugawa period, only physicians were able to achieve an independent stance from which to view the world in a new light. It was naturally they who brought modern universal science to Japan. But their independence was bought at the cost of alienation from the true sources of power in Japan.

The Science of the Physician and the Science of the Samurai

The pattern of response to Western science changed fundamentally when Japan was opened to the free flow of foreign ideas in the 1860s. The initiative passed from the physicians to the samurai. Among graduates in the class of 1890 at Tokyo University, the per centage of those who came from samurai families were as follows:[30]

Medicine	– 40.8
Agriculture	– 55.9
Law	– 68.3
Literature	– 75.0
Science	– 80.0
Engineering	– 85.7
(Total population)	– 5 (approximately).

Why was the proportion of students from the elite class so high in science and engineering, and so much lower in medicine and agriculture? As we have seen, it was customary for the sons of doctors to become doctors. They did not belong to the samurai class, unless they were employed by the government. Similarly, many agriculture students were the sons of wealthy farmers. The downfall of the samurai regime (which made the foundation of the European-style university possible) had little effect on the livelihoods of either doctors or farmers.

On the other hand, that transformation was catastrophic for the hereditary military elite, whose traditional occupations in the bureaucracy of the old regime, as well as their hereditary stipends, were lost. The fields of medicine and agriculture were largely occupied. The law was not considered a dignified occupation;

and until the Civil Service Examination Act of 1890, the law schools did not provide a route to upper-level civil service careers.

The only promising professions left for the samurai were science and engineering. In nineteenth-century Europe, the upper class tended to occupy the legal and medical professions, and science and engineering were largely shaped by the rising middle class. In Meiji Japan, on the contrary, scientists and engineers were drawn largely from the top 5 per cent of the population.

In the wealthier European countries the scientific and engineering professions drew on the ideology of the middle class rather than on that of the old aristocracies, whereas in Meiji Japan they were entirely subservient to the social and international aims of the imperial government. Scientists were indispensable to the policy of modernizing and Westernizing, and engineers played key roles in building the physical structure of a modern state. Much of the increased demand for engineers was for survey work and telegraph network construction. From the time these professions began to form on the Western model, those who entered them were government officials. The descendants of samurai, who valued public over private occupations, were thus attracted to engineering. The result was a group of public-spirited engineering professionals oriented towards civil service, in contrast with English engineers in Britain, for example, who came mostly from the class of skilled mechanical people and served private interests.

Before the Meiji period, Western-style physicians took up the physical sciences as intellectual pursuits; those who continued to study them did so largely out of an interest in natural philosophy. Although they legitimized their study, in the Meiji period the formation of scientific and engineering professions became the concern of the samurai and the doctors largely returned to their family practices. Samurai entered science and technology because of their contributions to the state. Their response was not so much intellectual as institutional.

MATHEMATICS

The Royal Society and Tokugawa Mathematicians

Many would consider the appearance in 1660 of such a disinterested group as the Royal Society of London to be quite unthinkable outside the sphere of Western civilization, but Japanese mathematicians of the Tokugawa period (*wasanka*) were similar in many ways. Members of the Royal Society secured a charter from the king for reasons of prestige and frequently studied subjects of no economic importance. Leisured gentlemen constituted the entire membership and dues financed most of the activities. Similarly, groups of mathematicians in Japan were purely private in nature, consisting of samurai, rich merchants and affluent peasants; they gathered solely for leisure activities.[31]

In its early period the Royal Society tried to realize certain Baconian ideals, but the activities of the amateurs declined in prominence after the first generation.[32] Learned societies in the British provinces also tended to become philosophical societies after the model of the Royal Society. That is, salaried officials – journal editors, directors of libraries and museums – became central in both while the socially more prominent members were relegated to support functions and retreated from front-line research.[33] In Japan, however, interest in *wasan* activities grew over time, spread outside urban centres and expanded until the nineteenth century. Participants were increasingly recruited from lower strata of society. Because mathematical activities lacked a significant occupational base, distinctions between professionals and amateurs did not arise. Mathematics was enjoyed by leisured groups in the same way as *waka*, *haiku* or the tea ceremony. In fact, modern historians of Japanese mathematics have commonly observed that *wasan* was more of an art form than a field of scholarly inquiry.[34] Here, however, I should like to raise the question of how art and scholarship differ from each other and to consider *wasan* in that context.

Scholarship and Art

While *wasan* may be considered an art form, it is by no means easy to distinguish art forms (*gei*) from scholarship (*gaku*). Activities that individuals consider scholarly in nature are not necessarily so regarded by society; such evaluations depend on the views of certain social groups in specific locations during a particular period. They may also depend on the value standards of intellectuals and be subject to influence by the presence or absence of official authorization as well as by popular impressions.

I shall not attempt conceptually rigorous definitions of scholarship and art here, but merely note that scholarship is usually thought to have some public function, while art forms are often regarded as private indulgences that may or may not have significant social value. This distinction was consciously employed by many writers during the Tokugawa period. For example, Seki Takakazu (Seki Kowa), often called the *sansei* or 'sacred mathematician' for creating the dominant *wasan* paradigm, wrote on a student's diploma in 1704, as a way of conferring legitimacy on his field: 'Mathematics, after all, is more than an art form.'[35] He thus refused to define mathematics as an art and tried, instead, to establish it as a dignified, prestigious form of scholarship. He even referred to mathematicians as 'scholars'.[36] Among themselves mathematicians acted as if they were pursuing an art form, but in the presence of nonmathematicians they tried to present their work in such a way as to give it prestige. Some kind of legitimation was necessary for mathematicians to succeed in this effort. While a leisure activity does not require public legitimation, scholarship does; and there had to be some basis on which to differentiate the one from the other. The evidence of this concern for social legitimation is best seen in Seki's introductions.

Introductions to mathematical treatises usually did not reflect the authors' personal views, since they were written by Confucian scholars according to a fixed, decorative formula in order to partake of the prestige of Confucianism. Confucian scholars were asked to insert hackneyed phrases into mathematical texts even when they confessed ignorance of the subject. One of these phrases declared that mathematics had been mentioned in the *Chou Li* as one of the six classical arts. Another alluded to its association with divination and numerology (esoteric doctrines about the nature of the universe). Mathematicians who wrote the introductions themselves said the same things in other words. These introductions, however, had no connection whatever with the highly technical matters discussed in the main text and were essentially empty, formalistic passages.

The basis for the contention that contemporary people should respect mathematics was its high status in antiquity, stemming from Confucius' esteem for it, and from its designation by the sages as one of the six classical arts. This Confucian form of legitimation derived from the conceptions of classical scholars. Tokugawa mathematicians, however, generally were socially marginal curiosity-seekers and thus did not care whether mathematics had been one of the six classical arts. During the more than 2,000 years since the time of Confucius, Confucianism had become securely established; but astrology, mathematical astronomy, medicine and arithmetic had come to be considered lesser crafts and were consigned to a peripheral, low status in the hierarchy of disciplines. Arithmetic, which was associated with such mundane matters as surveying and tax collecting, was assigned a status well below that of astrology and even further below that of mathematical astronomy, which described the principles governing the heavens and earth. Even so, unlike chemistry or other sciences, arithmetic had a guaranteed position in the Chinese bureaucratic system; and in Japan's prefeudal period there were doctors of mathematics and official arithmeticians.

During the Tokugawa period, however, mathematics was not formally recognized in the governmental structure. Astronomy had its hereditary doctors in the Tsuchimikado family of Kyoto; the more competent astronomers in the shogun's government were officially, if nominally, subordinate to them. The Tokugawa mathematical tradition, however, existed entirely in the private sector and had no link to the prefeudal tradition of the mathematical doctors. In fact, Tokugawa mathematicians had no interest in the tradition of court mathematics.[37] From the introductions to their writings one perceives instead a recognition of Seki Kowa as forebear or perhaps of Mori Shigeyoshi or Yoshida Koyu. Historical awareness of founding fathers, in other words, did not antedate the Tokugawa period. Consequently, Japanese mathematicians, unlike the school of such Ch'ing mathematicians as Mei Wen-ting, did not try to use the ancient designation that made arithmetic one of the six classical arts as a basis for defining their own identity.

Belief of the Pythagorean type that numbers permeate all objects in space, or constitute the basic principle of the cosmos, was certainly part of the Chinese mathematical tradition; it was specifically the creed of Chinese diviners and specialists in yin-yang cosmology. Even Kawakita Chorin, a mathematician at the end of the Tokugawa period, wrote: 'Numbers constitute the elements of the heavens, the earth and all of nature. Everything that happens is a result of their presence.'[38] This kind of belief was often used to justify the activities of mathematicians; but such an argument was more commonly espoused by cosmologists, astronomers and specialists in calendrical science than by mathematicians themselves. An attempt like the Pythagorean to explain the cosmos by numerical cycles and the cyclical world view as such existed in the Chinese tradition into the second century A.D.; but as astronomical observation became more precise, cycles came to be described algebraically in more complicated ways, and this cosmological view collapsed.[39] Experts in calendrical science in the T'ang era (seventh through ninth centuries) who became specialists in exact, empirical science did not consider expatiating on the cosmos to be a legitimate part of their work and they rejected cosmologizing altogether.[40] In the eleventh and twelfth centuries, cosmology once again became a subject of discussion among Confucian scholars; but the calendar makers were inclined to think that Chu Hsi 'talked nonsense because he did not understand mathematics' and generally refused to consider the problem.

During the earlier period Japanese mathematicians, especially their founding father, Seki Kowa, frequently studied problems in calendrical science. Seki's investigations, however, were confined to the technical aspects of the Shou-shih calendar. As scientific mathematical activity of a kind that might be called 'normal science' continued, the classical Pythagorean view of nature as based on mathematical principles disappeared from the *problematique* of Seki's followers. This was at a time when the essayist Nishimura Tosato was claiming that *sugaku* (the scholarly study of numbers) was an important subject while *san* (arithmetic) was a minor practical art.[41] Propounding a mathematical view of nature, he designated *sugaku* as a learned discipline that 'investigates basic principles, constitutes a major element of divination and was expounded by the Sages'. Nishimura also criticized the activities of the *wasan* mathematicians as 'arithmetic done by people of little consequence'. This mathematical view of nature had to conform to the Confucian values of a Confucian society. Consequently, mathematical studies were not considered scholarly unless they made a contribution to 'self-cultivation, husbandry and the pacification of society'. They were devalued if the practitioner admitted a 'desire to investigate mathematical principles merely for amusement'.[42]

It is said that mathematicians such as Wada Yasushi made a living by practising divination during the Tokugawa period, but this report may be based on a popular misconception deriving from the fact that diviners and mathematicians both used *sangi* (computing rods). In fact, mathematical calculations and divination based on calculating rods were entirely different.

Mathematics during the prefeudal period was completely practical and thus was at least socially legitimate. Even in the Tokugawa period appeals to practicality appear in introductions to mathematical works designed for such purposes as surveying, and these appeals apparently were accepted at face value. Practical mathematics, however, did not interest most mathematicians. *Wasan* lost even its practical character and became explicitly nonutilitarian, a situation that readily allowed Nishimura Tosato to dismiss it as simply an art form.

Let us compare this experience with that of the West. According to Alexander Koyré, Platonism had an important influence on early modern science because of its programme for the mathematicizing of nature.[43] Although Platonism dominated intellectual discourse during the Renaissance, its indispensability for the Galilean school may well be doubted. A close reading of Plato's *Timaeus* shows that it had almost nothing in common with Koyré's notion of Platonism.[44] Galileo did, however, use popular Platonist ideas to legitimize his mathematical methods, an effort that had precisely the social effect intended. Thus Galileo presented Plato as the defender of geometry against the Aristotelians, who emphasized logic; and thus Platonism had an important role justifying the inclusion of mathematics in school curricula, and eventually in the emergence of modern science.

Unfortunately, Japanese mathematicians had no charismatic figure to invoke in opposing the Confucian tradition. If Mo-tzu had not been forgotten for two millenia, this might have been possible; but Mo-tzu and the yin-yang natural philosophers had almost totally disappeared from intellectual discourse in Tokugawa Japan. The mathematicians would have needed something in their own tradition approximating Galileo's Platonism to have become full members of intellectual society. No reiteration of references to divination or the six classical arts in the introductions to their books could have raised their status significantly.

Social Position of Tokugawa Mathematicians

Whenever scholars demand legitimacy from society, they display a sense of mission that reinforces their commitment. This sense of mission is associated with the rise of professions that in Western society are intellectually based associations not explicitly connected with worldly gain. Theology, law and medicine were recognized as professions in medieval universities. In the early modern period, scientific researchers and technicians were also recognized as professional men. The largest profession in Tokugawa Japan was probably the class of Confucian scholars. One might also consider physicians, astronomers and specialists in calendar making as professionals. But whether the mathematicians could be called professional is a difficult question.

A sense of mission includes a desire to achieve a lofty objective beyond immediate personal interest. Scholarship or science for their own sakes seemingly represent an early modern form of consciousness that developed after the emergence of scholarly elitism, especially that of the universities. Before this modern attitude became established, a scholarly discipline could form only when learning proved to be useful to the sacred or secular establishment that monopolized universes of meaning.

But as society became more complicated, intellectual groups managed to secure autonomy as third parties between various powerful agencies and were expected to ignore matters of direct economic interest. Since the Middle Ages, universities have proclaimed their independence as professional bodies, meanwhile maneuvring between religious and secular authorities. Even in Tokugawa Japan, with its very strict regulations, officially sponsored Confucian schools, and even heterodox Confucian schools were generally recognized as authentic disciplines by the establishment. Nor was there any problem with medicine, even for doctors in the private sector, because practical utility assured recognition. Fields of knowledge related to commercial production or manufacturing were invariably supported on their own terms. It was difficult, however, to discover in either Confucianism or in Baconian pragmatism any grounds on which to legitimate the activities of the *wasan* mathematicians.

The popular image of mathematics was that of the abacus. The social position of mathematicians was probably based on demand for their services in teaching people how to use this device. But daily use of the abacus did not require anything like the elaborate technique of the mathematicians. In merchant families, abacus manipulation was considered a form of spare-time study. Apprentices were introduced to it by the head of the business. There was a saying to the effect that 'While use of the abacus is one of the most important things a merchant must learn, he should not take it too seriously. Excessive study will hurt business.'[45] Studying more advanced mathematics than was required in business generally was forbidden, being considered a form of dissipation. Some mathematicians managed to make a living by opening schools. The majority of such schools were run by masterless samurai. The mathematical training they gave usually stopped with simple arithmetic and the calculation of interest rates. Thus, according to Professor Oya Shin'ichi, the *enri* calculus, which included the most sophisticated problems studied by the mathematicians, was not generally taught in these schools.

From the government's point of view, mathematics was closely associated with simple calculation and land surveying. But in the *Ryochi shinan* ('Introduction to Surveying') one finds such statements as 'People who study mensuration say that not all mathematics is intended to be used by surveyors. According to them, there is nothing about mathematicians' theories that is contradictory to mensuration; but if you look at their work, it seems too much involved in mathematical theory and divorced from practice. And in general mathematicians' talk about surveying is all of this sort.'[46] Or 'Mathematicians' techniques are a distraction, with no utility whatever.'[47] The traditions of the academic mathematician and of the practical surveyor were quite distinct. Because of an attack on the *Sampo jikata taisei* ('Manual of Practical Mathematics', 1837), written by Akita Giichi and edited by Hasegawa Kan'ei (both mathematicians of the Seki school), surveyors subject to feudal authority were reluctant to publish significant mathematical works for fear of their lords' reactions. Even Seki Kowa, founder of the *wasan* mathematical tradition, was warned about this. He disregarded the admonition, however, and later wrote a text describing approximate or simple methods for solving problems in the style of Yoshida Koyu's *Jinkoki*. Seki's disciples, however, considered that text a disgrace to their school.

Astronomy offered mathematicians far more sophisticated problems than did surveying. Examples from traditional Chinese astronomy include indeterminate procedures for calculating multiple conjunctions, the problem in spherical trigonometry of transforming equatorial coordinates into ecliptic coordinates on the sphere, and interpolation procedures for handling the equation of the centre. The trigonometry and algebra that came to Japan with the Jesuits' later transmission of European astronomy may have opened up new mathematical vistas. There also were problems in navigational astronomy that the mathematicians Honda Toshiaki and Sakabe Hirono investigated. These topics were, however, considered astronomical problems and, as such, were separated from the mainstream of mathematical activity.

In the *han* schools, which were for the training of the sons of samurai, mathematics apparently became part of the curriculum about the 1780s.[48] Textbooks published by various domains seem to have included problems in applied mathematics taken from surveying, calendrical science and navigational astronomy. Some mathematicians served as Bakufu astronomers, and others worked as accountants or surveyors in the domains. They saw their public duties as separate from their research in mathematics, however, considering the latter a private activity. They apparently honoured this distinction to the point of not publicizing their research.[49] Astronomers and astrologers employed by the domains all used the algorithms of *wasan* mathematicians when they calculated but, aside from the writings of Nishimura Tosato, one finds no significant public comment on the *wasan* tradition in works by astronomical specialists. This may have been due to their bureaucratic consciousness. They gave no thought to the work of mathematicians, who belonged to the private sector. The result was that mathematics had no real place either in the government or in private life.

Nor did mathematical research have any base in the occupational system. Mathematicians' researches were separate from their occupations – which ran the gamut from warrior to farmer, artisan or merchant.[50] The orthodox Seki school was typical in this respect.

In general, the mathematicians' greatest achievements appeared in their early years, although Mikami Yoshio's research showed that the *wasan* mathematicians were most active in their later lives.[51] According to him, this was due to the shallowness of the roots of *wasan* mathematics in the educational system, the inadequacy of the available textbooks, and the extraordinary amount of time required to become proficient. All of these factors surely had an important impact, but an even more influential factor may have been that the time and money required to indulge in such a leisurely activity came only in later life. Anyone who gave up his regular occupation for mathematics encountered financial problems, including difficulties in paying publication costs. In particular, samurai employed in the government had scruples about participating in such activities and, for the most part, published mathematical works only after retirement.

Unlike astronomers or physicians, mathematicians did not have to deal with occupational inheritance. Because of the special ability required to produce original achievements in mathematics, the occupation could not be passed on – indeed, in the economic sense it was not worth passing on. Mathematics thus did not establish itself as a specialty of certain families. The status of *wasan* teacher had no economic implications (such as guild protection), despite the licensing system established by Seki Kowa. Nor, in consequence, was there any basis for giving academic autonomy to mathematics, such as the modern university has provided in the West.

The prestige of any field of scholarship is bound up with the social status of those engaged in it. From that point of view, there is little reason to think that mathematicians were particularly respected by society in general. In China, Chu Shih-chieh and Ch'eng Ta-wei, authors, respectively, of the *Suan-hsüeh ch'i-meng* and the *Suan-fa t'ung-tsung*, from which *wasan* mathematics derived, were itinerant teachers. In Japan there also were mathematicians who travelled from place to place and were supported by wealthy patrons.[52] In point of livelihood, these intellectual salesmen were no different from travelling artists.

The gentlemen members of the Royal Society were not professionals in the sense of earning their livings by research but, rather, in the sense that royal patronage gave them a degree of institutional independence. In a formative period in either scholarship or the arts, when institutional protection is lacking, a leader will assemble a supporting constituency and create an occupational base for his activity. When scholarship becomes systematized, institutionalization and professionalization occur simultaneously. The Royal Society, which was officially recognized yet lacked an occupational base, was exceptional; but its recognition made its dignity and status as a scholarly organization secure. *Wasan* mathematicians had no such recognition. There was no system for training or recruitment. The occupational base consisted solely of the inadequate patronage of a few wealthy individuals. As a result, there was no agency to press the claims of mathematics upon society. Because their organizations lacked status, *wasan* mathematicians could not defend the scholarly aspects of their work effectively. They received little or no social recognition. Nevertheless, in certain respects, they were more active than the Royal Society.

The artistic character of *wasan* mathematics helps to explain this fact. In publishers' catalogues of the Tokugawa period, books on *wasan* were classified with materials on the tea ceremony and on flower arranging, which suggests that *wasan* was mainly viewed as a popular art. Art, however, involves the pursuit of aesthetic pleasure, while purely intellectual matters are regarded as scholarly. During the Tokugawa period there were several pursuits that were not scholarly occupations and had no academic prestige – for example, Japanese chess (*go* and *shogi*). It was probably the aesthetic and playful aspect of mathematics which provided scope for the development of *wasan* and its diffusion down to the commoners.

In his book *Les jeux et les hommes* (published in English as *Man, Play and Games*), Roger Caillois defines recreation as activity that is free, isolated, indeterminate, unproductive, rule-bound and unrealistic. The activities of *wasan* patrons fit these six conditions perfectly.

Internal Logic of Wasan Development

What distinguished *wasan* from poetry, haiku and the arts in general? In what sense was it a scholarly rather than an artistic form? While it would not be quite accurate to say that the methodology of *wasan* was that of a modern scientific discipline, it definitely did come closer than any other field of inquiry existing during the Tokugawa period. It had, for example, a way of asking questions that was very similar to that of modern science. Thomas Kuhn states that all scientific traditions begin with a paradigm or model for raising and answering questions in terms of which scientific progress will necessarily occur.[53] *Wasan* conformed to this pattern rather well. It did not command a very extensive body of knowledge, but it stated its problems and questions in a precise way. Among the disciplines of the Tokugawa period, *wasan* and mathematical astronomy were methodologically closer to modern exact sciences than were the schools of moral philosophy or of clinical medicine. Since the questions raised in *wasan* were not limited by traditional structures (save those it evolved itself), it was free to move off in new directions.

In its formative period *wasan* mathematics developed the unique custom of *idai* (bequeathed problems). A mathematician would pose scores of problems of several kinds at the end of a book. Another mathematician would publish answers to these problems and present his own in the same manner. According to convention a third mathematician might propose answers to the second set of problems and issue his own, in relay fashion. This interest in mathematical puzzles greatly stimulated the formation of *wasan* groups. They were influenced by Chinese experience, but there does not appear to have been anything like the custom of posing *idai* in China. The tradition began with twelve problems from the *Shimpen jinkoki* (1641). A succession of mathematical lineages soon developed and reached a peak during the lifetime of Seki Kowa. Certain themes continued to appear in these problems, which passed through a number of phases. Practically all of the important problems in the history of *wasan* date from the period of these *idai*.[54]

Mathematics has a strong puzzle-solving character. Problems need not be constrained by physical or social reality. Mathematicians can freely create new intellectual worlds which may violate the logical forms of daily language.

During the early period there were two types of *idai* – problems used in daily computation and those relating to geometry. In the Chinese mathematical tradition that influenced *wasan*, practical problems were more common, but later more whimsical problems were added. The *Jinkoki*, which defined the popular image of *wasan*, also emphasized practical problems. As *idai* passed through many generations, a trend towards purely intellectual or recreational problems developed among the heirs to the tradition, enthusiastic puzzle-solvers uninhibited by utilitarian constraints. The puzzlelike character of *wasan* mathematics made this development entirely predictable. Given the sense of *problematique*, practitioners were not inclined to select problems with practical applications. Moreover, other factors reinforced the trend. Once a problem had been abstracted in the form of a diagram, people did not concern themselves with its utility and could enlarge or develop it freely. The purer the mathematical character of a problem and the greater its detachment from practicality, the greater was the enthusiasm with which it was received. We shall consider below a typical example, the *yojutsu*, which involved fitting various large and small circles into a triangle.

It also seems significant that the impracticality of *wasan* precluded links to mechanics or optics like those of mathematics in Western science. The seventeenth-century *idai* were probably responsible for the leisure-oriented character of *wasan*.[55]

Commerce, surveying and calendar making represented practical applications of mathematics. The first two, however, did not offer very sophisticated problems. The degree of precision required in their calculations was much less than that demanded in the theory of errors or in higher-degree equations. Calendar making, of course, was an exact science; but the major problems presented by the Shou-shih calendar had already been solved by such eminent mathematicians as Seki Kowa and Takebe Katahiro. In theory, the planetary motions should have generated problems fully as interesting as the theory of epicycles in Western astronomy. The Japanese art did not emphasize planetary movements, so they were not investigated very thoroughly. Ultimately the absence of kinematic and dynamic problems in Japan's scientific tradition handicapped and retarded *wasan*'s approach to analysis and proved decisive in its race with the Western tradition.

When *idai* were popular, from about 1650 to the early eighteenth century, the natural sciences did not develop to any extent. Science from the West had not yet been imported. It was during this period and from these *idai* that *wasan* mathematics created and established its significant problems, although in certain respects prematurely. The topics investigated were taken from such concrete problems as the volume of a rice bag or measuring cup. They were often focused on diagrams of circles or cones that had to be solved for a numerical value. As this trend progressed, the enthusiasts of pure mathematics gave little thought to the relationship between observation and its practical meaning; they simply invented fictitious problems whenever they wished. Once they discovered a strategic or paradigmatic problem, practical problems of astronomy, trigonometry or Western logarithms were no longer considered legitimate. The eccentric genius Kurushima Yoshihiro wrote: 'In mathematics it is more difficult to raise a problem than to give the answer. Only mathematicians who cannot invent problems borrow them from calendrical science.'[56] In short, the idea that looking for subject matter in society or nature was undesirable had already developed by the eighteenth century, when the notion of mathematics for its own sake emerged.

In the practical problems of applied mathematics, it is important to obtain an answer stated as a numerical value. Given the determination of topics in this field by social or natural conditions, it is entirely appropriate that obtaining numerical solutions should be considered more important than inventing problems. On the other hand, the asking of questions is essentially unlimited in pure mathematics, as Kurushima implied, and is necessarily considered supremely important. As with crossword puzzles, inventing the question is more difficult than supplying the answer. That one person paves the way by inventing a problem, while many others follow in trying to solve it, is part of the normal science tradition and further underscores the process by which normal science is conducted.

This was not, however, the way in which *idai* were developed and passed on in the early period (before Seki Kowa's time). The form of the problems at that time was not fixed; and when people invented problems, they did not simply adhere to those devised by predecessors. During that time there was a shift from practical to pure mathematical problems. Numerical solutions converged towards diagrammatic problems using the *tengen* algebra (based on the use of computing rods). *Wasan* did not yet have a characteristic type of problem.

As time passed, the problems became more intricate and multifaceted. Inventors did not simply present problems that they themselves had already solved, since other mathematicians considered that too simple – even foolish. There was an emphasis on solving problems by some unusual means or in presenting problems for which it was not known whether a satisfactory solution existed. People would take up a mathematical challenge and expend considerable energy trying to solve it. The way in which *idai* were presented gave an enormous stimulus to the competitive spirit of later mathematicians. Difficult problems constituted an enduring challenge. There have been, of course, similar examples in the history of Western mathematics – for instance, the unsolved problem of trisecting an angle, which has continued to engage the interest of mathematicians.

As problems became more complex, however, impossible problems appeared, and mathematicians began expending excessive energy to little effect. In the *Sampo kokon tsuran* the following comment appears: 'The shallowness of present technique and reasoning, together with the enormously complicated effort expended, lead one to think that the really interesting problems have all been exhausted.' Such problems could not constitute paradigms, could not lay the groundwork for normal science, and could do nothing but create confusion. Indeed, the confusion suggests that Japanese mathematics was at a preparadigm stage. Consequently, what brought Sekí Kowa his enormous reputation was his creation of the basic paradigm for posing and answering questions in an intellectual setting where almost total confusion had prevailed earlier.

After the importation of the *tengen* technique, changes occurred in the way questions were posed through the *idai*. In the earlier period, *idai* were highly diverse and multifaceted; now they emphasized higher-degree equations solved through the use of the computing rods. The confusion of the period was compounded by the popularity of *idai* deliberately designed to solve increasingly complicated problems by transforming systems of simultaneous equations into a single equation of increasingly higher degree. These problems were known as *handai* (troublesome problems). The tendency may have been an aberration, but it prompted Seki Kowa to introduce an important innovation – the *tenzan* algebra, a system for expressing unknowns in symbols similar to *A, B, C*, in order to use simultaneous equations freely.[57]

Seki also developed a theory of equations based on the existence of negative and imaginary roots, which he treated in his *Daijutsu bengi* ('Discussion of Problem Specification') and *Byodai meichi* ('Clarification of Impossible Problems').[58] From China's pragmatic mathematical tradition *wasan* had inherited the idea that an equation can have only one root.[59] Seki, however, introduced discussion of negative and imaginary roots, and tried to interpret their meaning. Taking an approach characteristic of *wasan*, he transformed and 'corrected' such problems to provide real positive solutions, rejecting the implications of the original problem setting. Thus his theory of equations ruled out further development towards imaginary and complex number theory in *wasan*.[60]

Seki's writings on the theory of equations perpetuated *wasan*'s orthodox way of asking questions. These writings were included in a seven-volume work transmitted esoterically by the Seki school. They were studied and passed on by pupils who were eminent enough to devise new problems themselves. Since Seki had previously laid down guidelines for solutions in the *tenzan* algebra, one may say that he fully paved the way for *wasan*'s later development.

Seki Kowa was not the only outstanding mathematician of the period. He certainly had the intuition of a genius; but if he had been an isolated figure too far ahead of his time, the paradigm he laid down would not have paved the way for normal science and there would have been no Seki school. During Seki's lifetime his school was extended by such eminent disciples as Takebe Katahiro and Matsunaga Ryohitsu. Seki was placed on a pedestal within the tradition. One may suppose that the special treatment accorded him was instrumental in establishing the diploma system. In the introductions to *wasan* books, the origin of the discipline is always attributed to Seki, not only in his own school but also in the Saijo school of Aida Yasuaki. Certainly the tendency of the *wasan* mathematicians to revere Seki limited their field of vision, yet it would seem from comparing such works as the *Suan hsüeh ch'i-meng* and the *T'ienwen ta-ch'ang kuan-k'uei chi-yao* with his own that Seki himself was more influenced and informed by Chinese writings than his later disciples ever imagined. Everything in the *wasan* tradition was, however, referred back to Seki, and the original Chinese mathematical works were ignored.[61] This was very different from the attitudes of Chinese mathematicians at the Ch'ing period, who turned their attention towards ancient Chinese mathematics as they mastered the Jesuits' astronomy.

Seki Kowa left answers to a large number of *idai* during the period of their greatest popularity. In fact, the theory of equations seems to have been born from the many different kinds of *idai* that he considered. There is no indication, however, that Seki himself left behind any *idai*; and after his time the practice of issuing them declined considerably. Their frequent appearance corresponded to a preparadigm stage in the delineation of mathematical problems. *Idai* were scrutinized and restated according to the principles of equation theory, and the significant ones were passed on. The paradigm emerged when the techniques for solving these problems by means of *tenzan* algebra were given; and at that point a way of asking and answering mathematical questions emerged that was quite different from what had existed in the preparadigm period. Creation of the paradigm was not the achievement of Seki Kowa alone; but circumstances led to his being given credit for it and he was, in fact, at the centre of the *wasan* tradition. Since the procedures for raising and answering questions were fixed, it was meaningless to set forth a large number of *idai*. With the issues about structure clarified, subsequent mathematicians readily resolved particular problems. *Enri* (circle theory) calculus was one of the areas that attracted attention. This technique developed from mensuration of the circle and led to the development of linear progressions and analysis. It probably began with either Seki or his leading disciple, Takebe Katahiro.[62] In any event, its inclination to the analytical approach of Ajima Naonobu is apparent. *Enri* calculus was applied not only to problems involving the circle, but also to curves and curved surfaces in general. Wada Yasushi helped to develop mathematical analysis by compiling tables of definite integrals and applying them to the mathematically infinite and infinitesimal. The Takuma school of Osaka developed a calculus that in some respects was superior to that of the Seki school.

Mathematical problems are not limited to those posed in nature or by society. But a pattern of development in mathematics will change according to the kinds of problems taken up; different choices of problems may create different mathematical worlds. The *enri* calculus, which developed from problems concerning the arc, an important topic in astronomy, coincided in its results with the Western-style calculus. The course it followed, however, was completely different from that of Western mathematics, which began with problems in dynamics. We should say, on the other hand, that dynamical problems offered mathematics greater scope for development than those concerned with arcs and circles, simply because the element of time was involved. There were greater limits to the problem development possible in *enri*.

Another stimulus to the development of *wasan* was the *ema sangaku* (pictured mathematical tablet) form that came after the *idai* tradition. On these wooden plaques were written both problems and answers; they were offered at shrines and displayed there. The best mathematicians made their accomplishments known through books, but it was largely the custom of *sangaku* that supported the activities of the *wasan* enthusiasts. That *wasan* was a hobby costing money to pursue is best shown by the elegant diagrams that embellished such work. In fact, the offering of *sangaku* had a strong attraction for local gentlemen over other art forms – drama, music, poetry – as a way of making their work known; and their frequent indulgence in it shows a desire to keep themselves more or less before the public at all times.

The most important kind of problem taken up in *sangaku* was called *yojutsu* (packing problems), which involved the attempt to inscribe the largest possible number of small circles in a larger circle. *Sangaku* also treated problems involving solar eclipse predictions in the Shou-shih calendar; but table of eclipses, with their arrangements of letters and numbers, could not attract attention even if displayed as framed pictures. Popular interest was limited to *yojutsu* decorated with designs of circles and squares. The pictures were intended, as far as possible, to produce the impression that the designer had obtained the solution to a difficult problem through a complex diagram. Thus, while *yojutsu* seemed to concern itself with very difficult problems, its emergence was not very significant mathematically.

In the early development of *yojutsu*, there was a tendency for different techniques to compete on the same problem. Later, however, the method of solving them became fixed. There was a tendency to devise new problems to which the standard technique could be applied. Problems became increasingly complicated while the technique scarcely developed.

Enri calculus and *yojutsu* both developed as normal sciences but in somewhat different ways. The *enri* calculus developed step by step: when one problem was solved, its result was used to solve a problem requiring deeper investigation; and in the process there appeared what might be called subparadigms. In principle *yojutsu* was also a problem-solving technique, but its pattern of development amounted to merely a series of transformations and variations. Moreover, its technique was not based on a demonstrational logic in the Euclidean style, and did not aspire to general principles of problem solving. During its 200-year history, *yojutsu* yielded several useful by-products, but its reliance on casual inspiration ranks it well below *enri* calculus in scholarly value.[63] One could have referred to *yojutsu* as a Japanese form of geometry only if it had gone beyond mere puzzle solving and had achieved a general methodology. In fact, it did not move very far towards analytical geometry. Because it was purely a puzzle-solving technique, it did become an acceptable recreation for amateurs, who were inclined to consider logical rigour in bad taste.

Wasan mathematicians apparently did not fully realize the importance of logical foundations; rather they valued insignificant, complicated and overly elaborate problems.[64] When Euclid's *Elements* first appeared in Japan, people said the simplistic, poorly developed and inferior character of European mathematics compared with *wasan* could be determined just from looking at its pictures – a reaction suggesting that *wasan* was not the kind of discipline to raise basic questions. One might even call it an art form that had as its major goal the refining of trivialities. On rare occasions someone like Takebe Katahiro, who respected precision, would come along; but because the *wasan* mathematicians did not generally raise basic questions, they made no revolutionary breakthroughs and confined themselves to refinements and improvements on the paradigm set forth in Seki Takakazu's time.

Mathematics, especially a pure form like *wasan*, differs from the natural and social sciences in the absence of checks imposed on it by the objects it investigates. It does not follow the same development as physics, in which the interpretation of a phenomenon can change completely during a scientific revolution. Non-Euclidean geometries can coexist with Euclidean geometry, and the replacement of the latter by the former is far from inevitable. Non-Euclidean geometries are not so much replacements as variations on Euclidean geometry. But unless one counts such trivialities as different assortments of diagrams in *yojutsu* problems, one would have to say that *wasan* had very few basic variations and, what is more, that almost none of its basic notions were conceptually deep.

Perhaps the conceptual poverty of *wasan* can be explained by the derivative character of the culture in which it developed. Most of Japan's basic cultural patterns were of foreign origin. Highly refined art forms evolved from these patterns, but the fact that they were borrowed precluded critical examination of them. By contrast, the Chinese were forced to reconsider their mathematical heritage during the intellectual crisis precipitated when the Jesuits introduced European exact sciences. The Chinese approach was characteristic of a society with strong autochthonous values; scholars used the new learning to resuscitate forgotten mathematical conceptions from China's past, and thus preserve the identity of traditional science. In Japan, *wasan* mathematicians raised neither the question of historical origin nor any other involving the theoretical foundations of mathematics.

Given that Japanese astronomers and physicians constantly compared China and Europe and took from either what they judged to be good, why did the mathematicians remain in their own world? In the first place, there was a difference in disciplinary structure between medicine or the natural sciences and mathematics. Practitioners of the former could readily determine what was better and what was worse by having both Eastern and Western examples before them. In natural science the search for a single truth could more or less readily lead to replacement of inferior Chinese conceptions by Western ones. In mathematics the belief that *wasan* and Western mathematics differed only in style allowed the two to exist side by side. This difference was rather like that between Japanese *shogi* and European chess. In this sense *wasan* and other forms of pure mathematics are closer to art forms than scholarship. The *wasan* mathematicians did not feel threatened by the importation of Western mathematics and remained in their own artistic world. Comparing the applied mathematics of China, which emphasized astronomical orientation, with the pure mathematical tradition of *wasan*, one perceives a clear difference in the extent to which practitioners of each considered the Western impact threatening.

According to the *Tokyo-fu kaigaku meisai sho* ('Survey of Schools in Tokyo'), a report on private academies in Tokyo prepared for the inauguration of the early Meiji school system, Western-style mathematicians trained at the Survey Office or Nagasaki's Naval Training Centre, and the more numerous *wasan* mathematicians of the Seki tradition belonged to sharply differentiated groups.

Contact or movement between the two groups seems to have been quite difficult. Modern Japanese mathematics really began when *wasan* mathematicians were superseded by those who had adopted Western styles. It would seem that differences in the ability of astronomers, physicians and mathematicians to restructure their disciplines and respond to foreign knowledge is explained primarily by the fact that the first two groups were professions and the last had no recognized occupational base. Occupational concerns of the first two groups made some awareness of Dutch studies inevitable.

It seems significant that *wasan* mathematics persisted longest in the north-east, where contacts with Europe and general cultural development lagged furthest.[65] During the Meiji period, Western mathematics came to dominate in the cities, following the establishment of the modern school system, and *wasan* was essentially banished to the countryside. It became, so to speak, an exotic flower blooming by the roadside of civilization.

Japanese Mathematics and the Pure Mathematics of the West

Could early modern mathematics in the West also be described as an art form? After all, people no longer accept the Pythagorean notion that cosmic mysteries can be discovered in the nature of numbers, nor do they believe that divine attributes can be deduced from the transcendental axioms of Euclidean plane geometry. J. W. N. Sullivan, author of *A History of Mathematics in Europe* (1925), asks why mathematicians enjoy greater social esteem than chess players, given that they no longer claim to be pursuing a single absolute truth.[66] The artistic spirit of *wasan*, moreover, seems to embody the essential ethos of mathematics; and one suspects that Platonism gave mathematics an excessively authoritarian aura that may not be essential to its nature.

The European public apparently viewed mathematics as the handmaiden of science. This handmaiden proved to be more competent in the West than did its Chinese or Japanese counterpart. It helped bring about the seventeenth-century scientific revolution and constituted one of the key-stones of the mechanical view of nature. The common image of the mathematical practitioner in that period was apparently that of a man who makes a living by producing and selling calendars or maps, or by conducting land surveys. The names of such people seldom appear in the histories of mathematics – in contrast with those who enjoyed royal patronage. Thomas Wright, a notable figure in the history of cosmological theory, was such a man. He came from a tradition separate from academic mathematics, and is thus excluded from histories of scientific astronomy.

Academic mathematics, however, retained its links with science well into the eighteenth century. *Wasan*'s weaning from science, by contrast, occurred quite early. In the eighteenth century, applied mathematicians sought numerical solutions. The more academic applied mathematics became, however, the greater was its tendency to seek elegant solutions of differential equations in preference to crude numerical calculations. This trend seems to have been somewhat similar to the dominant mentality in *yojutsu*.

Since applied mathematics dealt with issues posed in nature or society, numerical solutions were essential. In contrast, because *wasan* problems were recreational in character, practical issues and numerical values were beside the point. Nevertheless it remained characteristic of *yojutsu* that solutions were expressed concretely.

During the nineteenth century, pure mathematics in Europe tried to detach itself from various practical applications. Thus, even if numerical solutions were obtained, they remained nonessential and had little meaning. To this extent mathematics inevitably became more abstract. By contrast, *wasan* could be described as a form of pure but not abstract mathematics. The kind of numerical answers sought by *wasan* had no practical significance. Numerical solutions were sought only to determine the winner of what amounted to a sporting competition.

After the seventeenth century, the abandonment of Archimedean mathematical rigour in the West opened the way for the development of a normal scientific tradition in the application of calculus to mechanics.[67] In the nineteenth century, through a search for precision and logical rigour, mathematics became independent of science for the first time. Mathematicians of that era were very proud of this development and sought the origins of their discipline not in the Renaissance or the seventeenth century but in classical Greece. They even styled themselves the Greeks' successors.[68] Thus Greece was revived as the fount of 'modern' mathematics.

In this context the differences between the Euclidean tradition and that which developed from the *Jinkoki* would seem to be clear. Western mathematics was moving towards greater reliance on generalization and rigour, while *wasan* emphasized the solution of puzzles by finely honed intuition. Although *wasan* won its independence before Western mathematics arrived, it did so in a rather problematical manner. *Wasan*'s independence was merely separation from science and practical application. Separate though it was, computational mathematics did not extricate itself from the form of mathematics that seeks numerical solutions to problems. So long as the major emphasis was placed on quantitative calculation, questions about the quality of the underlying theory were difficult to raise.

Nor was Western mathematics prompted to raise basic questions through a concern with such quantitative problems as how to obtain an approximate decimal fraction. Potentially revolutionary developments were hidden in problems of quality. A new paradigm finally did separate itself from, and became independent of, science by returning to qualitative fundamental questions. From the fact of its independence or separation it developed the possibility of a new intellectual universe.

In trying to explain how this separation occurred, one must necessarily look to the institutional background. In the Western academic tradition, Platonism supported the recognition of mathematics as a legitimate and established subject. Thus, in the early modern period, calculus and analytic geometry, unlike other new disciplines, could be recognized immediately because there were posts for mathematicians in the university system.

In the history of science, the nineteenth century was a period during which the German university system occupied centre stage. Kant's *Der Streit der Fakultät* tells how the universities reconstituted, under the impact of modern science, an organizational structure that had existed since the Middle Ages. The newly constituted faculty of philosophy in particular acquired equal status with the traditional faculties of theology, law and medicine. As such, it was able to offer more than preparatory training for the three higher faculties. Elementary algebra and Euclidean plane geometry were introduced into the Gymnasium, while differential and integral calculus and analytic geometry were taught in the university. It is conceivable that mathematics might have had to serve as the exclusive handmaiden of such physical sciences as mechanics, physics and astronomy. In the nineteenth century there were disciplinary wars among the various specialized subjects, particularly over the constitution of chairs in the faculty of philosophy. Were mathematics to hold a subordinate position to physics, it might well be absorbed into the dominant disciplines. In the eyes of the public the value of a new field was recognized to the extent that it contributed to society; but in the universities, disciplines that raised fundamental questions had higher status than pragmatic ones, and were more appealing to students capable of responding to their professors' enthusiasms. Each speciality and every chair justified itself in this academic system.

For mathematics to become independent in the universities it first had to sever its ties to physics. Rigorously questioning the logical basis of mathematical assertions used by physicists elevated mathematics to be the queen of the sciences. A typical example is found in the career of Karl Weierstrass (1815–1877), preeminent figure in the German academic community and founder of the mathematics seminar at the University of Berlin. We do not know how far Weierstrass played academic politics as he developed his discipline. We may assume, however, that students who decided to prepare at the university for careers in mathematics were glad that he had enhanced the prestige of the mathematical profession.

Fundamental paradigms (such as that of Galois) saw their nature emerge outside the established line of development, and often outside the universities. They are eventually accepted there, developed there into disciplines, and follow the course of a normal science. All disciplines represented by chairs in the German universities were considered *Wissenschaften*. Even if a discipline had no stronger link to a more comprehensive system of values than that of the advancement of learning, its status, once it became part of the university system, was secure.

Since no tradition of university scholarship existed in Tokugawa Japan, it is unlikely that any of *Wasan*'s supporters would have rejoiced at (or even comprehended) the advent of a Weierstrass. There was no university environment to tolerate or even encourage questioning of fundamentals. As for the local devotees of *yojutsu* who patronized the Japanese mathematicians, they viewed the study as an embellishment, and were uninterested in close scrutiny of fundamentals.

As we have seen, Western mathematics, by precise and rigorous questioning of fundamentals, overcame its position as the handmaiden of science, and came to be widely considered the most basic and independent of disciplines. But Tokugawa mathematics merely separated itself from practical problem-solving and indulged in aesthetic pursuits. It would appear that the difference between European and Japanese mathematics is related to a difference in patronage. Through its recognition in a university system, the former shared state patronage redistributed according to academic criteria. The latter had only the direct patronage of individuals enthusiastic about an art form, with their own ideas about how their money was to be spent.

CONCLUSION: THE JAPANESE VIEW OF THE LAWS OF NATURE

Primarily through discussing Tokugawa practitioners, I have tried to describe the dominant view of science and of the laws of nature that existed during that period. What I have tried to argue in part is that aside from notions of law, the Japanese at that time had a conception of nature different from what we have today.

Our present conception of law in nature, stated in value-free terms, was produced in the nineteenth-century university. Earlier intellectual activity pursued much wider goals than the rigour of modern science allows. In that early intellectual world, modern concepts of the laws of nature were applied only in limited situations. Similarly, in Japan one readily discovers, from the tone of the introductions to books on calendrical science, medicine and mathematics, that the academic notion of science for its own sake did not exist in Tokugawa society.

In earlier times, science in the West was pursued on the assumption that its investigations would demonstrate the glory of God; in Japan the ideology of Tokugawa science derived from Confucian emphasis on individual moral cultivation and social harmony. Morality was the basis of law; laws of nature conformed to and were necessarily subordinate to it. Thus, astronomy and medicine in Japan ultimately had to subordinate themselves to an essentially Confucian set of priorities in order to guarantee respect for their status as disciplines. Pursuits like *wasan*, which diverged from moral values or had nothing to do with them, not only became isolated from the Confucian framework but remained a recreational art.

In its unconcern for moral values *wasan* was the discipline closest to modern science in the Tokugawa period. Takebe Katahiro consciously used the term 'law' in his methodological writings, saying: 'The establishment of laws provides the basis for technique; thus mathematics needs to have laws.'[69] Modern science's objectives consist in trying to establish or prove certain laws, not in defining the metaphysical realities that preoccupied scientists for centuries. Neither does modern science have such grandiose and far-reaching objectives as alchemy, which tried to prolong life indefinitely, or astrology, which sought to predict the course of nature and human affairs. It tries, rather, to achieve immediately foreseeable objectives. As Japan accepted modern science, astronomers, unlike Shibukawa Harumi, maintained a mechanistic framework, and avoided using historical changes in the celestial movements to explain changes in the heavens. In coping with disease, early modern physicians emphasized solidistic explanations rather than supposing that the human body is governed by temporal vicissitudes as doctors in the Chinese tradition had done. Attempts to prolong life indefinitely and to interpret celestial phenomena as portents were believed to impede technical progress, and thus were abandoned.

As science has restricted itself to universal and objective features of physical phenomena, the idea that values inhere in the physical world has been excluded; as the centre of activity shifted from one European state to another national tendencies have dropped out. These tendencies towards objectivity and universality would have been uncongenial to Tokugawa thinkers, to whom the ideals of modern Japanese scientists would make little sense.

Unlike Joseph Needham, I do not believe that the Chinese and Japanese scientific traditions were converging towards the same ends as early Western science. Perhaps because of Needham's great esteem for Chinese science, he stresses China's priorities in discovery and invention.[70] Some of these claims of priority are unquestionable, but when Needham compares and evaluates the contributions of China and the West, he uses proximity to today's standard of knowledge and technique as his criterion of value.

Today's standard, however indispensable to the working scientist, is too parochial to be useful in evaluating the past. It gives unwarranted normative value to knowledge much of which will shortly become outdated, and it accepts the research emphases of fields shaped and dominated by the educational, career and fiscal patterns of North American, Western European and Soviet institutions.

Treating today's understanding as more than transitory may seem to Needham unavoidable if he is to make China's great contributions intelligible to Westerners. Nevertheless there is ample room for doubt that, as Needham believes, Chinese and Western science had the same objectives.[71] Would Chinese science or that of Tokugawa Japan have developed in the same direction as that of Europe in the absence of influence from the latter? Needham can offer no convincing support, beyond a profession of faith, for his affirmative opinion. It seems to me more likely that East Asian and European science were diverging. The goals that were explicitly stated in China and Japan differed fundamentally from those articulated in the West, and the intellectual frameworks too were so different that the implicit goals we deduce from them also do not greatly resemble their Occidental counterparts. Many discoveries which seem to have been made on both sides of the world lose much of their similarity (especially similarity of significance) once they are examined carefully in context.

Japanese science developed within a framework of Confucian values. When Chinese culture was first accepted by Japan, science was included as part of the court culture and bureaucratic system. Japan was earlier practically a *tabula rasa* so far as the intellectual aspect of science was concerned. But by the time Western science entered Japan during the Tokugawa period, the ideological aspects of traditional Japanese science had been fortified by the recently developed Japanese Confucian system. To put it as simply as possible, the naturalistic and metaphysical aspects of Japanese science had been deemphasized, while its moral dimensions were stressed. The collision of Western science with traditional values was headed off by transforming the new science according to the dictum of Sakuma Shozan about 'Eastern morality, Western art forms.' The outcome in the early Meiji period was a highly pragmatic science, utilitarian and materialistic, quite lacking in moralistic connotations. Whether strictly speaking it was an art form was debatable, but its impact on the intellectual and spiritual transformation of Japan was delayed and minimized.

NOTES

1. At the end of the nineteenth century, Theodore Merz, in *A History of European Thought in the Nineteenth Century*, 4 vols. (London, 1903–1914), classified views of nature as astronomical, atomistic, mechanical, physical, morphological, genetic, vitalistic, psychophysical and statistical.
2. Shigeru Nakayama, *A History of Japanese Astronomy, Chinese Background and Western Impact* (Cambridge, Mass., 1969), 19.
3. Ibid., ch. 6.
4. Kawakatsu Yoshio, 'Shigaku ronshu' ('Treatise on Historiography'), in Asahi shinbun (*Chugoku bunmei sen* 12, 1973), 24–25.
5. Joseph Needham, *Science and Civilisation in China*, II (Cambridge, 1956), sec. 18.
6. Shigeru Nakayama, 'Educational Institutions and the Development of Scientific Thought in China and the West,' in *Japanese Studies in the History of Science*, no. 5 (1966), 172–179.
7. Shigeru Nakayama, *Rekishi toshite no gakumon* ('Academic Traditions'; Tokyo, 1974), 74–78.
8. I showed a table of contrasts between these two treatises in Nihon shisoshi taikei, *Kinsei kagaku shiso, ge* ('Japanese Thought [series], Modern Scientific Thought, II'; Tokyo, 1971).
9. Shigeru Nakayama, 'Accuracy of Pre-modern Determinations of Tropical Year Length,' in *Japanese Studies in the History of Science*, no. 2 (1963), 101–118.
10. Shigeru Nakayama, 'Cyclic Variation of Astronomical Parameters and the Revival of Trepidation in Japan', ibid., no. 3 (1964), 68–80.
11. Nathan Sivin, 'Cosmos and Computation in Early Chinese Mathematical Astronomy,' in *T'oung Pao*, **55** (1969), 1–73.
12. *Sorai sensei tomonsho* ('Queries and Answers of Master Sorai'; 1727).
13. Shigeru Nakayama, 'Edo jidai niokeru jusha no kagakukan' ('Confucian Views of Science During the Tokugawa Period'), in *Kagakusi kenkyu*, no. 72 (1964), 157–168.
14. The main source of information on this conflict is Koide Shuki, 'Rarande yakureki zenbun' ('Preface to the Translation of Lalande'), preserved in the Japan Academy.
15. Ibid.
16. Preserved in Kunaisho, Zushoryo.
17. Preserved in Naikaku Bunko.
18. Shigeru Nakayama, 'Diffusion of Copernicanism in Japan,' in *Studia Copernicana*, **5** (1972), 153–188.
19. 'Tengaku zatsuroku' ('Miscellaneous Records on Astronomy': n.d.), preserved in Naikaku Bunko.
20. Gregory Zilboorg, *A History of Medical Psychology* (New York, 1941).
21. Ogawa Teizo, 'Meiji zen Nihon kaibo gakushi' ('History of Anatomy in pre-Meiji Japan'), in *Meiji zen nihon igakushi* ('History of Medicine in pre-Meiji Japan'), I (Tokyo, 1955), 159–166. Also see his 'Kindai igaku no senku' ('Forerunners of Modern Medicine'), in *Nihon shiso taikei, yogaku, ge* ('Japanese Thought [series], Western Learning, II'; Iwanami, 1972), 506–509.
22. Joseph Needham, 'Science and China's Influence on the World,' in Raymond Dawson, ed., *The Legacy of China* (Oxford, 1964), p. 239.
23. Manfred Porkert, *The Theoretical Foundations of Chinese Medicine* (Cambridge, Mass., 1974).
24. Tsuruoki Genitsu, *Idan* ('Medical Critique'; 1795).
25. 'Shen Kua,' in *Dictionary of Scientific Biography*, XII, 369–393.
26. Ogawa Teizo, op. cit., 92 ff.; and Uchiyama Koichi, 'Nihon seiri gakushi' ('History of Physiology in Japan'), in *Meiji zen nihon igakushi*, II, 122 ff.
27. *Yakucho* ('Pharmacological Effects'), vol. II; repr. in Nihon shiso taikei, *Kinsei kagaku shiso, ge*, 256. Also see Otsuka Keisetsu, 'Kinsei zenki no igaku' ('Medicine in the Early Tokugawa Period'), ibid.
28. Shigeru Nakayama, 'Kindai kagaku to yogaku' ('Modern Science and Western Learning'), ibid., 460–461.
29. 'Scientific Productivity and Academic Organization in Nineteenth-Century Medicine,' in Bernard Barber and Walter Hirsch, eds., *The Sociology of Science* (New York, 1962), 305–343.
30. Amano Ikuo, 'Kindai Nihon ni okeru koto kyoiku to shakai ido' ('Higher Education and Social Mobility in Modern Japan'), in *Kyoiku shakaigaku kenkyu*, no. 24 (1969), 84.
31. Maruyama Kiyoyasu, 'Chiho ni okeru wasanka no shiso to seikatsu' ('Thoughts and Lives of Local Traditional Mathematicians'), in *Shiso*, no. 356 (Feb. 1954); repr. in *Kyodo shiso no bunkenshu*, no. 2 (1966), 13–31.
32. Robert G. Frank, 'Institutional Structure and Scientific Activity in the Early Royal Society,' in *XIVth International Congress of the History of Science, Proceedings* (1974), IV, 82–99.
33. Arnold Thackray, 'Natural Knowledge in Cultural Context: The Manchester Model,' in *American Historical Review*, 79, no. 3 (June 1974), 672–707.

34. For instance, Mikami Yoshio, *Bunkashijo yori mitaru Nihon no sugaku* ('Japanese Mathematics From the View-point of Cultural History'; Tokyo, 1947); and Ogura Kinnosuke, *Nihon no sugaku* ('Japanese Mathematics'; Iwanami, 1940).
35. Hirayama Akira, *Seki Takakazu* (Koseisha, Tokyo, 1959), 178–179.
36. For instance, *Daijutsu bengi* ('Discussions on Problems and Solutions'), reprinted in *Sekiryu sanpō shichibusho* ('The Seven Books of the Seki School'; Tokyo, 1907).
37. There is not much citation of traditional doctors of mathematics, but Aida Yasuaki wrote a few words on the Miyoshi and Kotsuki families in his preface to *Seiyo sanpō*.
38. Kawakita Chorin, *Ken'o sanpo* (Tokyo, 1863), preface.
39. Sivin, op. cit.
40. *Hsin T'ang shu* ('New History of the T'ang'), Li chih ('Treatise on Mathematical Astronomy'). Also see Yabuuchi Kiyoshi, *Zuitō rekihoshi no kenkyū* ('Researches in the History of Calendrical Science During the Sui and T'ang Periods'; Tokyo, 1944).
41. 'Sugaku, sangaku no hanashi' ('Topics on Mathematics and Arithmetic'), in *Sudo shodan*, I (1773).
42. Ibid.
43. Koyré's assertion that 'the allusions to Plato so numerous in the works of Galileo . . . are not superficial ornament born from his desire to conform to the literary mode of the Renaissance . . . nor to cloak himself against Aristotle in the authority of Plato. Quite the contrary, they are perfectly serious and must be taken at their face value' is groundless. Alexandre Koyré, 'Galileo and Plato,' in Philip Wiener and Aaron Noland, eds., *Roots of Scientific Thought* (New York, 1957), 174.
44. Shigeru Nakayama, 'Galileo and Newton's Problem of World-Formation,' in *Japanese Studies in the History of Science*, no. 1 (1962), 76–82.
45. Okumura Tsuneo, 'Kinsei shōnin no sanyo ishiki' ('Mathematical Concern of Modern Merchants'), in *Shuzankai*, **19** (Sept. 1969), 5.
46. Murai Masahiro, *Ryochi shinan, kohen* ('Introduction to Surveying'; II, 1754), preface.
47. Ibid., 4.
48. Kasai Sukeharu, *Kinsei hanko niokeru gakuto gakuha no kenkyu* ('Researches on Academic Tradition and Schools in Modern Fief Schools'), I (Tokyo, 1969), 9, 75.
49. Akabane Chizuru, 'Shomin no wasanka to hanshi no wasanka' ('Commoner Mathematicians vs. Samurai Mathematicians'), in *Kagakusi kenkyu*, no. 34 (1955), 25.
50. Hirayama Akira, 'Wasanka no shokugyo ('Occupations of Japanese Traditional Mathematicians'), in *Kyodo sugaku no bunkenshu*, no. 1 (1965), 192–198.
51. Mikami Yoshio. 'Sugakushijo yori mitaru nihonjin no dokuso noryoku' ('Japanese Creativity in the History of Mathematics'), in *Wasan kenkyu*, no. 11 (Oct. 1961), 7–8. Originally written in 1927.
52. Mikami Yoshio, 'Yureki sanka no jiseki ('Facts About Wandering Mathematicians'), in *Kyodo sugaku no bunkenshu*, no. 1 (1965), 166–184.
53. Thomas Kuhn, *Structure of Scientific Revolutions* (Chicago, 1962).
54. Hosoi So, *Wasan shisō no tokushitsu* ('Characteristics of Mathematical Thought in the *Wasan* Tradition'; Tokyo, 1941), 44.
55. For instance, Ogura Kinnosuke, *Nihon no sugaku* ('Japanese Mathematics'; Tokyo, 1940).
56. Ajima Naonobu, *Seiyo sanpō* (1779), postscript.
57. Hosoi, op. cit., 48–49.
58. Both repr. in *Sekiryu sanpō shichibusho* (see note 36).
59. Fujiwara Matsusaburo, *Nihon sugakushi yo* ('Epitome of Japanese Mathematics'; Tokyo, 1952), 127.
60. Hosoi, op. cit., 69–78.
61. Yabuuchi Kiyoshi, *Chugoku no sugaku* ('Chinese Mathematics'; Tokyo, 1974), criticizes the narrowness of Japanese mathematicians.
62. Mikami Yoshio, 'Enri no hatsumei ni kansuru ronsho ('Arguments on the Discovery of *Enri*'), in *Sugakushi kenkyu*, no. 47 (1970), 1–43, and no. 48 (1971), 1–42. Originally written in 1930.
63. Hosoi, *Wasan shiso*, 298–300.
64. *Yamaji Kunju sensei chawa* ('Table Talks of Master Yamaji Kunju'), repr. in *Rekizan shiryo*, 1 (1933).
65. Mikami, *Bunkashi* . . ., 77–80.
66. J. W. N. Sullivan, *A History of Mathematics in Europe* (London, 1925), intro.
67. Dirk Struik, *A Concise History of Mathematics* (New York, 1948).
68. Mori Tsuyoshi, *Sugaku no rekishi* ('A History of Mathematics'; Tokyo, 1970), 126.
69. Takebe Katahiro, 'Tetsujutsu' (1722), in Saegusa Hiroto, ed., *Nihon tetsugaku zensho*, VIII (Tokyo, 1936), 375.

70. Joseph Needham and Shigeru Nakayama, 'Chugoku no kagakushi o megutte' ('On the History of Chinese Science'), in *Gekkan Economist* (Oct. 1974), 86–87.

71. Joseph Needham, 'The Historian of Science as Ecumenical Man,' in S. Nakayama and N. Sivin, eds., *Chinese Science* (Cambridge, Mass., 1973), 1–8. Sivin has outlined an approach to Chinese science less bound to such assumptions in his preface to *Science and Technology in East Asia* (New York, 1977).

First published in *Fundamenta Scientiae*, Vol.2, No.1, 1981

14 The Future of Research – A Call for a 'Service Science'

Abstract – Viewed in terms of the operative mechanism, three types of scientific activity are distinguished: 'Academic science' is assessed by peer review, 'industrialized science' by higher level administrators and sponsors and 'service science' by local residents and the general public. The first one belongs to the past and second to the contemporary world; I try to clarify the function of the third kind which belongs to future.

Résumé – Nous distinguons trois types d'activité scientifique, considérés en termes opérationnels: la 'science académique' est évaluée par la critique qu'en font les pairs, la 'science industrialisée' par celle des administrateurs de haut niveau et des financeurs, et la 'science d'utilité publique' l'est par les communautés locales et le grand public. La première appartient au passé et la seconde au monde contemporain. Nous tentons ici de clarifier la fonction de la troisième qui appartient au futur.

Nearly a decade ago, Jerome Ravetz concluded his study of *Scientific Knowledge and its Social Problems* with a call for a 'critical science'.[1] Briefly stated, Ravetz's proposal was as follows. The years since the Second World War had brought major changes to the scientific world. What had formerly been an academic activity carried on in a relatively isolated university setting had now become so deeply involved with the military-industrial establishment that science itself could be said to have become 'industrialized'. Fuelled by abundant financial resources, this new-style 'industrialized science' moved swiftly from project to project, but in the absence of a satisfactory means of checking and guiding its advance, it had proceeded with little thought for the consequences of its research and its effect on human society. This had given rise to a situation in which science seemed at odds with society, engendered an anti-science atmosphere, and led to an erosion of public confidence in science that science could ill afford to ignore. To overcome this breach between science and society, Ravetz pointed to the need for a science that would take the side of the average citizen and review scientific research with an eye towards defending the interests and promoting the welfare of those likely to be affected by it.

Ravetz's proposal, however, drew surprisingly little response. This may be attributed in part to the fact that he was not very specific about who was to take the initiative in the creation of this 'critical science', about what strategies might be available for its purposes, and about how it was to be financed. But doubts about the efficacy of this critical science were also voiced by those who shared his basic concerns. Some wondered whether his approach could really succeed in curbing the runaway of industrialized science. Others suggested that the concept itself was too passive – that it would take a more positive and creative alternative to capture the imagination of youth and lure them away from the rewards of industrialized science. This proposal for a 'service science' represents an attempt to take account of and respond to this kind of criticism. What I intend by the term will, I hope, become clearer as I proceed, but to facilitate initial comparison with the more familiar academic and industrialized types of science I have prepared Table 14.1 below.

Table 14.1

	Academic science	*Industrialized science*	*Service science*
Assessors	Peer review	Sponsor	General public
Motive	Individual competition	Business competition International competition	Response to community needs
Referees and examiners	Fellow researchers	Higher level administrators Sponsors	Local residents General public
Rewards	Personal distinction	Promotion in the organization	Solidarity
Values	Pure scholarly interest	Fidelity to the organization (business principles)	Contribution to society
Evaluative criteria	Objectivity, universality	Practical use (business profits) (national prestige)	(Restoring science to society)
Form of presentation	Academic meetings Scholarly journals	Research reports Patents	Handbills General magazines
Promoting body	Elite unversities Professional associations	Industry research institutes National research institutes	Citizen movements Regional University Local research institute
Source of funds	Pocket money University budgets	Industry Government	Voluntary contribution Local government
Organizational configuration	Professorial chair	Centralization Integration	Decentrialization Dispersion
Research location	Laboratory	Research institute	Field
Spokesmen	Humboldt, Helmholtz	Technocrats, Bernal	Ravetz, Shibatani
Users	Fellow researchers	Business and industry State	Local residents Society
Ranking disciplines	Basic sciences	Engineering research and development. Military science	Social sciences Public sciences
Work-style	Interest-intensive	Capital-intensive	Labour-intensive
Political regime	Laissez-faire capitalism	Monopoly capitalism State Monopolies	???

To anticipate my argument, the crucial difference between these three types of science lies in the social mechanism by which each is assessed. Adopting a sociology of science point of view, I hold, in other words, that the question of to whom scientific research is addressed is the most important factor in shaping and defining its character. Other distinctions such as research location and ranking discipline are not necessarily hard and essential. They are typical characteristics, enumerated here in an attempt to depict an overall configuration.

ACADEMIC SCIENCE

When we say that modern science is the product of the scientific revolution of the 17th century, we are normally thinking chiefly of its intellectual and methodological aspects – the mechanistic view of nature, the Newtonian paradigm, etc. Yet this revolution also involved a change in the institutional and moral apparatus by which scientific activity was promoted and evaluated. This period witnessed, for instance, the birth of academic societies like the Royal Society of London. In these groups, members presented their research for debate and discussion by their colleagues, and saw their work recognized by publication in the proceedings of the society or its journal. Research was initiated out of personal interest and pursued for reasons of personal honour and distinction meted out through a referee system. These arrangements proved to be so effective that academies and amateur scholarly societies sprang up throughout the Western world, and the scientific journal became the vehicle for a voluminous stream of scientific research.[2] This mechanism has been an efficient means of promoting academic science.

In the nineteenth century, academic science invaded the university where it found a reproductive mechanism, acquired a market for its product and generally succeeded in establishing itself on a sound institutional basis. This pattern was best exemplified by the philosophy faculties of the nineteenth century German university. According to Joseph Ben-David, the German academic world benefitted greatly during the first half of this century not only from competition among individual researchers but from the competitive conditions that existed at the time between universities in the many small German states – conditions that prompted the introduction of new positions and chairs for new fields of science and gave rise to competition for top-notch researchers.[3,4]

Chemistry is the classic example of a discipline that developed under the impetus of this inter-university competition. No sooner had Liebig succeeded in establishing an experimental laboratory for students at Giessen University than chemistry labs began to appear elsewhere in Germany and throughout Western Europe. Until that time, scientists had conducted their experiments with their own pocket money in makeshift laboratories they set up in their homes. Hereafter, the experimental sciences – first chemistry, and then biology and physics – were to flower in university laboratories. To the general public, the academic scientist soon became someone who wore a white lab coat and did his research in a red brick laboratory.

Advocates of academic science have commonly drawn their arguments from the ideals of the German university as well, from Humboldt's conception of the unity of education and research or Helmholtz's attempt to promote science for its own sake, for science as a self-justifying, value-free activity possessed an objectivity that transcended secular interests. They also argue for the academic freedom necessary to advance this vision of science and the autonomy required to defend that freedom.

While these ideals state the case for an established academic science against competing claims in society, it is quite unlikely that they had much to do with its formation. As Ben-David has argued, the real driving force behind the rise of academic science was the free competition which existed under the decentralized conditions that prevailed in Germany at the time – though one might also add that it shared this competitive mechanism with the burgeoning *laissez faire* capitalism of the nineteenth century. Yet even today, these ideas continue to influence the values and standards of university men and women and are still used to promote research in the basic sciences.

INDUSTRIALIZED SCIENCE

Academic science, however, is not the dominant form of contemporary research. In the 20th century, and especially in more recent decades, the value of scientific research in furthering the goals of industry and the purposes of the state has been widely recognized, and the world of research seems to have been restructured along these lines. In Japan, research today is typified by what goes on in the central research institutes of major industries. Thus the term 'industralized science' would seem to describe the Japanese situation rather accurately, more accurately perhaps than it describes the situation in other advanced countries where the majority of research is funded by the State, and even the research institutes of business and industry depend heavily on government contracts. In this sense it might be better to refer to contemporary research more generally as 'incorporated science'. Still, the term 'industrialized science' intends to indicate not simply science conducted by or for the purposes of industry, but all scientific research that is patterned on the formulas employed in the production of industrial commodities. Whether the research is conducted in a university or in a national research institute, insofar as it is conceived along these lines, it can be called industrialized science. Indeed, this is the reason the term 'industrialized' rather than 'industrial' is used.

In contrast to academic science, industrialized science is reviewed by a sponsor who stands outside the scientific community and tends to have little or no interest in or respect for science for its own sake. The result is a science that cannot be sharply distinguished from technology.

The motive for industrialized scientific research is furnished by competition with business competitors or among states, while profits, military needs or national prestige determine how its results will be evluated. The researcher's work is assessed wholly within the organization by management or administrative personnel and decisions about an individual's abilities are made only in the context of the larger purposes of the organization. Research reports are not publicly presented at scholarly meetings but in internal reports and applications for patents. Although a clear-cut merit system and a standard form for the presentation of research are not as well-defined in the research institutes of business and industry as they are in academic science, and though there are many individual differences among enterprises, one thing is clear; research is not reviewed or evaluated by anyone from outside the organization or unconnected with the project.

Supported by industry and government resources, big research-institute science can engage in large projects which require an investment of talent and money on a scale far beyond anything that can be attempted in a traditional university setting by an ordinary university professor with a few assistants and graduate students to help him. At the same time, in these projects the researcher must place the project team's goal above any desire for personal recognition. While those at the top may well get the same kind of satisfaction from their work that comes to anyone who successfully completes a task, this kind of research setting can give rise to alienation at lower levels. On the other hand, the sense of belonging to a large organization may compensate the individual for his self-effacement.

In any event, the large research institute is not the setting for the kind of competition among individuals that has done so much to promote scientific research in the past. In addition, in the world of big science where research is carried out under State auspices, such huge amounts of human and financial resources are required that one cannot afford to create two or more competitive teams. Who will do the research is determined at the funding stage and there is nothing with which to compare the results. There may be competition among industries and States, but assessment occurs only at a political level that lies beyond the reach of the scientific community. In this sense industrialized science corresponds to the age of monopoly capitalism or State monopolies.

A list of proponents of industrialized science would include men like J. D. Bernal, who argued over forty years ago that the systematic organization of scientific research would greatly enhance its contribution to mankind, as well as all advocates of technocracy.

THE SOCIAL ASSESSMENT OF SCIENCE

The most problematic aspect of industrialized science is that it goes on behind closed doors.

Large sums of the taxpayers' hard earned money are invested, but industrialized science proceeds along policy lines laid down by a handful of bureaucrats, or – in the case of private industry – in the direction that promises the greatest profits. There is great danger that research so totally removed from the average citizen may ignore his interests and proceed, as it often seems to be doing today, in directions that lead to pollution of the environment or bring some other social harm. The scholarly societies and associations of academic science are powerless to check these excesses. For one thing, academic science has been caught up in the powerful mechanisms of industrialized science and conformed itself to them as a subcontractor. Moreover, the scientific community is isolated from the man in the street and does not speak for him.

This being the case, there is only one way to deal successfully with the dangers contemporary science has brought to society: the creation, at the popular level, of machinery for the review and assessment of science that will represent the concerns of the average citizen. In the spring of 1978, the Science Policy Committee of the International Union of the History of Science held a symposium at Bielefeld University to explore this question. Here it was urged that all social assessments of science must include an examination of the research process as well as a review of research results.[5] In the past technology assessment has been chiefly concerned with large-scale developments and things whose consequences have been fairly predictable. But this symposium focused on a type of problem new in the annals of science history: the technology for changing the genetic make-up of the human race.

Armed with the experimental method, modern science has advanced by learning from its mistakes. But with the prospect of producing an unforeseen aberration capable of reproducing itself at will, the researcher is no longer in a position simply to acknowledge his mistake and try again. There is a real danger that the scientific community may create a problem that cannot be controlled. We have, in other words, come face to face with one of those rare moments in the history of modern science when the traditional freedom of academic research, the freedom to make mistakes, must accommodate itself to social rules and regulations.

Adequate protection aside, the experimental facilities required to change genetic composition do not amount to much. This means that the industrialized science of pharmaceutical interests is not the only potential source of danger in this field. Academic science also has the resources to do it. Moreover, attempts are being made to develop protective installations capable of preventing any immediate danger to the researcher and the residents in the vicinity. The larger question, however, is where this type of research will lead in the long term and whether it should be permitted to continue idefinitely. And such a major decision should not be left entirely to the scientific and business communities. It must be based on a broad social consensus. Yet the channels of communication between science (both academic and industrialized) and society that might help build this consensus are missing. The question of changing human genetic make-up has prompted a renewed awareness of the need for such channels and led to serious discussions on 'science and society' in public forums and university courses in several advanced countries. On these occasions the idea of a critical science that would conduct an assessment of scientific practice from the standpoint of the interests of society as a whole has attracted considerable attention.

Nevertheless, insofar as the function of critical science is simply to check on the excesses of industrialized science in much the same way that consumer protection groups review and examine the quality of products on the market, it is still too passive to accomplish what the situation really requires. Since it is traditional academic and industrialized science themselves (and not merely the products of scientific research) that are causing social problems, it is necessary to go one step further. We must, that is, adopt a sceptical attitude towards the mechanisms and structures which promote them and work towards the development of an alternative concept of scientific research.

SERVICE SCIENCE

A contemporary scientist who looks at his profession as an outgrowth of man's inherent desire to know and sees its institutions as an indispensible subsystem of human society may be inclined to believe that science as we know it today will exist forever. But the structures of modern academic science have only been fully developed for about a hundred years and they are already being eclipsed. Industrialized science came into its own only in the post-Second World War period, but it has been seen as problematic for the last decade. That either should enjoy its place in the sun in perpetuity is quite unimaginable.

Let us, then, turn our thoughts to the future, and more specifically to two of the most important factors in the promotion mechanism of any research system: motive and value system. Service science does not reject either the incentive afforded by individual distinction or the vitality engendered by individualistic competition in academic science. Nor does it deny the efficacy and efficiency of industrialized science with its organized cooperation to achieve organizational goals. Our aim has been to delineate and describe certain general types. We do not by any means suppose that all researchers engaged in these two types of science are either no-holds-barred competitors or wholly dependent on and loyal to the organization. There are surely some, perhaps many, who feel themselves involved in the lives of those around them, who find satisfaction and meaning in their relationship with ordinary men, and who would rather do research that pleased the old lady down the street than their superiors in their research organization or their colleagues in the groves of academe. Some scientists no doubt value the specialized scientific knowledge they possess because it sets them apart from others, but there must also be those who feel isolated because the research they are doing is neither meaningfully related nor comprehensible to their families and loved ones. It should not be necessary for such men to sacrifice their natural feelings of solidarity with their fellow men in order to maintain their standing as respectable research scientists. If they find that this kind of sacrifice is actually necessary, then it is the researcher's code and the way in which research is evaluated that are out of line.

'Service science' is an attempt to put these motives and values to work. The word 'service' has a variety of meanings, but in our case it stands for 'public service'. Service science, then, may be defined as science sustained by a sense of solidarity with and sympathy for local residents and responsible to their standards. To underscore its concern with the common welfare, I initially considered using the term 'science for the people'. But afraid that some might find this too facile and others too radical, I finally settled upon the somewhat more modest 'service science'. Though less dramatic, it will, I hope, make my proposal more widely acceptable.

I am not interested in rejecting either academic or industrialized science outright. The fact is, however, that both these types of science have a well-established and highly respected position in contemporary society while service science has yet to be properly appreciated by or acquire legitimate status in the scientific community. This I find all the more disconcerting because this service science may very well be the trump card that will overcome the dangerous breach between science and society and open the way to the future. It is out of this concern and frustration that I am making a plea for the legitimacy of service science and for the creation of structures that would promote its advance and make it a viable alternative for the practising scientist.

A 'FIELD SCIENCE' APPROACH

The first characteristic of service science is that it does its research in the field. The field sciences are expected to play a major role. The first thing one must do in dealing with things that are of concern to local residents and affect the human ecological system such as environmental pollution is to get a firm grasp of what has been and is happening. Bringing materials back to the university laboratory and subjecting them to high level laboratory technology is a secondary, even a tertiary matter. The laboratory science approach with its isolation and analysis of experimental phenomenon under ideal laboratory conditions has been one of the most effective instruments of both academic and industrialized science. Yet once one has correctly grasped the phenomenon in question and analysed each of the causal factors involved, the scientific problems associated with these factors are resolved in most cases and hence of no further interest to the academic scientist. The scientific problems that occur in ecological systems, however, are commonly the product of a complex of factors and conditions that make them very sticky and hard to deal with, and limit the effectiveness of the laboratory science approach. The field science, ecological way is really the only way.

By the standards of laboratory science, the ecological approach is not equipped to give clear-cut answers. It simply tries to stick closely to the phenomenon at hand, shun easy generalizations and consider all possible factors. Still, service science would revive the kind of natural history, field science approach that has been overshadowed by the laboratory centred research of academic and industrialized science for so many years. One cannot, of course, simply equate academic with laboratory science, industrialized with research institute science and service with field science. The real point I want to make here is that the laboratory approach has only a supplementary role to play in service science.

Ecologists say that if one is going to study, for instance, monkeys, he must begin by becoming thoroughly familiar with their individual worlds. Unless one lives with monkeys and shares the feelings of each individual monkey, they suggest, one cannot fully understand their behaviour. How much less then can a service science that deals with the environmental problems faced by particular groups of citizens hope to understand the situation by treating victimized human beings as experimental animals and subjecting them to tests under isolated laboratory conditions. Human beings cannot be dissected alive; nor can one simply pull them out of their life environment and bring them to the lab or hospital for analytical tests. In a laboratory atmosphere where the individual being examined is at the mercy of the examiner, it is not possible to come to an understanding of the conditions that actually prevail in his environment. After the Second World War, the Atomic Bomb Casualty Committee (ABCC), backed by Occupation authorities, conducted a fact-finding survey of survivors of the Hiroshima blast. But those who were brought to the lab for tests were silently outraged at finding themselves regarded purely as experimental guinea pigs rather than as persons in need of treatment, and they misrepresented the facts and concealed the truth. There are in fact a whole host of problems which the researcher cannot grasp unless he stands on the same ground as the victim and shares his feelings. In these cases going out into the field and making contact with the people is an essential first step, a step without which the researcher stands very little chance even of being able to identify the problem. Here compassion is more important to success than competition, fellowship more effective than rivalry.

Service science also requires forms of publication that differ from the scientific papers of academic science or the reports of industrialized science. This need arises directly from the fact that service science addresses its findings neither to fellow scholars (as in academic science) nor to research administrators (as in industrialized science) but directly to local residents. Moreover, since neither originality nor the accumulation of know-how is involved, it can make its points in handbills and appeals, in a style designed to secure as wide an audience as possible. It also makes maximum use of journalism and the broadcast media. Finally, inasmuch as service science research is not published to lay claim to a new discovery, one need not be afraid of duplication or repetition. Freed from inflexible adherence to the formalities of academic science with its quotations from reliable sources, one can make use of modes of expressions more effective in speaking to the general public to say what he wants to say about a wide variety of subjects.

DECENTRALIZED, DISPERSED STRUCTURES

Inasmuch as service science is closely tied to particular local areas or regions and emphasizes sensitivity to the individuality of the people and research materials with which it is concerned, centralized institutions are not appropriate vehicles for its promotion. It would not be at home either in the elite universities of metropolitan centres or in the research institutes of major enterprises. It emerges first as the work of vanguard movements of volunteers who must support their research themselves with the aid of contributions from friends and from the general public. Eventually, the research institutes of local governments as well as the municipal and prefectural institutions of higher education might be able to provide a regular budget. The citizens' institutes for scientific research envisaged by Shibatani Atsuhiro[6] and the science and technology centres of the prefectures also come to mind as potential means of institutional support.

This tendency to think in terms of regional dispersion is international. In a March 1971 report, the Science Policy Committee of the Organization for Economic Co-operation and Development dramatically reversed its advocacy of investment in research and its all out support for industrialized science in advanced capitalist countries, pointing to the need for a change in science policies still based on the old idea of the unrestricted development of science and technology, it called for a science that was guided by social needs and dedicated to the pursuit of a better quality of life. The report also observed that in order to respond to localized social needs such as environmental or pollution problems, the need was for research based in regional or local universities and carried out with the cooperation of local residents rather than for more luxurious big science projects designed by government agencies. Noting that these problems were not of a kind that could be successfully dealt with through a major, one-shot investment in a single research institute, it argued, furthermore, that local governmental machinery could be more responsive to local requests for scientific investigations. Finally, it expressed the hope that research organizations would be more open to society than they had been in the past, enlisting the participation of the average citizen where circumstances permitted and making an effort to stay closely in touch with social needs.[7]

THE PLACE OF RESEARCH IN THE MASS UNIVERSITY

Under service science the education and training of scientists would also undergo a change. Today the mass university is a fact of life. In such institutions, the Humboldtean idea of the unity of education and research is obviously anachronistic. Finding their research lagging behind that done within the framework of industrialized science and their educational efforts frustrated by the large number of students, university scientists are being pulled in two different directions and are unable to do either task well.

Academic science has gone after solvable problems and solved them one by one, in accordance with individual interests and the internal logic of the various disciplines. But many of the problems that society really needs to have solved remain. The lay man or woman who lacks an understanding of the structures that promoted the growth of academic science, still entertains the illusion that the university scientist has the capacity to solve them. But this is an excessive expectation that can only lead, first to despair at the academic scientist's apparent ineffectualness, and then to a disillusionment that breeds an anti-science mood.

Lord Ashby has observed that while the need for new discoveries at the frontiers of scientific research has declined, the 1970s have brought an increase in the relatively more difficult type of problem that combines scientific and social dimensions. The task of scientists in the mass university, he suggests, is to develop a new generation of concerned scholars and scientifically literate citizens who will be able to bridge the gap between science and the scientific community.[8] University scientists continue to be evaluated by the standards and values of academic science, but teachers at mass universities might be able to produce work that would meet these standards by taking advantage of the sea of human resources in their classrooms and organizing students for manpower intensive research projects.

Still, the mass university will always be less a research than an educational institution. The OECD Science Policy Committee has recognized this in its basic guidelines for future university research,[9] Calling for 'research for education', they suggest that in selecting research themes, framing problems and devising projects respect be shown for and priority given to student ideas. Once a project has been worked out, they envisage research teams composed of students from different disciplines going out in the field to investigate accompanied by a teacher who serves as an adviser. For example, a team set up to study dye pollution of the Rhine might include students from such fields as chemistry, biology, law, international relations, etc.

The approaches taken by such student teams will no doubt seem naïve to adult society. Nor is it at all reasonable to suppose that students have the capacity to solve enigmatic problems that even adults find baffling. Yet if the fresh social consciousness that is born anew in every student generation but which seems to have disappeared from so much of academic science – if this social consciousness is confined within the regimen of academic science for a long period of time, the student's emerging awareness of himself as a responsible citizen will also degenerate. Students learn much from field research. Sometimes the experience changes their lives. Research for education should not be judged by whether or not it produces, within a limited period of time, an article in an academic magazine, but in terms of the long range effects it has on the future lives of students who participate in it.

As a model for this 'research for education', the OECD cites the newly-conceived Roskilde University Centre in Denmark. In this institution lecture style instruction is kept to a minimum and most of the student's time is spent in the field. In the USA, the National Science Foundation has started a programme to fund research projects submitted by students. In England, student unions take the lead in conducting research that calls to mind the settlement house movement among Japanese students prior to the Second World War. Recently I have also learned of an Indian university which includes field research in local villages in its regular curriculum. Students move into rural villages and, while sharing the life of the local people, test their water, study their economy, and explore ways to improve village life. Reports written by these students seem not only more moving but also more true to life than any government survey I have ever read, for their contacts with the people (including the lowest untouchables) enable them to depict village life in sharp relief. Thus in developing countries student research surveys can actually serve to improve local living conditions.

THE NEED FOR NEW MEANS OF ASSESSMENT

These are just some of the forms service science might and does take. As I have suggested, however, the fruits of these kinds of projects cannot be adequately evaluated within the research systems of traditional academic or industrialized science. Most of the elements listed on the chart under one of the three types of science I have been discussing can actually be found in all three to a greater or lesser degree. For example, individual researchers have all three kinds of motives and values; it is the mix that differs. Again, service science is not all field science. It makes use of laboratory methods just as there are fields within academic science that emphasize the importance of field research. In fact, if there is any one element that distinguishes these three types of science from one another it is the system by which research is reviewed and evaluated. Yet service science still does not have any clearly defined apparatus for this purpose. The institutionalization of these procedures will inevitably dilute the motive, but this cannot be avoided. Inasmuch as there is likely to be an increasing shift to service science type research as time goes by, it is important that steps be taken to deal with this problem.

CONCLUSION

For instance, work done by a local pollution laboratory should not be evaluated by the criteria of academic science but in terms of the contribution it makes to popular awareness of public problems and issues through such things as public forums and journalistic writings. Academic scientists whose writings are confined to the conventional scientific journal article often apologetically acknowledge that they are not addressing themselves directly to the general public, and express the hope that their scholarship will ultimately be made popularly available through science journalism. The fact of the matter is, however, that in most cases the work of academic science never gets outside academic circles. Even when an academician's ideas do exert a broader influence, there is no feedback from the populace. Hence those who seek solidarity with the people have no alternative but to appeal directly to the general public.

In the West, the professions have enjoyed a special status ever since the days of the medieval university with its faculties of theology, law and medicine. This special status and the prestige which has accompanied it was conferred upon the professions because of their obligation to serve all men in accordance with self-defined ends and values that transcended the powers and interests of this world. When science emerged as a new profession in the 19th century, it inherited this status and prestige. Thus far, however, it has largely served its own profession and, more recently, the military-industrial establishment. This kind of activity has not traditionally been considered professional service, and unless the scientific community makes the adjustments that are required to meet the needs of client taxpayers and the general public, the scientist's reputation could conceivably decline to the point where he would no longer enjoy a status comparable to that of the lawyer and the physician.

But what are the broader social implications of the idea of service science? What kind of future society will accommodate its concerns and give it a proper place? As yet, this society has no name, but we can think of a society in which science and technology are fully assessed by the general public. As is most obvious in the military technology developed by the military-industrial complex, the central problem with incorporated or industrialized science is that both the process and the products of research and development are the effective monopoly of a special sector in the establishment. Insulated from popular assessment, they are utilized to oppress and to thwart the general interest.

Western democracy has thus far proved unable to control runaway science and technology. Indeed science has never been a major election issue. The situation is, if anything, even worse in socialist countries despite the exaggerated official claims for popular channels of assessment made by their science planners. Thus whether the society be socialist or capitalist, service science for the people is called upon to be their advocate, and to work against all those who would use science to enhance their power over others and their capacity for social control. The need for the development of established criteria for a citizen assessment of science is urgent.

REFERENCES

1. Jerome R. Ravetz, *Scientific Knowledge and its Social Problems*. Oxford University Press, Oxford, 1971.
2. Nakayama, Shigeru, *Rekishi toshite no Gakumon*. Academic Traditions, Chuo Koron, 1974, see chaps 4 and 5.
3. Joseph Ben-David, *Kagaku to Kyoiku* (a collection of essays translated into the Japanese by Shinbori, Michiya). Fukumura Shuppan, 1969.
4. Joseph Ben-David, *The Scientist's Role in Society; a Comparative Study*. Englewood Cliffs, New Jersey, 1971.
5. E. Mendelsohn, D. Nelkin and P. Weingart, eds., *The Social Assessment of Science Proceedings*. Bielefeld, 1978.
6. Shibatani, Atsuhiro, *Han-Kagakuron*. On Anti-Science, Misuzu, 1973.
7. OECD, *The Research System*, OECD, 1974, Vol. 3.
8. Lord Ashby, Science and anti-science, *Proc. R. Soc.* B **178**, 29–42, 1971.
9. OECD, *The Structure of Studies and Place of Research in Mass HigherEducation*, OECD, 1974.

First published in *Occasional Paper* No.23, Centre for Studies in Higher Education, University of California, 1982

⑮ The Transplantion of Modern Science to Japan

What happens to a paradigm and the scholarly tradition that has grown up around it in one culture when it is introduced into another? Newly situated among groups who employ a different scholarly jargon and work under different social conditions, the paradigm confronts new challenges that occasionally affect its very structure as a mode of learning. The removal of Greek learning to the Islamic world, the introduction of Chinese culture into Nara-Heian Japan in the ninth and tenth centuries – history does not lack examples of this phenomenon. Here I would like to take a look at what is perhaps the most dramatic instance of them all, the entry of modern science (that most excellent product of the modern West) into a non-Western country (Japan) for the first time. More was involved in this case than the gradual spread of a cultural achievement from one region to another. The influx of modern science into Japan was intimately bound up with a newly-risen element – the policy of a modern state.

A scholarly tradition that emerges from within normally follows the following course: (1) generation of the paradigms, (2) paradigm elaboration through the formation of disciple-advocate-support groups, (3) canonization (incorporation into text books or manuals), and (4) institutionalization. The level on this spectrum at which a foreign paradigm is transplanted determines the type of acculturation thereof.

First, inasmuch as the academic traditions of East and West have grown out of different paradigms and present different spectrums, the reception of Western scientific thought in Japan could not have taken place at the paradigm level. The paradigms of modern science, that is, could not have been subsumed under traditional paradigms. This had already proved impossible even in the West – indeed this is why there had been a scientific revolution – and had they been successfully taken in under traditional paradigms, they would, by definition, have ceased to merit the designation paradigm in the basic sense. The introduction of the basic paradigms of modern science involved the transformation of old concepts and the creation of new ones. When Shizuki Tadao first introduced Newton to East Asian in the late eighteenth century there were no words in the Sino-Japanese vocabulary that could adequately express Newton's notion of particles or his concept of mechanics. In attempting to ground Newton in the East Asian concepts of *i* (Change) and *ch'i* (matter-energy), he may well have been taking a first step in the direction of a new natural philosophy, but in the absence of a mature support group, Shizuki's work was not expanded into a new scholarly tradition.

In the contemporary world, where professional groups share common internationally accepted goals that transcend particular cultures, like the physics community, a paradigm that appears in one country quickly finds groups of supporters in many countries (step 2), groups that have the capacity to put the paradigm to work in a normal science programme . In this sense, the international physics community today may be said to share a common culture, and the transmission of a paradigm follows the course outlined above.

But when a paradigm is transplanted from another cultural area, where there are no established support groups sharing common philosophies or outlooks at its reception (and in the case of modern science this precludes the existence of a normalized research tradition), the sequence is rather different. In the absence of participation and feedback in the paradigm-elaboration process (step 2) the paradigm brought in from abroad is treated as an established canon to be faithfully translated (step 3). This pattern is commonly observed where contact with a foreign culture has led to a gradual osmotic acceptance of its cultural achievements by various individuals in the private sector, as exemplified in Japanese history by the ancient arrival of the Buddhist classics and the advent of Dutch learning in the late-eighteenth to mid-nineteenth centuries.

There is also another pattern. When foreign learning is systematically introduced by the state, the establishment of an educational system under government auspices comes first (at step 4). Paradigms and canons are then inserted in the school curriculum and advocate-support groups are created as a part of national policy. Though this has become typical wherever scholarly paradigms have been imported on a national scale by modern states, Japan's introduction of Western science after the Meiji Restoration (1868) remains the classic and earliest example of this pattern, and this will be examined later in this paper.

WESTERN INFLUENCE ON TRADITIONAL SCIENCES

But first let us turn to the earliest influence of Western science in Japan. In the mid-sixteenth century Jesuit evangelists 'discovered' Japan; however, soon after, the Japanese government, considering the influence of Christianity detrimental to the cohesiveness of their culture, banned all foreigners from the country, but sanctioned trade with the Dutch only through the port of Nagasaki, on the island of Kyushu. This relatively effective, self-imposed containment remained in effect until the mid-nineteenth century. Western ideas did, nevertheless, seep through, directly and indirectly.

There were in the Tokugawa pariod (seventeenth to mid-nineteenth centuries) four traditional sciences, each with its own paradigm and professional groups; (1) calendrical astronomy, (2) medicine, (3) materia medica and (4) mathematics (*wasan*). A spectral analysis of these fields reveals that Western learning was accepted somewhat differently in each case, depending upon the nature of the field, the character of the support group, and the strengths and weaknesses of the native tradition and its classics.

(1) Astronomy

In astronomy the influx of the new Western paradigm did not spell destruction of the old. Data generated by the new were simply incorporated into the old framework.

Traditional astronomy was Chinese-style astronomy for calendrical calculations. Its purpose was to investigate the apparent movement of the sun and moon, to put them together to construct the luni-solar calendar, and lunar eclipses. The many systems for computing the ephemerides incorporated in the *Lu-li-chih* (chapter of calendar) of official Chinese dynastic histories served as the paradigms for this astronomy. The Yuan calendar *(Shou-shih li)*, which was considered to be the crowning achievement of the traditional Chinese astronomy, was particularly important. Its mastery was the final step in obtaining a licence in traditional mathematics (*wasan*).

Western astronomy had arrived in China with the Jesuits. Calendar making was a vital function of the Chinese bureaucracy, and the Jesuits hoped that by demonstrating the superiority of Western astronomy they would succeed in persuading the bureaucratic elite that Western culture and Christianity were superior. Astronomy was well suited to their purposes, not only because it was so important to the Chinese bureaucracy but because the phenomena with which it dealt possessed a universality that transcended East and West and the objectivity of celestial phenomena that did not permit of much human manipulation. Furthermore, it was quantitatively precise. Thus when the tally sheet was in, Chinese astronomers had to acknowledge that Western astronomy had superior features. All that the Jesuits made available from Western tradition, however, were peripheral data and methods of calculation. The structure, style and purposes of Chinese calendrical astronomy remained unchanged. As Hsu Kuang-ch'i, a high Chinese official who collaborated with the Jesuit Matteo Ricci on several projects, put it, 'We melted down their materials and poured them into the *Ta-t'ung* (the traditional Chinese calendar) mould.'

In Japan, the eighth Shogun Yoshimune (1684–1751) attempted to use this refurbished Chinese astronomy as the basis for calendrical reform, but he was effectively opposed by conservative court circles in Kyoto. The revision of Horeki that came into effect in 1754 was still modelled upon the traditional *Shou-shih* calendar. Yoshimune's desire was fully realized, however, in the Kansai revision of 1797.

The Tempo revision of 1842 made use of Western astronomy learned directly from Dutch books. Yet even here the framework of Chinese-style astronomy was not broken. Japanese calendar makers, like their Chinese predecessors, were chiefly interested in the more precise numerical data of Western astronomy. A man like Takahashi Yoshitoki (1785–1829) who worked in the tradition of Asada Goryu could dabble in questions of planetary motion out of intellectual interest, but the subject was peripheral to the official task of calendar making. Indeed, the new Western paradigms of modern astronomy – theories of the cosmos and celestial mechanics – had nothing to contribute to traditional paradigms. No attempt was made to fit them into the Chinese mould.

Calendrical astronomy was official learning and government managed science. When Western learning entered the country it initially became the work of the official astronomers of the Tokugawa Shogunate. With the fall of the Tokugawa regime in 1868, the office and the routine observational work its occupants had conducted were abolished. Even if it had survived it is unlikely that the Tokugawa astronomers would have become modern scientists. A remnant worked on the solar calendar reform that came into effect in 1872, but given their training this routine computation was about as far as they could have gone. Thus the field of astronomy passed into the hands of modern professionals, newly graduated from the university. Between the two generations there were no teacher-disciple relationships, no continuity. In summary, the traditional paradigm is this field proved so strong that it kept the incoming paradigms of modern science at arm's length, and Japanese astronomers made only supplementary use of the data they generated, using it to strengthen the fabric of their tradition.

(2) Medicine

In medicine, Chinese and Western paradigms coexisted peacefully, for there was little overlap in their areas of strength and hence, little room for competition.

Western medicine first came to Japan in the *Namban Igaku* (Medicine of Southern Barbarians) that entered the country with Christianity in the sixteenth century. Before long, 'barbarian surgery' was being welcomed to supply deficiences in medicine. Yet what was picked up at this point was not an organized body of knowledge but techniques such as rubbing salve on wounds and the use of alcohol as a disinfectant.

When we say that science is governed by paradigms, we mean that paradigms determine the shape of the questions posed and the answers to be delivered by normal science. But there are some fields in which their dominion is far from total. This is particularly true in medicine where the mechanistic paradigms of modern science were not readily applicable to the many intricate and varied techniques used in dealing with the human body and all its complexities. Older paradigms peculiar to medicine and biology have continued to have meaning. The relationship between theory and empirical techniques was even more diffuse in Chinese medicine where physiological and pathological theories were based on doctrines of *Ch'i*, drawn from *Yin-yang* and the Five Forces natural philosophy. These theories were indeed little more than window-dressing in Japan, where they were, after all, imported from China, and much less coherence between theory and practice was maintained than in China. Medical techniques governed by *Ch'i* theory have survived on their own, without the support of the old paradigm.

Scientific medicine, in contrast to bodies of medical techniques accumulated through experience, approaches treatment on the basis of physiological and pathological theories that are grounded in anatomy. Chinese medicine also had its theories of physiology and pathology, but anatomy was considered to be rather unimportant so they did not trace the source of disease in any particular organ but in the disharmony of the *Ch'i* permeated and circulating in the body. To the comparative eye they appear even weaker after the European Renaissance, when great strides were made in anatomical studies, and the seventeenth century, when contact with mechanistic ideas opened the way to modern physiology with its attempts to explain the functioning of the human body in mechanical terms.

Seventeenth-century Europe also witnessed pioneering work in the chemical treatment of disease.

Nevertheless, it is extremely doubtful whether this kind of basic research contributed much to actual medical care before the nineteenth century. Half-baked theory could on occasion even lead treatment astray, and once embraced, serve to deprecate and discourage the use of 'popular' empirical remedies. All medical theory of the time regarded the quinine treatments of the Peruvian Indians as popular superstition, but there were doubtless many patients who were thankful to them. Much medical theory was little more than academic trapping, and one may say that the paradigms of modern science had yet to become accepted as a definitive guide to practical medical care.

Under the circumstances attempts to argue for the superiority of Western medicine were not very persuasive. Western superiority might be accepted in surgery – undertaken as a last resort, since this at the time often proved fatal – and such demonstrably effective procedures as vaccination, but these were exceptions. In so far as general internal medicine was concerned, there seemed to be little to choose between China and the West. Since regardless of theory, whatever worked was best for the doctor as well as for the patient, the average Japanese physician adopted a syncretic approach, treating patients with a combination of Chinese and Dutch medicine.

Still, the theoretical paradigms of the two traditions were scarcely compatible. It is at this level – and not because of any contribution it made towards advances in medical treatment – that the appearance of the *Kaitai Shinsho* (New Treatise on Anatomy) can be considered an epoch-making event. This 1774 translation of a Dutch version of German work on anatomy represented the first real evidence upon which the advocate of Western medicine tried to make the case that Western medicine's much greater knowledge of anatomy meant that its methods of treatment must also be superior. Although anatomy was not relevant to actual medical treatment, it gave Japanese physicians an opportunity to become acquainted with the anatomical basis of the localized approach to treatment that, after about 1800, was the distinctive feature of Western medicine. In East Asian medicine, physiology rested on the operations of a *Ch'i* that filled the universe and the human body, and medical treatment centred on assisting the whole body to return to a state of harmony within and without and recover its normal functions. Early nineteenth century Western medicine, however, looked at the body as a machine and saw pathology in 'solid' terms. The problem was to locate the part of the body that was ill and treat it in a concentrated fashion.

In a scientific revolution the revolutionary process ordinarily reaches completion when the new paradigm succeeds in encompassing the empirical knowledge that had been explained in a different way under the old paradigm. At least this will be the objective. But modern, chemotherapy-centred medicine has found Chinese herbal medicine and techniques such as acupuncture and moxabustion difficult to cope with. In China, Chinese and Western medicine continue to co-exist and are thought of as complementary – but are integrated only at the level of treatment. In Japan, the creation of a modern medical system occasioned a political struggle between Chinese and Western support groups. The older group of physicians, who were the chief backers of Chinese medicine, were defeated. It was excluded from the new licensing system of 1876, after a century of attack by spokesmen for 'Western' medicine, and came to be regarded simply as a bag of popular remedies.

(3) Herbology

The encounter of East and West in this field led to the growth of a new branch of normal science through the addition of empirical data from the East to the Western paradigm or vice versa.

In the Chinese tradition, there was a field of study known as *Pen-ts'ao*. This field was centrally concerned with the medicines used by Chinese medical practitioners, but it also included identification of animals and plants mentioned in the Chinese classics with local Japanese flora and fauna, and 'product study' which compiled information about useful animals, plants and minerals found in various localities. It was chiefly concerned with collection, classification and description. It had developed the artificial-classification system that appears in the *Pen-ts'ao kangmu*. This system was based on use and external appearances, and distinguished mountain herbs, tropical herbs, poisonous herbs, medicinal herbs, fragrant plants, trees and shrubs.

Japan was introduced to the natural classification system of Linnaeus in 1829, primarily through the work of Ito Keisuke. The *pen-ts'ao* and Linnaean paradigms had little to say to each other but Keisuke had studied both. By explicating Linnaeus' classificatory scheme and identifying Latin names of Japanese plants, he made possible the addition of Japanese data to the Western mode of natural history. But Ito Keisuke's methods differed little from the collecting, classification and description of the traditional Chinese herbologist. He does not seem to have understood the modern method of examining structure, properties and patterns of growth. Though his contributions were acknowledged and he lived to see the dawn of the twentieth century, he became increasingly out of touch with students trained after the 1870s. Since Chinese herbology was traditionally carried out mainly by physicians as an adjunct to medical practices, and had no independent group of professional practitioners comparable to the pharmacists of the West, it was destined to share the same fate as Chinese medicine.

(4) Mathematics

In mathematics, Japanese and Western paradigms were self-contained systems, mutually exclusive in structure and constitution. When the Meiji political revolution and institutional change came, there was a complete change from one to the other.

The *wasan* (Japanese mathematics) had roots in the Chinese mathematical tradition, but it represented an independent development. Taking the work of Seki Takakazu (?-1708) and Takebe Katahiro (1664–1739) as paradigmatic, it emphasized the algebraic solution of geometric problems. Although this development involved a turning away from practical concerns in favour of mathematical puzzles for amusement, it represented something rare in the Japanese academic tradition – a mode of learning whose protagonists pushed forward to new scholarly frontiers without consulting foreign authorities, a discipline which moved from paradigm creation to the formation of a support group. Having a vitality all its own, it experienced substantial normal science growth in relative isolation, largely unmindful of developments in Chinese or Western mathematics.

Western-style mathematics was introduced in the mid-nineteenth century in conjunction with practical matters such as navigation and surveying. The notation of the Western system was completely different from that of the traditional system – a factor that has frequently had paradigmatic significance in dividing one tradition from another. The change from Roman to Arabic numbers and the differences between English style Newtonian mathematics and the continental, differential notation of Leibnitz had a substantial influence on subsequent development, and the ability of Japanese mathematics to establish its independence from China owes much to its system of algebraic notation. In turn, those who had need of and studied Western mathematics – naval officers and surveyors in the late nineteenth century, who had come up through Nagasaki Kaigun Denshusho (Naval School) – Sokoryoshi (Office of Surveying) route – were very different sorts of men than the practitioners of *Wasan*. Having found it a nuisance to convert all the mathematical notations they encountered in their study of Western science into *wasan* forms, they determined to use the Western system of notation as was. This choice was made on the basis of political convenience and before anyone had asked themselves which of the two systems was superior or thought about blending them.

An examination of the backgrounds of mathematicians listed as teachers in private academies in Tokyo in 1873 reveals the radical split that existed between the two traditions. Almost none of the teachers had had training in both Japanese and Western-style mathematics. After the Ministry of Education decided for the latter in 1872, some *wasanka* (practitioners of traditional mathematics) attained a solid grasp of basic theory before going on to applied problems. In terms of personal as well as notational system, it was by leaving the tradition of *wasan* behind that Japan was able to embrace Western methods of calculation.

SUCCESSIVE SUPPORT GROUPS OF WESTERNIZATION

Let us now turn to the occupational support groups who successively played a leading role in the introduction of Western science to Japan:

(1) Astronomers and Interpreters

The adoption of the Western 'barbarian' astronomy in the seventeenth century by the Chinese dismayed conservative elements in Japan; however, it did promote the desire in others for a similar change in the Japanese calendar, which was modelled on that of the ancient Chinese. As has been mentioned, the eighth Shogun Yoshimune and his court astronomers clearly recognized the superiority of Western over Chinese astronomy, but these men were primarily government bureaucrats or technicians whose scope remained limited to their assigned duties of drawing up the official calendar; their interest in Western science was limited to the precision of astronomical data and methods of calculation, and they made no attempt to jeopardize their hereditary posts by entertaining revolutionary paradigms such as were developing in Europe at that time.

Professional interpreters at Nagasaki were the most well versed in the Dutch language, through which they must have become acquainted with the concepts of Western science. But again, they were also hereditary government officials who remained within the boundaries of their duty of faithful translation and nothing more. Neither official astronomers nor interpreters published their work for general audiences.

(2) Independent Scholars and Physicians

From the late eighteenth century on, a sizeable number of Dutch books containing the term *Natuurkunde* (study of nature) found their way into the country which aroused the interest of various independent scholars who set about translating them, although their foreign language skills were much inferior to the Nagasaki interpreters. The majority of these 'Dutch scholars', as they came to be called, were avant-garde physicians who were primarily free-lancers, with no strict subordinative links to the governing elite or subsequent interest in maintaining an existing status quo. Thus they were not inhibited to step out of their line of work and extend their interest to anything Western (except perhaps the Christian doctrine). Astronomy was the first area in which men sensed that the West was superior, but the notion that the West was superior in other fields of scientific endeavour first spread among these independent physicians.

As they inched their way through Dutch texts, they realized that Western science was more than a variant of the natural history line of their own tradition. As they saw it, its essence could be translated as *kyuri* (literally, 'investigating the principles of things', a neo-Confucian term), or 'natural philosophy', being a systematic and fundamental investigation and consideration of the nature of things. Thus, although at first there was no clearly established tradition for the term, *kyuri* was later to become the most common, and given the vocabulary of the period, it was an informative translation. In recognizing an enquiry into principles at the bottom of such traditional practical studies as medicine and calendar-making, in becoming aware, that is, of natural philosophy or physics, they also grasped the hierarchical structure of modern science, from basic to applied. Above and beyond the culturally-bound achievements of Western science, they seemed to have sensed that it contained a revolutionary paradigm – since the belief in underlying laws in Nature which could be formulated was weak in Japanese traditional thought. This was in marked contrast to the official astronomers who looked upon science as something that could be handled at the technical level and assigned it only a supplemental role.

The physicians recognized the importance of physics and chemistry as the basis of medical work, and several founded schools for the teaching of Western medicine. Their students, during their apprenticeship and internship days, moved from one school to another in major medical Centres like Nagasaki, Osaka, Kyoto and Edo (Tokyo), and played a role of diffusing knowledge of Western culture as well as medicine. While some physicians remained in the practise of medicine, others turned to the teaching of the Dutch language and even the discussion of international knowledge and politics and also founded schools for such. Thus, the physicians' cultural influence was more considerable than that of the astronomers who limited their activities to the translation of technical works.

But these support groups for Western science were by no means adequate preparation for the reception of the contemporary science. The 'Dutch scholars' were still only amateur supporters of uninstitutionalized paradigms, comparable in this respect to members of the Royal Society. Their schools were few and regarded on the whole as avant-garde. Had they flourished in the seventeenth and eighteenth centuries, however, their work might have had greater consequences. By the mid-nineteenth century, however, modern science in the West had entered the systematically structured world of the university and reached new levels of normal science advance. Reception of learning that was now systematically pursued in Europe within the structures of the university required the creation of an institutional system. It had to be met not by self-taught amateurs but with professional scientists who had received a modern education. These early scholars were not aware of the institutionalized aspect of Western science up until the mid-nineteenth century, remaining bookish translators of the *kyuri*, paradigmatic aspect of Western science.

(3) Samurai

The Opium War in the 1840s between the Chinese and the Western colonialists had created concern among the cognoscenti of the samurai class, but the event of Commodore Perry's visit to Japan in 1853 and subsequent threat of war to the Japanese government unless the country was opened to foreigners caused all samurai to realize that the Westerners' science, in the form of their superior technology, was needed for the sake of national defence. As a result, young samurai flocked to the schools of Western learning established by physicians. As they were traditionally the ruling class, and in time of war the warriors, it was natural that samurai interest in Western science and technology was based on more real and pressing needs than reception of and support for particular scholarly paradigms.

First, they tried to learn the Western art of manufacturing firearms, but soon they realized that it was impossible to catch up and compete with Western forces by a crash project of manufacturing cannons. They could purchase hardware, but what was really needed was the software of Western military training and tactics. Moreover, the acquisition of Western military disciplines was one of the contributing factors in the overthrow of the ruling Tokugawa family by anti-Tokugawa samurai and the subsequent establishment of the Meiji government in 1868. The samurai recognition of Western science was therefore political rather than cultural.

After the revolution, convinced that Western scientific learning was essential, the ruling class set about the dissemination of their conviction through the establishment of an educational infrastructure. With Meiji effort and initiative, modern science was fully assimilated in a wholesale introduction through governmental institutions rather than a piecemeal cultural infiltration, as in the previous era, and modern scientific and technological professions became the artificial creation of the new Western-oriented government.

The main practitioners of these new professions were former samurai. In the past they had received hereditary family stipends in exchange for their loyalty to the feudal powers, the Shogunate or local feudatories. But in the 1870s, efforts were made by the Meiji government to curtail the inherited family stipends of the samurai class as a step towards social modernization While other classes, farmers, artisans and merchants, could continue to be engaged in their inherited vocations, samurai completely lost their vested source of revenue. Consequently, the samurai had to find new ways of living independently. Since the samurai could not compete with other classes in the fields of traditional occupations like agriculture and medicine, science and technology was one of these new fields into which jobless samurai were attracted and invited. Almost all of the early graduates of the engineering colleges were samuari. Even as late as in 1890, the per centages of Imperial University samurai graduates in engineering and science were 86 and 80 respectively.

Thus, Japanese modern science and technology professions were, in the beginning of their formation, very much 'samurai-spirited'. The samurai were the class long accustomed by mental habit to think in terms of public affairs and by behaviour patterns to playing the game of public office. Unlike the European pattern in which science and technology was one experience of the rising bourgeoisie, the new Japanese scientific and technological professions in the last quarter of the nineteenth century were dominated by the proud old samurai class, comprising the top five per cent of the total population.

Table 15.1 summarizes what has been said so far.

Table 15.1: Sequence and summary of support groups' thrusts of Westernization in Japan

	I	*II*	*III*
Occupation	Astronomers & Interpreters	Physicians	Samurai
Leading period	18th century	late 18th – early 19th century	Mid-19th century
Interest	Technical	Cultural	Political
Status	Technicians	Free-lancers	Planners & administrators
Role	Referees of superiority	Diffusion & popularization	Institutionalization

In the first stage, Western science was merely confined to technical knowledge by technicians in the government sector. In the second, Western knowledge in the form of its revolutionary paradigms was garnered by vanguard physicians but they were alienated from the central power structure. Only in the third stage was Western science fully admitted into the power structure, and, after the Meiji Restoration, through institutionalization, it was recognized to be a most legitimate study for the youth of the samurai class. Let us look now in more detail to this remarkable third stage.

DEFINITION OF SCIENCE

The Meiji encounter with Western culture was in many ways an unprecedented experience. The many academic disciplines of the modern West suddenly burst in upon the country, creating confusion in the intellectual world. A former Dutch scholar, Nishi Amane, sought to provide translation for the array of disciplines and develop a classificatory scheme for all knowledge in his *Hyakugaku Renkan* (Links of a Hundred Sciences), but it was the image of diversity that was to prevail. If Japanese scholars had initially been struck by the fact that Western learning probed the basic principles of things, by the 1860s and 1870s they were marvelling at the degree of specialization in which it presented itself. In the Chinese tradition, astronomy and medicine had been recognized as independent professions, but the main current of scholarship – study of the histories and the classics – was the work of Confucian scholars among whom specialization was minimal. In Western learning, however, the Japanese not only discovered an array of fields – chemistry, natural history, physics, etc. – but noted that each had its own group of practitioners.

It was against this experience that the current Japanese terms for science, *kagaku* (literally 'classified learning') gained currency. To my knowledge, the first appearance of the word is in the 'Opinion on the School System', drafted early in 1871 by Inoue Kowashi, who proposed 'to invite selected students teach them the Western languages and then let them specialize in *kagaku*', with the help of foreign professors whom he would employ. In this case, *kagaku* did not necessarily imply natural science but any kind of particular discipline, the 'hundred specialities', successfully capturing the salient feature of the specialized and institutionally differentiated phenomenon that was a characteristic of science of *Wissenschaft* in the late nineteenth century.

The suggestion that *kagaku* is a translation of the German *Fachwissenschaft* has also been made. Yet whatever its origins, Japanese of the Meiji period clearly used the term less with reference to the unique methods and paradigms of modern science than to the configuration of differentiated special disciplines in which it manifests itself. Although *kyuri* had represented an understanding of science at the cognitive level, *kagaku* was an attempt to handle it in institutional terms.

Moreover, in Ito Hirobumi's *kyoikugi* (Proposals on Education), drafted by Inoue Kowashi and presented to the Emperor in September 1879, we read: 'Upper level students should be schooled in the science (*kagaku*); they should not be drawn off into discussions of politics.' Science, was, in other words, saddled with the task of curbing the Movement for Liberty and Popular Rights and training politically docile, narrowly specialized professionals. Thus the very same 'science' that, as *kyuri*, had been considered as a philosophic paradigm during the Edo period (up until the mid-nineteenth century), was perceived in Meiji eyes as an array of non-philosophical, technical disciplines.

THE UTILITARIAN IMAGE OF SCIENCE

In Europe in the eighteenth century science and technology were distinct activities with different social origins. In spite of the effort made by the Encyclopaedists to liquidate this social interface, the dual structure of science and technology was still maintained, even in the nineteenth century, by socially separated groups. This was exemplified by such institutional separations as between the German *Universitat* and the *Technische Hochschule*. However, there was no particular reason for the mid-nineteenth century Japanese to distinguish between science and technology when facing the impact of modern Western military aggression. To the Japanese it appeared that modern science and modern technology grew in a single Western tradition. It was not the science-versus-technology dichotomy but rather the tradition-versus-Western dichotomy over which the Japanese were seriously concerned.

While science in nineteenth century Europe was still in the main a cultural activity, well illustrated by the issue of the theory of evolution, rather than a practical means of achieving economic growth, the Japanese image of science in the late nineteenth century was pershaps the most modern. It was exclusively utilitarian and pragmatic, planned for national interests if not purely for profit-making, specialized and compartmentalized. Emphasis was laid on physical and applied science rather than on biological and hence the style was closest, for that period, to our contemporary scientific technology.

Neither did the debates that were raging in the 1880s in Europe on science versus philosophy and science versus religion have any significant impact in Japan. Since the philosophical and religious dimensions of the Japanese intellectual and academic tradition were attenuated and the sciences were imported in a non-ideological and non-philosophical fashion as already established, compartmentalized edifices, the conflict between science and religion never became a major source of controversy in Japan. Western philosophy was considered only one of the newly imported 'hundred disciplines'. In other words, Japanese modern science was characterized as being unrelated to, or free from, its European philosophical roots, accepting the paradigms developed in Europe as self-evident and concerned only with mastering them technically.

THE PRIORITY OF INSTITUTIONS

The introduction of Western learning as a system left little room for a comparative examination of paradigms in particular fields. If the learning of the West was advanced, it was entirely due to the fact that the West had undertaken to pursue scholarship in an open, regularly organized, institutional way. The first imperative then was the adoption of the Western style of scientific institution. This was the conviction and the commitment that underlay the introduction of the sciences en bloc and inspired the enthusiasm with which men greeted the age of wholesale Westernization booming in the 1870s and 1880s.

The modernist samurai leadership soon eatablished a policy of Westernization in which the establishment of Western-style educational institutions occupied a central position. They did wide research on the Western system, country by country, and in 1870 drafted, perhaps on the suggestion of G.F. Verbeck and other foreign advisers to the government, rules for the dispatch of students for study abroad. The following subjects and countries are listed:

Britain:	Machinery, geology and mining, steel-making, architecture, shipbuilding, cattle-farming, commerce, poor-relief.
France:	Zoology and botany, astronomy, mathematics, physics, chemistry, architecture, law, international relations, promotion of public welfare.
Germany:	Physics, astronomy, geology and mineralogy, chemistry, zoology and botany, medicine, pharmacology, educational system political science, economics.
Holland:	Irrigation, architecture, shipbuilding, political science, economics, poor-relief.
USA:	Industrial laws, agriculture, cattle-farming, mining, communications, commercial law.

In view of the history of nineteenth-century science, the above assessment was, by and large, correct and objective.

EMPHASIS ON PHYSICAL SCIENCE AND SPECIALIZATION

Japanese pioneers in Western science during the late Kokugawa period were impressed by the Western process of inquiry into natural laws. They came to feel keenly that, although there was no great gap between East and West as far as the classificatory knowledge of natural history was concerned, Chinese and Japanese cultures seriously lacked the belief in the underlying regularity in Nature and the 'investigation of its principles', namely natural philosophy. This *kyuri* aspect was now under the Meiji *kagaku* scheme narrowed down to mean the physical sciences and scientific technology. The primary school curriculum, therefore, prescribed for the first time by the Ministry of Education, was not natural history and biology oriented, like that of American primary education, but physics-oriented.

During the 1870s and 1880s greater emphasis was placed on science and technology in the Japanese educational curriculum, from elementary school to university level, than in any other nation. For instance, mathematics and science occupied about one-third of the school curriculum at the lower grades (first four years) and two-thirds at the upper grades of the eight-year elementary education, though because of the shortage of qualified teachers available, it is somewhat questionable to what extent these ideal plans were put into practice. At the university level too, the emphasis on science and technology was evident in the high per centage of graduates in scientific disciplines of Tokyo University (85 per cent in the 1880s as compared to 40 per cent in the 1920s).

OFFICIAL AND PLANNED CHARACTER

At the frontiers of newly forming disciplines in nineteenth-century Europe, the scientist was free to take up for his research any problem that interested him, and the scientific communities were formed by individual scientists drawn together by a common interest. The voluntary activities of scientific or professional societies usually preceded their inclusion in the university curriculum. In the case of Meiji

Japan, however, and as is generally the case when a foreign discipline is artificially transplanted under state sponsorship, this process was reversed. The government first created institutions for training specific personnel and only then did university graduates in each discipline form their scientific societies, not purely for academic purposes, but mainly with common interests in their new and still very weak scientific community in Japan which had a 'planned character,' planned by and for the set purpose of catching up with the Western standard of science as quickly as possible.

In the nineteenth century, however, it was uncommon, even among the advanced nations to find established precedents or formulas for a national science policy. Thus, the Meiji government had to find its own way by trial and error, Unlike today, it was simply unimaginable to have some sort of international aid or technology transfer from a developed country to a developing country. In Europe and America science and technology were not yet state-supported but largely private activities within the private sector. The Japanese government was simply trying to purchase and procure the scientific facilities and manpower available in the European and American free markets.

On the practical level, first of all, it urgently needed qualified teachers and engineers in building up a modern state. Here the new government, in pursuit of its set aim, built schools and factories, trained scientists and engineers in short and intensive courses, and sent them off to their posts. Rather than having every scientist following his own research interest, as was often the case in Europe and America, priority was given in a collective way to certain basic tasks, the accomplishment of which was indispensable for the operation of a modern state: matters such as geographical and geological surveys, weights and measures, meteorological observations, sanitation, printing, telegraph and telephone, military works, railways, survey of natural resources, etc. All these activities were carried out by the Ministry of Technology, the Ministry of Interior, the Ministry of Finance, the Commission for the Colonization of Hokkaido, the Army, the Navy and other functioning governmental agencies under the supervision of many foreign engineers working in Japan. To conduct such ambitious nationwide projects, these agencies had to have their own short-course training programme to provide field assistants to foreign employees. These agencies did not then exclusively depend on the Ministry of Education, which was responsible for regular long-term education programmes. An example was the Telegraph School of the Ministry of Technology.

We may label these activites as 'public science' initiated by the government. This step in science for public service was the indispensable prerequisite for the industrialization by the next generation.

Besides this, the government entered into private entrepreneurship constructed and managed pilot plants and subsidized new kinds of industries. The Ministry of Technology and the Commission for the Colonization of Hokkaido were the two major institutional innovations, which carried out new experimental programmes and were also the centres of Westernization and modernization.

Their enterprises were from the beginning exposed to financial risk. Many of the projects of the Ministry of Technology eventually proved to be too far ahead of their times, as they intended to introduce the technology of an industrialized society into a pre-industrial environment. For instance, their railway construction enterprise was economically unsuccessful at the time, and only paid off commercially after 1880, in the next phase of industrialization.

Thus, Y. Fukuzawa, Japan's foremost exponent of industrial revolution, concluded: 'We should not blame them too much for their financial failure. After all, it was a costly tuition for the Japanese to learn civilization.'

In giving priority to the construction of an institutional system within which to transplant Western paradigms, Meiji Japan paid more attention to the configuration and format of learning than to its content. Men troubled themselves little over how new scholarly paradigms were being born. Neither did they entertain the notion of participating in the ongoing advance of normal science. Their first preoccupation was the creation of an institutional framework to house knowledge previously canonized and accepted as standard in other traditions.

First published in Everett Mendelsohn (ed.), *Transformation and Tradition in the Sciences*, Cambridge University Press, 1984

16 The American Occupation and the Science Council of Japan

The creation of the Science Council of Japan was an epoch-making event in the history of science. The council is a representative organ for scientific researchers. Its members are elected by the entire body of professional, working scientists – not only by the established groups but by rank and file researchers as well. Council members, like those in the American National Academy of Science, are recognized and given official status by the government. The council was undeniably, however, a product of America's occupation of Japan following the Second World War. Had it been created independently, separate from America's occupation policy, there is no doubt that the council would have taken on a substantially different character. This essay is an attempt to examine the influence of occupation policy on the birth of the Science Council of Japan. The analysis is from the vantage point of the occupation forces and is based on the documents of the Economic and Scientific Section, Scientific and Technical Division (ESS ST) of General Headquarters (GHQ).[1]

HARRY C. KELLY AND THE JAPAN ASSOCIATION OF SCIENCE LIAISON

Harry C. Kelly (1908–78) was a former professor at Lehigh University, Lehigh, Pennsylvania. As a physicist, during the war Kelly worked on radar research at the Massachusetts Institute of Technology. Shortly after the war he joined the American occupation forces in Japan as a civilian chief of the Fundamental Research Branch of ESS ST and later became the associate director of the division. Kelly can be regarded as one of the New Dealers active in the early years of the occupation. His aspiration was to contribute to Japan's social and economic development by helping to organize its scientific research. His first act was to create the private Japan Association of Science Liaison as a conduit to Japan's scientific community. This organization later evolved into the Preparatory Committee (Sewaninkai) of the Scientific Research Organization Renewal Committee, finally emerging as the Science Council of Japan. Kelly was, therefore, the true parent of the Science Council, the driving force creating Japan's post-war scientific organization system. Consequently, of the GHQ documents related to the recreation of Japanese science, those concerning Dr Kelly are among the most significant.

The first task undertaken by GHQ in academic and scientific matters was the dissolution of arms research. Beginning with atomic energy, aviation and naval research, the occupation authorities ultimately banned research on television, radar and other radio-related activities as well. A well-known story has it that GHQ went so far as to have Japanese cyclotrons destroyed.

Japanese researchers had no idea whatever about what to expect for the dismemberment of wartime scientific mobilization or about the extent to which scientific activity would be authorized under the occupation. It was widely believed that GHQ authorization had to first be obtained to carry out research plans, so researchers flooded ESS ST with letters outlining research projects and requesting permission to pursue them. The authorities responded to each individual, pointing out that there was no restriction on normal, routine research. GHQ in time came to feel the necessity of employing measures to correct misunderstandings about occupation policy.

The question of permission was not the only problem. Researchers also faced at that time the problem of research materials. In the early years of the occupation even tickets for railway travel were best obtained by having the authorization of the occupation forces. As a result, Japan's scientists harboured the vague hope that given the absence of even daily necessities, if they could only get the signature of the occupation authorities by petitioning them directly they could avail themselves of the usual supply routes, or even receive materal from GHQ itself and not have to rely on the black market. Although Japan's scientists were interested in material assistance, as scientists their first desire was to bridge the information gap that had cut them off from foreign developments during the war. Researchers, consequently, flocked to the Scientific and Technical Division seeking, most frequently, information on the American scientific world. Their primary objective then was to pursue the latest developments that had taken place outside Japan. To do so they would like to have travelled to the United States themselves but as this was very likely not possible they hoped to forge scientific ties by inviting American researchers to Japan.

Japanese scientists had many and varied problems for which they sought solutions. The occupation authorities – particularly Kelly and Dr G. W. Fox – recognized that to respond to these problems was a matter of some urgency. They therefore had scientists create a liaison group of Japanese researchers that later evolved into an organization known as the Japan Association of Science Liaison (JASL). GHQ for its part intended to use Japan's scientific organization to further the objectives of the occupation. To that end it requested that a report be drafted on wartime research activities and ordered a monthly submission of reports on research-related activities. A joint Japanese-American community was also established for atmospheric and oceanic surveys, oil exploration, and similar projects.

Projects of this kind, however, were the fruits of individual, personal contacts between GHQ scientists and their Japanese counterparts and not the products of institutional initiative. As a consequence, those researchers who sought relations with the Scientific and Technical Division represented that part of Japan's scientific community that was competent in conversational English. The concern over whether this group was representative of the scientific world as a whole was a question that dogged Kelly. In order to build a more reliable and meaningful linkage with Japan's academic world Kelly conceived of and attempted to put together JASL as a more rational means of fostering contact. This idea is expressed in GHQ documents as early as April 1946.

From that time until the birth of the Science Council in 1949 Kelly remained the occupation's point man for issues related to the rebuilding of Japanese science. Kelly's superiors, including Brigadier General O'Brien, chief of ST, allowed him a great deal of freedom, generally supporting Kelly's positions and protecting him from outside interference while to a certain extent controlling his activities. In the early years of the occupation, other branches of GHQ harboured the misplaced notion that if Japan's research institutions were strengthened, the wartime mobilization system would be resuscitated and disarmament dead. To deal with important problems of this kind Kelly convened a meeting of various GHQ branches to decide on an initial general policy. More important, however, Kelly used this meeting to win the approval of his proposals from other GHQ divisions and was thus able to proceed immediately after his negotiations with the Japanese.

In his travels around Japan, Kelly developed friendly relations with a number of Japanese scientists. One of these men, Horiuchi Juro* of Kokkaido Imperial University, introduced him to Professor Tamiya Hiroshi of the botany department, science faculty, Tokyo Imperial University.[2] With Tamiya, Kelly began discussions on the organization of Japanese science. The living standards of an Imperial University professor had fallen to such an extent that Tamiya later recalled that he had been drawn to Kelly and his plans and was induced to accept responsibility for the reorganization task Kelly envisaged by the mere offer of American cigarettes.[3]

Through Tamiya's efforts a nucleus of mainly Tokyo University scientists, beginning with Kaya Seiji and Sagane Ryokichi and several others, began to be assembled. Quickly Kameyama Naoto and Yukawa Hideki were included as the circle of participating researchers expanded.

Kelly may have originally intended to create only a liaison committee of Japanese scientists to aid in the execution of the occupation's policies, but in the preparation process it became clear that this modest objective was not sufficient. In his meetings with important Japanese scientists Kelly saw that the existing academic associations had become an impediment to scientific development and concluded that a wholly new organization had to be devised. GHQ, particularly Kelly, must therefore be identified as the catalyst in the ultimate fulfilment of this task: the creation of the Science Council of Japan.

Three academic organizations in post-war Japan were directly affected by this discussion about reorganization: The Japan Academy, the National Research Council of Japan, and the Japanese Society for the Promotion of Science. Since its founding in 1906 the Japan Academy, despite being frequently referred to as an old folks' home, had been an eminent institution. The Research Council, which had been created following the First World War to foster international technical exchange, during the Second World War guided the mobilization of scientific activity. The Society for the Promotion of Science was a government-endowed

* Japanese personal names here follow the Japanese style of family name first.

nonprofit foundation whose purpose was to disperse research funds. Along with eliminating the wartime mobilization system and disbanding the Research Council, questions arose about reorganizing the Japan Academy and the Society for the Promotion of Science and dividing between them the functions of the Research Council. The reorganization activities were moved forward by the Japanese themselves under the direction of Nagaoka Hantaro, the director of the Japan Academy.

On the fifth and sixth of June 1946, Kelly held the first meeting of the Japan Association of Science Liaison. Kelly seems to have used this organization and this opportunity to embark on his major objective of reorganizing Japan's technical and research structure. Kelly, in a letter of 22 June, indicated the extent of his interest in this reorganization. He suggested to Tamiya that he should

> . . . write a letter from the Japan Association of Science Liaison to the National Research Council in America telling what [the] functions of the organization are, what the problems are confronting the organization and what contact you would like to have with the National Research Association and what kind of aid you need from the American scientists. Also if you would like to have a member from NRC come over here not for just a few months but so that there will always be one member in Japan.
>
> I don't think we in America appreciate the problems here. If we could get some indication of the eagerness of the Japanese scientists to help your country and cooperate. I think American scientists would probably be in a much better mood to help out in this situation if they have this information. They know nothing except what is read in the papers. I will go back and see the NRC and ask them if they will help.[4]

Meeting the same day, Kelly and Tamiya discussed the proposal that had been outlined in the letter. The following is a portion of the transcript of their discussion.

K: Do you think it is a good idea?
T: Yes. But there is the National Research Council of Japan and also the Imperial Academy.
K: I would not worry about that. There is too much confusion now and I will tell them that they need help to get out of the confusion.
T: You see, we have stated our interests as messengers and our duty is only to transport the message.
K: I disagree with you.
T: A letter from our organization would not be representative of the Japanese scientists as the Japan NRC and Imperial Academy of Science are representative of Japanese science.
K: They are not doing the job right. I don't think they are doing very much good.
T: Then I will try to write a letter as you said.
K: How about writing it from the Japan Association of Scientific Liaison and telling them the group is representative of Japanese scientists?
T: We cannot do that, we are not intended to represent Japanese scientists.
K: I think the time has come when we have to come out in the open – we are trying to be that representative. That is really what we are driving to do.

T: Well, this group, the NRC and Imperial Academy has accepted our association with limitation: that it should not intend to reform Japanese science.
K: Those organizations are not good, they need reorganization. Therefore, it is up to your group to help reorganize this group.
T: Our present group is far from it. We have the hope, confidence and worth to be called active scientists but among such a political situation, it is not yet time to express this thought. It is too idealistic.
K: If anything is good, it is not too idealistic. Do you think it is a good idea?
T: Yes, but it takes time, but the name is a poor one if it is this idea.
K: It is necessary today that this be done gradually – not too quickly because if done quickly, you will do a poor job.
T: It will be easier if you yourself grasped the Japanese scientists' leg and organized them. I do not think this group is representative of Japanese nor does it have history – only 40 members.
K: Even if there were five members it could represent a nation.
T: Yes, many of the members of this new association do represent Japanese science – for example, Yukawa, Horiuchi, but then others are not.
K: Why do you think there is no history? This organization has history and is going to make history.
T: We started only one month ago. Frankly speaking, you Americans always judge only a cross section of these people and culture.

Kelly the New Dealer is clearly in evidence in this conversation. Secure in his authority Kelly moved in impatient pursuit of radical, rational renovation without regard for existing institutions.

Tamiya felt obliged to produce the letter to the American NRC as Kelly requested. By the following day a draft was completed. During a discussion at the JASL meeting of 9 July, however, the question of the organization's scope and significance was, once again, a point of contention between Kelly and Tamiya. Tamiya's concern was that by overcommitting himself to Kelly and his proposals he would lose the support of Japanese researchers. He feared that if he courted GHQ privately and strengthened JASL's position without its having obtained the authority or right to represent Japanese scientists as a whole its actions would be criticized and its negotiations repudiated.

Kaya Seiji's view of the issue was quite different. In contrast to Tamiya's cautious attitude at the 9 July meeting, Kaya aligned himself with Kelly and expressed his agreement with Kelly's demands. Kaya, as one observer has noted, possessed no unified, coherent body of thought. He was a pragmatist. He acted on the basis of a cogent analysis of existing conditions and possessed the flexibility necessary to achieve compromise.[5] Among those scientists and engineers who had been a part of the wartime research mobilization programme, Kaya became a representative post-war technocrat attempting to find a voice for Japanese scientists at the policy-making level of government and he ultimately achieved the greatest prominence in this effort.

The year 1946 was the most severe period of post-war Japan's economic and food crises. Like David Lilienthal, the first head of the Atomic Energy Commission and the planner of TVA, who after the depression urged economic rebirth through science and technology, Kelly suggested at least as early as April 1946 that GHQ 'encourage only those research programmes which are directed towards improvement of Japan's economy'. In an internal policy memo Kelly stated that:

> One of the immediate objectives of the Occupation is to insure that the Japanese secure for themselves the necessities of food, clothing, shelter and education. The scarcity of raw materials requires that the maximum use be made of such materials as are available and that substitutes be developed. Research should be directed towards these ends.

Fundamental research under such circumstances as Japan was then experiencing could only be considered, in Kelly's view, a 'luxury'. Kelly did not propose that basic research be forbidden but rather that Japanese scientists be encouraged 'to accept their obligations in this emergency'. Later, when these views were put before the JASL the members were generally apathetic. One reason for this response was that Kelly's initial contact with the Japanese scientific community was limited to the elite group of pure scientists at Tokyo University who were incapable of such an approach. A second reason was that Japanese researchers in general had lost their vitality and self-confidence in the post-war period, making such a course impossible. Among them, however, the views of Kaya and Kameyama Naoto, who later became the first president of the Science Council, were closest to the New Deal line laid down by Kelly, and they and their group later became the founding nucleus of the Science Council of Japan.

THE DEPARTURE OF MEIJI ACADEMISM

Kelly returned to Japan from the United States in late September 1946. On 28 September he convened a meeting of JASL members representing the existing academic organizations along with the relevant GHQ sections and from each group demanded the presentation of a proposal for the reform of Japan's scientific and technical infrastructure. Kelly, however, from early on, had decided to push for the evolutionary development of JASL itself.

It quickly became apparent, however that a movement had emerged that sought to defend and indeed broaden the functions of the existing research organizations. Members of JASL led by Tamiya, Sagane, and Kaya met with Kelly on 15 November to alert him to this problem. Kelly, determined to check such sentiments, met with the head of the Imperial Academy, Nagaoka Hantaro, on 18 November. Kelly attempted to impress upon Nagaoka the urgency of the problems facing Japan and Japanese science and the need to quickly and effectively reorganize scientific and research institutions. The following discussion then ensued:

N: The problem is how to organize the centre for directing research and we hope that the Academy would be the best place and we are now going to have new young members – 150 in number.

K: But that may not be the complete answer.

N: We are going to hold conventions and discuss the matter. We had a conference with NRC and also with the Society for Promotion of Science. The NRC wishes that 300 new members should be elected to the Academy but that was too much. We discussed in the Academy that the number should be limited to 150 and not for whole life term but [a member's] term would be four years.

K: That doesn't mean anything to me. Have you thought this way – What is the function of organization of science? Have you talked about it? Now is the time to discuss that. Nowhere else in the world do they have the opportunity to do something about it that you people have.

N: Since the establishment of the Society for the Promotion of Scientific Research, the number of research items is about 300; that involves all the researches that can be undertaken at that time which was most useful to the country.

K: That still doesn't mean anything to me. That is still not worrying about the basic problems which are facing Japan and the world – not the number of research items or papers – the question is what the function of a research organization is and what is the responsibility to other countries, to your people, and to the world.

N: That is national and international.

K: I have seen no discussion of that; you are just shuffling things around and nothing will happen. Science today has a frightful responsibility in the world and we can't just talk about it in this way.

N: It is a rather difficult problem and everybody will have his own opinion and they must be connected together and restudied.

K: But you're doing nothing about it.

N: Well, we are always considering that problem.

K: Your proposed changes have no consideration of these problems.

The difference between the two men stemmed largely from the fact that to Nagaoka the conduct of research was not controllable and was best left to the spontaneous creativity of individual scientists. Kelly, on the other hand, opposed the individualistic attitude of Japanese scientists and criticized them for their inability to cooperate to solve larger social problems. He was therefore clearly warning Nagaoka of the reorganization of Japan's scientific community by GHQ – or in Kelly's words, with GHQ as a 'catalyst' – as a means of dealing with the problems that confronted Japan.

At this time Nagaoka, who was more than eighty years old, and Kelly, who was not yet forty, were separated in large measure by a difference in generations. Nagaoka had studied in Germany and was the representative in Japan of the highest tradition of German academic scholarship. He was also one of the first Japanese scientists to be involved in and compete in international research front activities. Japan's Nobel laureate physicist Yukawa Hideki is reported to have said that 'without Nagaoka Hantaro we would not have amounted to anything'. It is for his role as a pioneer of Japanese theoretical physics that Nagaoka is today remembered.

Nagaoka, however, also typified the Japanese legacy of nineteenth-century German academism. It is erroneously believed that the Japanese national universities were modelled after the German university system. German universities' philosophy faculties, however, possessed in general institutional freedoms and organizational flexibility absent in the Japanese system. What Japan's universities did resemble most closely was the Prussian bureaucratism rooted in law faculty graduates. From around the 1880s the upper reaches of the Japanese governing bureaucracy came to be dominated by graduates from the Faculty of Law at Tokyo University. As a result Japan's national universities and its scientific community resembled the national government's organization and were subordinate to it.

Nagaoka disliked bureaucratism and formalism and as a result was continually uneasy about the strong bureaucratic strain in the academism of Japan's universities. He had participated in the highly elite late nineteenth-century German academic world, which at that time was on the eve of the quantum theory and which was an academic structure he felt to be ideal. To Nagaoka Japanese academism, while differing in time and place, was not very far from this ideal type.[6] During the war Nagaoka criticized the National Research Council of Japan, which conducted academic research under the auspices of the scientific mobilization programme. At that time he emphasized that what is called research is less a function of organization and faculties than of human beings, the originality of individual creativity. Such views were the very essence of the scientific ideals of German academism. Given Nagaoka's ideals it is quite natural that he held opinions at variance with Kelly's New Deal approach to scholarship and science. Even such approaches as those taken by J. D. Bernal's *The Social Function of Science* to Nagaoka represented a strategy identical to that of the wartime mobilization system.

FROM THE PREPARATORY COMMITTEE TO THE RENEWAL COMMITTEE

The model for Japanese scientific organization considered by Kelly was the American National Academy of Science (NAS). Like Japan's Imperial Academy, NAS was an honorary institute. But unlike its Japanese counterpart NAS was more a working institution. Responding to Woodrow Wilson's call NAS organized the National Research Council for the purpose of mobilizing wartime research activities during the First World War. NAS's members were limited to natural scientists, excluding engineering, the social sciences and humanities.

Following the war Japan's established scientific organizations – the Imperial Academy and other established scientific groups – had lost much of their strength and as a result their ability to resist overhauling. Kelly was convinced he could build an organization of unprecedented strength and effectiveness without interference or intrusion from existing power bases. It was felt, therefore, that if a complete organizational renovation were to take place this was the best time. Given the occupation's role as a 'godfather', Kelly developed the ambition to leave a lasting legacy in the history of Japan's scientific policy. For Kelly and for the other New Dealers of the occupation it was an opportunity to grapple with and bring to a successful conclusion the great problems – such as agrarian reform – that had become impossible in their own country. It is this sense in which Kelly's activities too must be seen.

Kelly's first task was to try to eliminate the existing scientific organizations. The organizations for their part, recognizing the circumstances that prevailed under the occupation, attempted to come up with their own plans for internal reform. These proposals were essentially hollow and were put forth in an attempt to forestall GHQ intervention. Kelly, however, took no notice of the organizations' considerations. He wanted to give official sanction to JASL and make it over into a more public forum. JASL began, however, as a private group centred on Kelly. Nishina Yoshio and others were very concerned by the complaints that had arisen in the scientific community about JASL's symbiotic relationship with GHQ and its reliance on the prestige and the power of the occupation forces. On 16 October 1946 they visited GHQ and warned the occupation authorities once again about this matter.

The Imperial Academy, despite its geriatric character, did possess the machinery for electing new members and it operated under formal rules. JASL had been provided with no such authority. JASL was criticized as being an undemocratic organ. To offset these complaints the more publicly visible Preparatory Committee (Sewaninkai) was created. This group acted as a supervisory board overseeing the election of members to the Scientific Research Organization Renewal Committee, ensuring that the election of members was conducted democratically and that the elections were representative of the Japanese scientific community.

The task of scientific reorganization moved step by step from JASL to the Preparatory Committee to the Renewal Committee, and in so doing became more public and formal. Kelly, who had originally conceived of JASL as a private intermediary organ, was unable to control the more public Renewal Committee. Japanese scientists had pushed for this opening of the reorganization process and Kelly could do little but leave its details in their hands. At the time of JASL, for example, it was Kelly's intention to broaden the organizational base from basic research to applied sciences such as engineering, medicine and agriculture. A JASL expansion of this kind Kelly could control.

With the rejection of the reorganization proposals drafted by the Imperial Academy and the Society for the Advancement of Science, however, organizational reform plans were devised in which the humanities and social sciences would be included. Such a conception was quite contrary to Kelly's American Academy model and represented an institutional expansion he was unable to control. At the same time the Ministry of Education's Scientific Education Division Chief, Shimizu Kinji, called for intermediaries representing the seven faculties of Tokyo University: law, letters, economics, science, engineering, agriculture and medicine. On 17 January 1947 the Preparatory Committee was launched with forty-four members selected in accordance with the Tokyo University seven-faculties model. The development of the Preparatory Committee based on Shimizu's 'Scientific Organizational Reform Plan' was a situation with which Kelly was not wholly content. He expressed his dissatisfaction in a handwritten memo of 18 August 1947 (presumably to be addressed to the visiting Scientific Advisory Group).

1. Sewaninkai [Preparatory Committee] created contrary to our ideas and members appointed by Mombusho [Ministry of Education].
2. [Sewaninkai] never submitted election procedure until too late to alter.

3. Procedure sounds good on paper but wrong in practice for following reasons: The members of the Executive of the societies are mainly old school, nominated old school people as electors, continuing process of old school ties until Renewal Committee [is] comprised mainly of them.
4. Mombusho, having such a hold, made representation of social sciences strong.
5. The Renewal Committee being comprised of a group Tokyo University men, it is inevitable they will fight to get an organization in which they retain their privileges and power.
6. The younger scientists and those from the universities other than Tokyo, not having respect for the Renewal Committee, will not abide by its findings (no matter how good they are) although they will pay lip service.
7. If we have to apply pressure to ensure findings of Renewal Committee are in accordance with our ideas we lose all the force of our policy to let the Japs decide for themselves: better to ensure that Renewal Committee is truly representative by assisting with the procedure for election – give the Renewal Committee some basic principles and let them ride, merely keeping a watch briefly to see they keep within the scope of the enunciated principles.

The Preparatory Committee began the process of electing members of the Renewal Committee in early April 1947. The election was completed by 10 August when 108 members had been selected.[7] On the American side the Scientific Advisory Group, which was sent to Japan by NAS to 'advise General MacArthur's staff on the decentralization of Japanese scientific research and to evaluate the proposals considered by Japanese scientific organizations', arrived in Japan on 19 July. The arrival of this group represented the fruit of Kelly's visit to the United States the previous year.

After its arrival the advisery group visited universities and research groups in various parts of the country. The group submitted its report on 28 August and on the following day returned to America. Kelly, who acted as the group's guide during its stay in Japan, explained to it the activities of the Preparatory Committee. Inevitably, Kelly's criticisms of the committee – which were expressed so clearly in his memo – found their way into the advisery group's report.

There was also criticism from a number of Japanese about the absence of democracy in the Preparatory Committee's activities. Comments on the preservation of the influence of the Tokyo University clique on the Renewal Committee appeared in various newspaper editorials of the time, which shows that Kelly's opinions were not unique to him.[8] It might be pointed out that since the adoption of the seven-faculty organization by the Science Council of Japan it has been very difficult to win support for newly developed areas of scholarship. The possibility of an organizational flexibility that would allow for the creation of new divisions to accommodate new knowledge was absent from the Preparatory Committee's thinking on the issue.

Both Kelly and the American Advisory Group reluctantly authorized the inclusion of social science in the incipient scientific organization. Kelly's reorganization efforts began with the sciences and broadened to include the applied sciences. This in large measure explains the organization and structure of JASL.

The representatives of the humanities and social sciences on the Preparatory and Renewal Committees were not as interested in the work of the committees as were the natural scientists. As a result, representatives from the science area attended committee meetings in the greatest number followed by engineering, agriculture, medicine, humanities, law and economics. Moreover, later at the first meeting of the Science Council the per centage of humanistic area representatives who voted was low compared to the natural sciences. Several factors may account for this difference in interest. For example, unlike the natural sciences, there was no national learned society in the humanities or social sciences. Also unlike the physical sciences, which moved towards reorganization at the time of JASL, discussion of organizational reform of the humanities was largely absent. Finally, the natural science division, which was the birthplace of JASL, was most reform minded and the average age of its representatives was the lowest.

GHQ RESPONSE TO THE RESEARCH RESTORATION COUNCIL

Of the various organizations of left-wing scientists the most well known is the Association of Democratic Scientists, an umbrella group for leftist scientists and science organizations. The association was launched as early as January 1946 but does not appear in ESS ST documents on scientific organization until much later.

On 10 June 1947, members of the Preparation Committee (Junbi Iinkai) for the Research Restoration Council met with Kelly and General O'Brien, head of the Scientific and Technical Division. The chairman of the Imperial Inventions Association, Ono Shunichi, acted as the group's spokesman. Ono was a skilled engineer – he had graduated in electrical engineering from Waseda University – and was fluent in Russian. According to friends of that time, his pleasant personality enabled him to deal with people of all ideological stripes.

The creation of the Research Restoration Council was sought as an organ designed to complement the Economic Restoration Council, made up of labour and popular groups. It was to comprise scientists' and technicians' unions in private, governmental, university and industrial research laboratories. Its intention, announced in a petition, was to put pressure on Kelly and O'Brien, under whose authority the reorganization of science, it was believed, was being carried out. Shortly before the meeting between ESS ST leaders and the representatives of the Preparatory Committee the new socialist government of Katayama Tetsu had been formed. To Ono, the opinions of the Economic Restoration Council were not listened to during the former Yoshida Shigeru cabinet. With the change of government, however, he believed that the council's views would be given their proper weight. Being a Socialist Party member himself, and having a number of friends in the new cabinet, Ono felt inclined to apprise Kelly and O'Brien of his potential influence. Thus for Ono the inauguration of the Katayama cabinet was a welcome event and represented a propitious time for his group to force itself on GHQ.

GHQ for its part was greatly concerned with the activities of the Economic Restoration Council and in the preparation process of the Scientific Restoration Council. It was, however, bogged down with conferences on the Science Council organization issue. So it was Ono who first approached Kelly. He pointed out the Preparatory Committee (Sewaninkai) was merely a collection of Tokyo University professors. It did not, he argued, reflect the opinions of researchers in other universities, government and private laboratories nor did it represent the views of younger scientists. In response Kelly stated:

> One of our wishes is that the younger scientists and technologists be represented and if they are not, there are two things to do, find out how they can be better represented in this group, Sewaninkai, than they are now. We would like to hear your suggestions. But I should also like to point out that if you want to set up a group like this you would have identical problems. You too would be criticized for not being representative, and I think you can help most by finding the best way of amalgamating and finding the best way to see that these younger scientists are represented. I am surprised that they do not have a vote in the Renewal Committee.

Ono then proposed a union of the preparation committees of the Renewal Committee and the Research Restoration Council. Giving O'Brien and Kelly a list of organizations that participated in the Restoration Council preparation process he introduced the members of the Preparation Committee who represented those bodies. The following discussion then took place:

K: We don't know why you need this group as there is already the Sewaninkai and the Renewal Committee.

O: One point is clear that when the ultimate organization of science will be established, the Council for Research Restoration will amalgamate so you can think that this Council is of temporary nature, but instead of establishing the final organization of science in Japan, it is clear we must prepare and endeavour in such manner that Sewaninkai will not eliminate the necessary younger strata and minor elements who are very essential for the final.

K: Your job is to see that Sewaninkai and the Renewal Committee do a good job.

From Kelly's point of view consultations with other organizations would simply cause confusion. As a result, despite their limitations, he believed there to be little choice but to work through the Preparatory Committee and the Renewal Committee. Although he had complaints about the direction of the Preparatory Committee and its ability to fulfil its functions, Kelly bureaucratically rejected the appeal of the Research Restoration Council. He then left the meeting and the conversation was carried on by O'Brien and Technical Division chief Colonel Allen. To Ono's group the Preparatory Committee's expressed methods of carrying out the election were not democratic but oligarchic and underlined the committee's lack of responsiveness to younger scientists and those in industry in particular. Beyond criticizing the Preparatory Committee, however, Ono and his group derived no satisfactory results from the meeting.

The quoted conversation between Kelly and Ono perhaps gives the impression that GHQ regarded the Research Restoration faction lightly and that it felt it could deal with the group as it wished. In fact, however, if occupation documents are examined carefully we see that Ono and his group made a very strong impression on Kelly, the result being that from this time research was started and a Research Restoration Council file began to be assembled.

Even within the Scientific and Technical Division, assessments of the Research Restoration Council differed. A memo written by Allen and circulated in mid-June suggested that if the Preparatory Committee (Sewaninkai) were likened to an employers' group GHQ should then provide appropriate proportional representation on the Renewal Committee to the Ono faction. If this were done the Research Restoration Council, for which Ono spoke, would renounce its idea of trying to control the Preparatory Committee. Moreover, the memo argued, such a course could serve to create a mediating organ for the two sides within the Economic Restoration Council.

A memo authored by Kelly countered, saying there was no reason for confusing matters by comparing academic reorganization with a labour-management contract. He stated that GHQ's goal was the creation of an organization of scientists to facilitate research, not the establishment of a political organization. To deal with the issue as a political question, he argued, would not be very useful in prodding Japanese scientists into action.

Both Allen and Kelly were invited to attend a meeting of the Research Restoration Council held on 30 June in the Science Museum in Tokyo. Kelly's speech to the assembly emphasized the necessity of conducting scientific reorganization through the Preparatory Committee. Pursuing other approaches at that point, he claimed, carried the danger of disrupting the reorganization process. Allen in his speech was more sympathetic to the Restoration Council.

Ono himself recognized that it was too late to reorganize the Preparatory Committee. Indeed, the new Renewal Committee, chosen in accordance with the original Preparatory Committee formula, met for the first time on 25 August 1947. On the following day a departmental memo penned by Kelly stated, 'We have diplomatically and tactfully announced to the science Restoration Council that we support the Renewal Committee alone.' On 5 September, General O'Brien finally spelled out the division's policy concerning the problem of Ono and the Restoration Council and in doing so affirmed the role of the Renewal Committee. He said that:

> The questions brought by Mr Ono and the Research Restoration Council are ones that must be decided by the Japanese themselves. The inaugural message of the Renewal Committee was endorsed by both the committee and Prime Minister Katayama. It should, therefore, henceforth begin its work. Opinions on the scientific reorganization issue which are presented to the Renewal Committee are not the province of GHQ. GHQ will involve itself only in situations from which the Renewal Committee abstains.

In October 1947 the Association of Democratic Scientists submitted a plan for the renewal of scientific organization to the Renewal Committee. The plan demanded the creation of a high-level conference of scientists whose advice and opinions the government would be obliged to follow. According to ESS ST documents (particularly a 23 February 1948 report from Paul Henshaw to Kelly) this conference plan was Ono's way of responding to the Renewal Committee. But, the report said, the plan 'has been completely ignored and has in no way influenced reorganization activities so it need not concern us'.

The Renewal Committee, which had been formed in August 1947, held a number of general conferences. GHQ, however, received only intermittent progress reports on its reorganization deliberations and had no opportunity to influence them. In late January 1948 Prime Minister Katayama sent a note to Renewal Committee chairman Kaneshige Kankuro proposing that the reports be terminated until the end of the fiscal year on 31 March, when preliminary investigations of the issue had been completed.[9] The Renewal Committee sought GHQ's approval for such a plan. At this final stage GHQ held a series of staff section meetings beginning from 1 March to study and finalize GHQ's position. In a memo of 19 March General O'Brien informed ESS chief General Marquat that agreement on the broad principles of scientific reorganization had been obtained from all GHQ sections. These principles were formally approved by the various sections on 23 March.

Concerning this period, Tetu Hiroshige has argued that 'between the seventh (23–25 February) and eighth (25–27 March) conferences of the Renewal Committee there must have been powerful pressure placed on the Renewal Committee by GHQ'.[10] This assertion is borne out in GHQ records of the staff section meetings of that time. One of the clearest expressions of this problem can be seen in the meeting of 10 March. The main subject of this particular meeting was the relationship between the government and the Science Council.

The first to speak was the government section representative, Mr Porter. He complained that the structure of the scientific advisery body as proposed by ESS ST was becoming unnecessarily complex and that what it seemed to desire was an organization that would coordinate Japanese scientific activity. Porter suggested that an advisery organ appointed by the prime minister alone might be a simpler and better alternative. Kelly and General O'Brien of the Scientific and Technical Division, however, felt it necessary to preserve the organization's independence from government and to maintain the dignity of scholars. To achieve these ends they expressed their wish to build an advisery body similar to the British Royal Society or the American NAS. Thus, it was necessary to avoid government appointment of members and government-paid salaries for them. If not, the members would be unable to criticize the government or resist its actions. Moreover, they suggested, the organization's dependence on government funding might be unconstitutional.

Colonel Johnson from the public health and welfare section brought up another question. He wondered what the objection was to having the prime minister appoint members from a list prepared by the scientific electorate as Porter had proposed. Rather than reducing; he asked if such a procedure might not in fact enhance the organization's prestige. On this question the following discussion took place.

O: No, you have 500 members elected. You have 50,000 people electing the members. You then present them . . . qualified people should pick the best 210, not the Prime Minister. [Cites Ono and his group of political ideologists. If PM is to retain party support he would have to elect Ono's group, if they were in power] . . . let them be elected by the scientists themselves and not have 210 members culled out for political reasons.

K: I understood you to say we would like this to have scientific prestige and not political prestige.

P: You say represent Japan in international science . . . it has both the weight of the scientists and the [government] back of it. Would it not have more prestige?

K & O: Political pressure groups like Ono's would influence the Prime Minister. They are interested only in political ideologies and trade union activities which do not fall into this category. I think you would lose [a] tremendous amount by taking [the] Prime Minister into this purely for the securing of funds, when it may be unconstitutional.

On 10 March, when discussions on the nature of the scientific organization were taking place, the new Ashida Hitoshi cabinet was formed. On February 10 the Katayama cabinet fell, creating a political vacuum and a long interregnum imbued with considerable confusion. It was under such circumstances that the GHQ debate on government-science council relations was taking place. Kelly's and O'Brien's greatest concern was that Ono, who had earlier flaunted his relationship with the Katayama cabinet, might influence the government, pressuring it into accepting a left-wing takeover of the Science Council. To prevent this they sought to restrict the government's ability to intervene in the Science Council's activities. The draft law for the Science Council of Japan, presented to GHQ by the Renewal Committee following its seventh meeting, included the provision that the 'government should refer the following matters to the Japan Science Council'. On the ESS ST copy of the draft law the words 'should refer' were underlined in pencil and in the margin a question mark was written. In a revised copy of the draft, this passage was changed to read, 'The government shall seek the opinions of the Science Council of Japan on the following. . . .' The word 'shall', however, was later crossed out and the word 'may' written above it. This correction appeared in the final draft and was later adopted in the approved Science Council Act.

The results of this process are well known to Science Council members. Council recommendations, having no binding authority, have steadily come to be ignored by the government, a development that defines the course of the Science Council of Japan.

CONCLUSION

In the early GHQ files on academic reorganization a name that frequently appears is Watanabe Satoshi. Watanabe bitterly denounced the old Japanese establishment, giving forth views highly critical of Tokyo University, the Ministry of Education and the existing scientific organizations. Watanabe's opinions can be considered to have influenced Kelly's views of the Japanese scholarly establishment. Thus Kelly paid little heed to the desires of the established scientific groups. Instead he believed it necessary to eliminate them. Some other Japanese group must have assumed a central role in the process of reorganization. At the outset that role was played by the Japan Association of Science Liaison (JASL), Kelly's private, scientific advisery group. Later, the principal position was expanded to include the Preparatory Committee (Sewaninkai) and the Renewal Committee, and was finally assumed by the Science Council of Japan. The most important activists in the Science Council were among others Kaya Seiji, Kameyama Naoto and Kaneshige Kankuro. It was these men who would become the spokesmen and technocrats for the later rapid growth of Japanese science and technology.

Kelly and other members of ESS ST had a keen interest in the activities of Japan's scientific community. If we are to judge from the data in GHQ files, however, their sources of information were surprisingly limited. Of the very large and influential Association of Democratic Scientists they were largely ignorant and they were unprepared to accept recommendations from Ono's group, which they perceived as threatening to the reorganization process they sanctioned.

Of course, the threat of Ono's pressure alone was not sufficient to force Kelly to reduce the government's role in the formation of the Science Council. Kelly must have had other sources of information. Moreover, the Japanese Renewal Committee strongly opposed the 'popular front' line of the Association of Democratic Scientists societies. Nevertheless, the appearance of the Ono group at the same time as the Socialist Party cabinet of Katayama Tetsu had a profound influence on Kelly's thinking. The consequence was, ironically, that the Science Council, freedom of thought and research autonomy were protected from the government while at the same time they were also insulated from leftist influence. In this way Kelly effectively isolated the council and reduced its significance.

NOTES

1. This study is based on documents found in the Science Reorganization and the Scientific Renewal files of the Scientific and Technical Division, Economic and Scientific Section, GHQ. These materials can be found in the National Recording Centre, Virginia.
2. Tamiya Hiroshi, 'On the Tenth Anniversary of the Founding of the Science Council of Japan' (Japanese) *Gakujutsu Geppo*, no. 5 (May 1959), p. 5.
3. Ibid.
4. This was shortly before Kelly's return to the United States for summer vacation. He hoped to use this trip to get in contact with American research institutions. He had earlier argued that it would have been preferable if the Japanese case were brought to American research groups by Japanese scientists themselves. As a result, to strengthen the appeal, he tried to persuade Tamiya to accompany him. Tamiya, however, proved unwilling.
5. Aochi Shin, 'Kaya Seiji' (Japanese) *Chuo Koron* (January 1958).
6. Itakura Kiyotsugu et al., *Nagaoka Hantaro* (Japanese) (Asahi Shimbunsha, 1963), p. 540.
7. Nihon Gakujutsu Kaigi, *Twenty-five Year History of the Science Council of Japan* (Japanese) (Nihon Gakujutsu Kaigi, 1974), p. 265.
8. *An Outline of Scientific and Technical History*, vol. 5 (Daiichi Hoki, 1964), pp. 137–9.
9. Nihon Gakujutsu Kaigi, *Science Council of Japan*, p. 269.
10. *Scientific and Technical History*, vol. 5, p. 128.

First published in *Higher Education*, 18: 31-48, Kluwer Academic Publishers, 1989

17 Independence and Choice: Western Impacts on Japanese Higher Education

Western models of higher education generally serve two purposes when they are adopted by non-Western countries. The first is the 'window-shopping' mode in which complete freedom is retained on the part of non-Western recipients in selecting any one of a number of Western models. The second is the 'involvement' mode in which a Western model is appropriated by a non-Western country, whether it be on a voluntary or involuntary basis.

Inherent in the window shopping mode are certain limitations in understanding and replicating Western institutions, because colleges and universities are so deeply rooted in tradition that many of the culture-specific elements are filtered out of newly created non-Western institutions. Consequently, the adopted model often exhibits indigenous traits as well as the most up-to-date trends, which have not yet found their way into Western institutions. Such is the case at some Latin American universities, which have unsuccessfully attempted to introduce a radically new curriculum in the wake of political revolution.

In the involvement mode, ties are very close between Western and non-Western countries as evidenced by the exchange of faculty members and students, similar degree programmes and even common employment opportunities for graduates. The recent internationalization of research has promoted a further exchange of information and personnel.

In the history of Japanese higher education, the first example of the window-shopping mode occurred in the late nineteenth century, while the involvement mode is best illustrated in the post-Second World War Occupation period, in which reforms based on the American system were carried out. In contrast to other non-Western countries, where a colonial involvement mode came first, followed by a window-shopping mode in the post-war period, the Japanese experience was exactly the opposite and therefore quite unique. The following sections will present an account of both the window-shopping and the involvement modes in Japan.

WINDOW-SHOPPING BY THE MEIJI GOVERNMENT

Absence of a Single Model

Western models of Japanese higher education used to be a favourite topic of discussion among Japanese historians of higher education. Some discussed the early American influence in structuring the entire educational system, of which higher education is a part. Others emphasized the predominant German influence in medical and later law schools. Still others appreciated British contributions in engineering education, while French law was taught in the early law schools. All in all, one can never come to any clear-cut picture of a single dominant model. This mixture of multi-country influence was an intentional one, as the Japanese government, consciously or unconsciously, preferred to strike a balance of Western power and influence not only in education but in all other fields as well. Thus, it is impossible to analyse the attempt to establish a Western-style modern university in terms of a single country model.

Governmental Initiative

Since Japan was neither colonized nor dominated by a single Western power, there was no colonial-type authority to enforce adoption of a particular existing foreign model, which contrasts sharply with the post-war Japanese experience during the American occupation period of 1945–1952. Western influence in the nineteenth century was exercised only on an individual basis by employed Western consultants rather than through a well-formulated national or corporative assistance programme as is often the case in the contemporary Third World context.

Instead, a single agency was responsible for model selection: namely, the new Meiji Government established in 1868. Unlike the continuous and spontaneous development of Western universities, whose medieval origin preceded that of modern nations, the prototype of modern Japanese universities is a purely artificial product created by a Western-oriented modern government. There existed a number of schools of Western language and science supported by the Shogunate as well as the local fief governments, but these were all summarily closed. In the 1870s, the new government attempted to reorganize them with a systematic educational policy.

A product of the new government policy of the 1870s was the creation of the University of Tokyo, the prototype of modern Japanese national universities, as an indispensable sub-department of modern Japanese bureaucracy. Thus, we may be able to say that the model of modern Japanese universities is derived from bureaucratic institutions, whose ultimate origin can be traced back to ancient or medieval Chinese institutions, rather than to the medieval Western university from which all Western universities originated. All teaching and administrative staffs of the university were government employees of the Ministry of Education. Its faculty-department structure was modelled after the departmental hierarchy of bureaucratic machinery, which has been more rigidly and rationally developed in China than in the West.

One of the exemplary features of bureaucratic character in Japanese higher education is that students do not form a body independent of faculty organization but rather subordinated to the departmental structure of the faculty. A student belonged to only one particular department from his university admission to graduation or even beyond, making interdepartmental mobility extremely difficult. The merit of this system might be that it has been a very efficient and systematic way to create the strategically necessary manpower for modernization and to acquire existing bodies of Western knowledge in a short period of time.

Another feature of the Japanese university bureaucracy is that those employed by universities are all civil servants so that all members have to observe the Civil Service Regulation, which often conflicts with the free and self-generating intellectual activity of faculty members.

Model of Private Colleges

Being outside of government initiative in higher education policy, private schools had long been denied official recognition. They remained, at least in official accreditation, in academy status. Their status was raised to the collegiate level, the equivalent of national universities, only in 1919. However, they have maintained a uniquely liberal tradition, especially among the older ones.

Some of the private colleges bear a strong resemblance to American liberal arts colleges, but they are, in the final analysis, a minor sideline of the total national educational structure. And in the course of time, they have had to assimilate towards the national system by sacrificing their independent and unique characteristics and subordinating themselves to the government regulations enforced by the Ministry of Education. Especially the newer private colleges are modelled after the government universities, since most of the faculty members are graduates of old national universities who 'colonized' the newly-founded private colleges.

Western missionaries, Catholic and Protestant alike, were not able to establish a stronghold in higher education, largely because the Japanese government took the initiative in nearly all areas of modern institution building. Missionary schools had a good share only in the education of females from kindergarten to the junior college, which was relatively neglected by the government up to the end of the Second World War.

Guidelines for Westernalization: List of Western Models

In the draft rule for sending students for study abroad, prepared in 1870,[1] the following subjects and countries were listed:

Choice of countries to study

Britain: machinery, geology and mining, steel making, architecture, shipbuilding, cattle farming, commerce, poor-relief;

France: zoology and botany, astronomy, mathematics, physics, chemistry, architecture, law, international relations, promotion of public welfare;

Germany: physics, astronomy, geology and mineralogy, chemistry, zoology and botany, medicine, pharmacology, educational system, political science, economics;

Holland: irrigation, architecture, shipbuilding, political science, economics, poor-relief;

USA: industrial law, agriculture, cattle farming, mining, communications, commercial law.

In view of the history of nineteenth century science, the above assessement was largely correct and objective. Presumably, the policies for science and technology, including the above recommendation, were drafted mainly on the suggestion of G. F. Verbeck and other foreign advisers to the government.

The government-sponsored students sent abroad seem to have followed the above guidelines. Parallel to this, the government hired a substantial number of foreign teachers to provide instruction in every Western discipline. As a result of these measures, Western science was successfully imported during the 1870s and 1880s. Towards the end of the 1880s, the process was nearly complete when those returning students replaced foreign teachers in the University of Tokyo faculty.

Language Prerequisites

Early in 1871, a proposal was presented to the government by Inoue Kowashi, who later played a most important role in the formulation of Japanese higher education policy. It states: 'Let students master the Western languages first; then they can automatically speed up in mastering each of the specialized sciences by use of the languages.'

At a time when neither native-speaking teachers of Western disciplines nor textbooks written in their native language were available in higher education, it was natural enough that the foreign teachers taught in their own language and that students were required to have excellent aural comprehension in that particular language. Thus, preparatory courses in language training were prescribed for all students. In view of the scarcity of qualified foreign teachers and the urgent needs of native personnel for administering the modernization and Westernization programmes, courses were often divided into 'regular' courses taught by foreign teachers and 'special' courses in which instruction was provided by Japanese graduates in the Japanese language. Most of the private schools were also of this type. The latter students fulfilled the immediate needs of society, while the students in the former category played leading roles in each area of specialization throughout their careers.

In the nineteenth century, when English was not yet established as the major international language, students focused on one of three European languages: English, French or German, depending on the language of the visiting foreign professors or of the predominant scientific work in particular disciplines. This linguistic diversity soon proved unworkable and the government moved to entrench teaching in Japanese in all fields of study.

Students also found it a great burden to have to master several different European languages in order to pursue advanced studies and in 1872 the Ministry of Education decided to abolish instruction in French and German, concentrating only on English in preparatory schooling. In the process of reform, the government had to disband the polytechnic department, in which French was the language of instruction and the mining department, which was taught in German, thus giving birth to temporary departments of French physics and German chemistry in 1875. At the same time, the language instruction requirement in the preparatory stage was uniformly limited to English. Its courses of instruction followed the model of the American high school.[2]

In the University of Tokyo, founded in 1877, efforts were made to systematize language requirements in preparatory education into two languages: German for the Medical School and English for the other schools. In 1884, the Ministry of Education suggested to the University that it would be preferable to teach primarily in the Japanese language. Finally, with the establishment of a more comprehensive higher education system in 1886, language education was assigned to secondary education. English and German were taught in the Middle Schools and Higher Schools (kotogakko) respectively.

Plurality and Diversity of Western Models

Around 1870, government officials eagerly examined various Western educational literatures with the idea of adopting the best elements of each model. On the basis of these investigations, they initially formulated a crude plan for imitating Western educational systems. In early 1870, the government issued 'university regulation' as well as 'primary and secondary education regulations', in which students in secondary and higher level education were supposed to specialize in any one of five departments. These included theology, law, science, medicine and humanities. These arrangements are reminiscent of the faculty structure in contemporary German universities, though science and humanities (philosophy) had not yet been separated in nineteenth century Germany. The Japanese, however, failed to recognize the meaning of creating a discipline in Christian theology. Instead, some traditional scholars tried to introduce Confucian studies into the modern university curriculum. These moral and ethical subjects could not win the support of political leaders in the Meiji government, who were utilitarian materialists. Later, in the same year, the 'college regulation' enumerated the disciplines of rhetoric, logic and Latin in the humanities curriculum, but in actual practice these were all neglected as the Japanese did not comprehend the content and significance of such a medievalistic course of studies. Out of the above subjects, only the discipline of logic was later recognized as a modern discipline to be studied in school.

In the early 1870s, American advisers such as Verbeck and David Murray were influential in the formation of school systems under the umbrella of the Ministry of Education. In order to meet the immediate needs of each governmental service department, such as the Army, Navy, Department of Justice, Department of Works and the Hokkaido Development Agency, each department had to have its own system and training programme in the higher education sector. For example, in the Department of Work, Scottish engineers, headed by Henry Dyer, were brought in. According to Dyer, they tried to introduce not a British but rather a Swiss system of synthesizing science and technology as practised in the Technische Hochschule of Zurich. The Japanese Navy followed British practice, while the Army adhered to the French model. The Hokkaido Development Agency adopted the American land-grant college as the exclusive model (in particular, the University of Massachusetts).

The Ministry of Education incorporated into the *Kaiseigakko* (the preparatory school or college level predecessor of the University of Tokyo) an American liberal arts college model suggested by Verbeck, David Murray and his colleagues from Rutgers College in New Jersey. German influence also came to the Medical school, where as early as 1870 German medicine was officially adopted to the exclusion of English and other styles of medicine. In the law school, French influence was predominant in the early years, while the German model replaced it after a political event in 1881 when the government decided to switch its institutional model from other Western countries to Germany. Subsequently, more Germans were employed while the number of other foreigners gradually decreased. Traditional fields of study such as Chinese studies and Japanese studies later found their place in the Department of Literature at the University of Tokyo, giving proof of counterbalance, and self-reflection on, the process of rapid Westernization.

Amalgamation and Inclination to the German Model

The early attempt at modernization conducted by each service department following its own Western model was concluded after a decade of effort. Their teaching institutions were now unified and reorganized into the Imperial University, founded in 1886 under the jurisdiction of a single agency of the Ministry of Education. By this time, most of the faculty positions were held by natives who had studied abroad. They formed a new type of a national university, amalgamating various Western models and casting them into their own Japanese mould under governmental auspices.

The characteristic of the Japanese University which is perhaps most revealing is its school-department structure. The basic structure of six schools, including law, humanities, science, technology, medicine and agriculture, was completed in 1890. In comparison with Western universities, there are noteworthy differences in that science became independent of humanities (philosophy). In addition, traditionally vocational subjects of the modern applied disciplines, such as technology and agriculture, were elevated to university status. In spite of opposition to the university status of agricultural science among faculty members who had been trained in Europe, they were obliged to implement the government's policy.

In the initial phase of modernization in the late 1870s and early 1880s, departments of science and technology were encouraged by the government to train the manpower and develop the hardware requisite to modernization, such as building a domestic telegraph network, conducting geographical and geological surveys and so forth. Once these basic tasks were completed, the government turned its attention to the software of controlling a modern nation: it built a modern bureaucracy by inviting selected elite students to serve in the administrative offices. For that purpose the government raised the status of the hitherto relatively neglected law school by granting certain privileges to its graduates such as exemption from the Civil Service Examination, which was formulated in 1887.

For purposes of academic research, the nineteenth century German universities, particularly the philosophical faculty, provided the world's most highly regarded model. However, the Japanese government had different reasons for adopting the German model. It intended to copy the modern German bureaucracy, which was dominated by law school graduates.

At the same time, the bureaucratic character of the University was further strengthened. The founding ideology of the Imperial University was designed 'to meet the urgent needs of the nation', a phrase which appeared in the beginning of the University constitution. This was not merely rhetoric for it was manifested in the career patterns of the graduates.

In spirit, it is often said that the pre-Second World War Japanese national universities were quite close to German universities. On closer inspection, the predominant German influence came after the 1880s. Previous to that time, American, British and even French models were being incorporated into uniquely Japanese institutions. In retrospect, German influence was felt rather belatedly. This made it institutionally difficult to completely replace the pre-existing amalgam with a purely German ethos. For instance, uniquely German concepts such as *Abitur, Privatdozent* and mobility of students and faculty from one university of another, have never been introduced into the Japanese system of higher education. Still, the intangibly authoritarian and elitist atmosphere of the Imperial University faculty was said to be derived from German academia. The notion of academic freedom, presumably along the lines of the German model, was advocated among university faculty members at the time of its infringement by the government.

POST-WAR INVOLVEMENT WITH THE AMERICAN MODEL

New Dealers in Spirit

Japan experienced a second phase of Westernization in the post-Second World War period. Its intensity was certainly comparable but its impact was qualitatively different, from the first phase.

By the mid-twentieth century, Japan had firmly established a Meiji-type institutional paradigm (which spread later to new Japanese universities and colonial universities in Korea and Taiwan). Its replacement with a new American model could not be done without the application of extraordinary pressure by the Occupation Forces.

One of the initial purposes of the Occupation was the eradication of ultranationalistic and militaristic ideology in Japan. The Allies decided that in order to replace that ideology with 'democratic' thought, it was imperative that a comprehensive educational reform should be instituted. Even though the Occupation Forces consisted of representatives of several Allied powers, the occupation of Japan, in contrast to the occupation of Germany, was administered almost exclusively by the US. Hence, the model for post-War reform, of not only educational but various other institutions, was purely American. In other words, for those members of the American Educational Missions who came to Japan as advisers to the Occupation Forces, it was simply inconceivable to propose any model other than the American one, with which they were the most familiar.

The American model is, however, not quite monolithic. Some Americans attempted to promote their own ideal model, which may not have existed at that time in their country. Especially in the early phase of occupation, a number of New Dealers joined the Occupation, introducing land reform and encouraging labour movements. Educational reform was also carried out, at least in the initial phase, in a similar spirit by the Civil Information and Education Section (CIE) of the Occupation Forces by way of indirect control and guidance using Japanese official and private agencies.

The reform began in the primary schools and then gradually extended to the secondary and finally to the postsecondary levels. Before the wave of reform reached the graduate schools in 1953, the Peace Treaty was concluded in 1952 and the US withdrew its forces from Japan.

Educational Reform from 6.5.3.3 to 6.3.3.4

The number of years of schooling at each level of education are arbitrary and hence differ from one country to another. There was much discussion among Japanese educational planners before the suggestions of the American advisers were outlined. The Japanese preemptively resolved to move towards the American system, in which the old system of 6 (primary), 5 (secondary), 3 (*senmongakko*, junior college level professional school or *kotogakko*, preparatory Higher School for the Imperial University) and 3 (university) was abandoned in favour of 6 (preparatory), 3 (compulsory junior high school), 3 (senior high school) and 4 (college).

Higher Schools were viewed by reformers as elitist and moves were made to abolish them. They were in some ways similar to the British public school, the German *gymnasium*, and the French *grandes écoles* but quite different from the American public high schools. The emergence of these schools was regarded as a failure of educational planning during the Meiji era by the administration, but students created and enjoyed a uniquely liberal elitist culture. These schools were abolished not because they did not fit into the Japanese society and school system but because of their elitism and the fact that they did not fit into American plans for educational reform.

Democracy – From Elite to Mass Education

In pre-war Japan, college graduates comprised fewer than 7 per cent of the age cohort and hence their education was naturally elitist (the prestigious Imperial Universities enrolled about 1 per cent of the age cohort). The Americans strongly suggested that the Japanese expand their higher education sector to the extent that each prefecture (46 in number) would have its own university. In spite of the economic hardhip that prevailed after Japan's defeat in the war, a significant expansion of expensive higher education was proposed by the Ministry of Education with pressure from the Occupation Forces, which generated numerous complaints and a great deal of confusion among the Japanese. This move certainly created a great opportunity for the *senmongakko* and normal schools to raise their status to university level. They took advantage of the opportunity and most were successful in obtaining accreditation at the university level.

These newly created colleges were very poorly equipped in terms of both teaching staff and facilities. The traditional elites criticized these new universities as being second-rate, but these new institutions prepared the way for the dramatic expansion of higher education that took place in the post-war years. One of the causes of the student revolts of the late 1960s in the Western industrial countries was interpreted as being due to overcrowding in universities. In this sense, the Japanese universities suffered less than their European counterparts because the expansion had started earlier and facilities were more adequate.

Egalitarianism – Plural Track to Single Track

Before the Second World War, different tracks of education were designed for those who terminated their training in the primary, secondary or *senmongakko* (junior college) levels, largely depending on students' social class origin. Hence, there were three levels of engineering, secondary, *senmongakko* and university graduates. Even in the field of medicine, practitioners were divided into two classes, university graduates and *senmongakko* graduates, despite the fact that both received the same medical licence. This complicated plural track of education was changed to a single track under the egalitarian policy of the American advisers. In the same spirit, the status of private universities and women's colleges was enhanced. Normal schools were given college status.

In reality, however, in view of the short period of preparation and the lack of facilities, it was impossible for the new universities to obtain the resources necessary to rise to the standard that had existed under the old system. Thus, in spite of the creation of a new university system, the unequal pattern of resource distribution that characterized the old system remained essentially unaltered.[3]

Decentralization

Americans viewed the power of the Japanese pre-war educational system as being too centralized. They seem to have had the idea of disbanding the Ministry of Education, the hub of the power structure, but in order to maintain an indirect means of governing, they had to use the existing bureaucracy. The relationship between the American CIE and the Japanese Ministry of Education was an ambivalent one.

The creation of a national university in each prefecture, mentioned previously, was primarily aimed at the decentralization of higher education. Like the state universities in the US, the CIE proposed, in 1947–48, to delegate supervisory power of universities to local prefectural educational commissions.

Opposition on the part of members of the Japanese Educational Renewal Committee was so strong that the CIE gave up its original intention. The main reasons for the opposition were that local governments were not accustomed to handling matters of higher education and that they lacked resources.[4]

Lay Control

In order to avoid bureaucratic control and to promote the participation of the citizenry in higher education, the second American Educational Mission in 1950 made a recommendation to the Japanese government and the CIE followed it up by introducing the American-style 'Board of Trustees' as the governing body of Japanese national universities. They also found the autonomy of and self-governance by Faculty Conferences (as practised in major Japanese universities) too self-complacent, self-righteous and isolationist. This move was met with still stronger opposition from the Japanese. Professors simply could not conceive of allowing lay people to exercise control over academic matters. The student body, led by leftist leaders, expressed its opposition to this policy change by means of a strike. They resented the external pressure and identified it with a capitalistic, and a symbolically militaristic, invasion of academia. This was persuasive especially in light of the 'Red Purge' in academia carried out by the Occupation Forces during the Korean War. The debate was continued in the post-Occupation period and finally the proposal was withdrawn.

Consequently, the Japanese universities could successfully resist outside pressures from industry and remain aloof from cooperation with industrial and particularly military research, as Harry C. Kelly, an adviser in science and technology in the Occupation Forces, witnessed in a comparison with American universities.[5] On the other hand, haughty Japanese national universities have turned out to be obsolete in equipment and facilities of scientific research because of their isolationist attitude.

General Education

In the pre-war period, the university curricula were all designed for the education of specialists while liberal education was conducted at lower level schools or was self-taught by students themselves. The CIE was enthusiastic in introducing the American ideal of general education. The Harvard style of general education, as advocated by J.B. Conant and illustrated in *General Education in a Free Society* (Report of the Harvard Committee, 1945), played the most important role as a model, in which the trinity of course arrangement (that is humanities, social sciences and natural sciences) was adopted.

In translating this into actual practice in new universities, courses were mostly taught to students in the first two years by teachers drawn from the Higher Schools. In the pre-war Japanese educational system, courses were divided into two classes; humanities and natural sciences. Hence, it was difficult to locate the social sciences in the new curriculum due to a scarcity of adequate teachers.

There has always been confusion between general education and preparatory or premedical courses for higher specialized courses. In the natural sciences, the demand for foundation courses was considerably higher than for general education. Later in the post-Occupation period, industry and the School of Engineering demanded more vocational education at the expense of general education.

Graduate Schools

Japanese graduate schools were founded as early as 1886. Their model might have been an American one, given the fact that such a system existed nowhere else at the time. Japanese pre-war graduate courses, however, were not structurally well-developed throughout the pre-war era. Doctoral degrees were conferred by the Faculty Conference, without having had any institutional relationship whatsoever with the graduate school.

After the war, the American Scientific Mission, rather than the Educational Mission, strongly urged that graduate schools be reformed.[6] The CIE of the Occupation Forces had formulated the model of the American graduate school system and provided guidance through a Japanese agency, the Japanese Association of University Accreditation.

There was not much resistance on the part of Japanese universities, because the old graduate school did not have any structure to conflict with the newly introduced American one. Thus, the three major aspects of the new graduate school were accepted: (1) The direct linking of the graduate school with degree programmes; (2) the introduction of the three degree systems (Bachelor's, Master's and Doctoral) and; (3) the introduction of graduate level education.

Since the system is not the product of spontaneous growth but rather the importation of an external model, there remained a certain administrative rigidity in regulations such as the prescribed residence of two years for the master's course plus the three years doctoral programme along with a *numerus clausus* for each discipline.

Graduate programmes are not well funded and graduate fellowships and assistantships do not adequately cover living expenses. Hence, the graduate school in Japan is not as efficient as the American one. Furthermore, it may be partly a matter of culture that the Japanese graduate school has failed to create a highly competitive atmosphere, as was the case in the US, even after more than three decades of its existence.

Though nominally linked to degree programmes, professors in the humanities and social sciences do not often confer doctoral degrees, simply because they themselves did not have the opportunity to earn a doctorate under the old system, in which doctorates were given to those older scholars who had reached the end of their research career. In the fields of science and technology, where greater internationalization has occurred, new graduate schools based on the American model are more readily accepted.

Process of Internalization

After the withdrawal of the Occupation Forces in 1952, the Ministry of Education gradually strengthened its control over higher education. However, after seven years of model change, it was utterly impossible to return to the old model that had existed before the Second World War. The process of internalization (in other words, domestication of the external American model) has followed.

The prototype of pre-war Japanese universities was established in the late nineteenth century and the minor changes that evolved were not enough to meet the new demands of the mid-twentieth century. Reform was desperately needed, regardless of its source. The American model was not necessarily new but the newer elements were taken up by the Japanese: one of these was mass higher education. If the elitism of the Imperial University and the Higher School still remained, it might appear to be grotesquely obsolete now.

Other elements which were incompatible with Japanese culture were, for the time being, excluded from the reform process. Hence, decentralization was discouraged in the period of internalization that followed and layman control was totally dismissed.

In higher education, academic decision-making power which resided in the Faculty Conference of each school remained intact. The efforts of reformers had been focused on the introduction of new models in general education and graduate school, new areas where old faculties had no vested interest. In the process of internalization, however, a realignment of the new arena gradually took place. For example, the University of Tokyo created a new faculty in the College of General Education where a new programme of general education was designed and new experimental interdisciplinary disciplines like area studies were located. In graduate school, a new division (humanities, social, mathematics, physical, chemical and biological sciences) was introduced to promote interdisciplinary research in contrast to the old faculties of law, economics, literature, science, technology, medicine and agriculture. But because the major decision-making power still resides in the Faculty Conference of the old faculties, the structures of both the College of General Education and the Graduate School are now subordinated to the old faculties division. The ideal of general education has fallen victim to the suspicion and neglect of the old faculties. On the other hand, the Graduate School will continue to expand, since it can meet the future demands of science and technology as well as the desire of raising the status of old faculties which control degree conferring authority.

CONCLUSION

As stated earlier, Japanese higher education has followed a unique course of development, as far as development of an organizational structure and the adoption of a variety of foreign models. It has been seen that Japan has moved from 'window-shopping' and experimenting with a number of Western models in the period prior to the Second World War to the direct involvement of a foreign power, the United States, following the war. In this sense, Japan has followed a bifocal approach to higher education, looking both to the American model and to earlier patterns of university development.

It is possible that another type of modelling will be more useful in explaining the contemporary situation in Japanese higher education – the transfer from a bureaucratic model to an industrial model. We have pointed out earlier that while many of the Western universities were created on a pure academic model (or ultimately on a professional guild model of the European medieval universities), the modern Japanese universities in the late nineteenth century were created by a modern bureaucratic state with specific goals in mind. In the course of time,

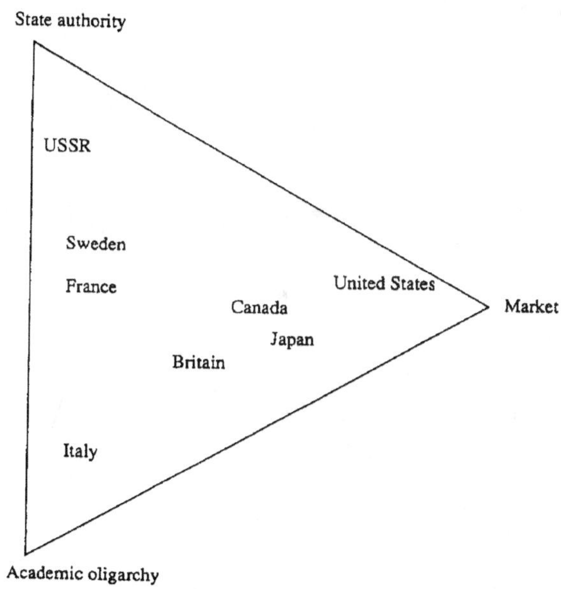

Figure 17.1 The triangle of coordination

Source: B. R. Clark, *The Higher Education System* (Berkeley: University of California Press, 1983), p. 143.

however, the Japanese universities attained academic freedom in a limited sense and self-government through the Faculty Conferences.

Since the late 1950s, a new external pressure has been imposed on Japanese higher education as Japan has recovered her industrial capacity and has started to promote it more than at any other time in her history. The influence of the industrial community became visible at the beginning of the 1960s, as it demanded from higher education the creation of abundant scientific and technological manpower.

Around the same time, most of the major Japanese industrial laboratories were founded and have already raised the level of provision for academic research in terms of facilities and equipments, as evidenced by the fact that industrial R&D resources now comprise more than 80% of total national expenditure for R&D, far outstripping those of national and academic laboratories. The impoverishment and obsolescence of university research is beginning to be scandalous. Leaders of industrial laboratories often boast, 'We do not have to depend on academic science for basic science. We can make basic science by ourselves, if needed.'[7]

The state of affairs of 'academic coordination' is internationally compared and nicely illustrated in a triangular representation by Burton Clark.[8] We shall partially borrow it in order to show the academic decision-making power in Figure 17.1 and make another triangle of budgeting as an indication of research expenditure in Figure 17.2. While Clark's Figure 17.1 is based entirely on

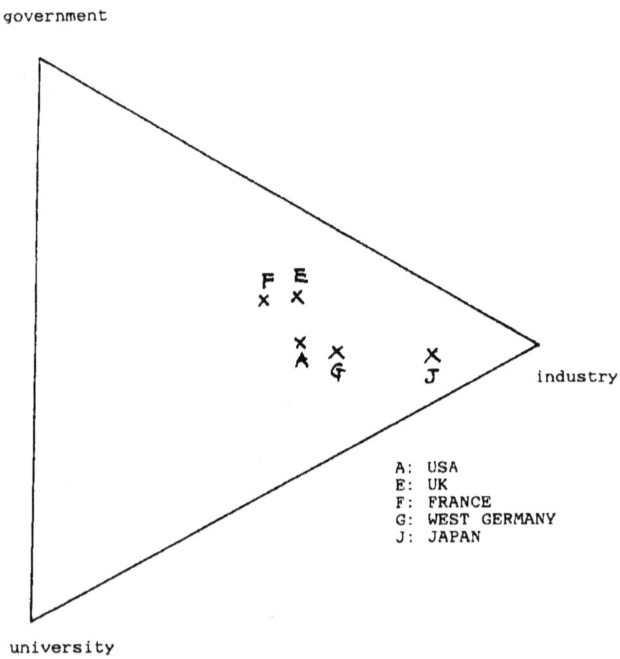

Figure 17.2 R & D Coordination triangle

Source: *Kagaku gijutu hakusho* (Science and Technology White Paper, 1986).

impressions without having any quantitative measure, Figure 17.2 has some numerical basis, as all the data come from the OECD statistics. While the location on Figure 17.1 has a close correlation with academic decision-making power, I feel that Japanese academic decisions are influenced less by the market and more by faculties than the figure indicates. In Figure 17.2, the share of private and public sources out of the total national R&D expenditures is shown on the horizontal scale. The vertical scale represents the portion between governmental (largely defence) laboratories and academic research of the amount spent for R&D.

If we superimpose Figures 17.1 and 17.2, we find a discrepancy between the location of academic decision-making power and universities' R&D resources. This distance is suggestive of the direction in which future higher education is heading. A comparison with the corresponding OECD data for 1965 shows that whilst both the US and Japan have during that period moved towards the right on the figure (i.e. towards private rather than public), on the vertical axis, the US has moved towards university research whilst Japanese expenditures have moved towards government laboratories. Another point which is not shown on Figure 17.2 is that while some public funds go to private laboratories in the US, some private funds go to university research in Japan.

There will be two distinctly different conceivable cases based on the dominating influence of Japanese industries:

(1) Higher education will switch its model to industry, so that it organizationally assimilates towards the industrial sector and produces quality-controlled manpower and efficient research findings which the industrial community demands. Most of the science and technology faculties, as long as they remain competent in their professional performance, are inclined to the industrial model and its value system. In such a case, their academic performance in both education and research is assessed by or subordinated to the value standards of industry. Consequently, academia becomes industrialized.

(2) Higher education will refuse to adopt the industrial model and preserve its autonomy and identity by keeping a respectable distance from the industrial world. While liberal and general education are maintained on campus, the outer world will no longer depend on higher education for any vocational training, for which industry will develop in-house training or some system other than existing institutions. Then higher education will turn into an enjoyable playground to spend youthful life perhaps more meaningfully or simply remain a leisure land.

NOTES

1. The original Japanese text is reprinted in S. Nakayama et al. (ed.), *Nihon kagaku gijutsushi taikei* (Source books of the history of science and technology in Japan) vol. 8 (Tokyo, Daiichi-hoki, 1967), pp. 35–36.
2. *Tokyo daigaku hyakunensi* (Hundred years of the University of Tokyo) (Tokyo: University of Tokyo Press, 1984), pp. 305–311.
3. Ikuo Amano, 'Continuity and Change in the Structure of Japanese Higher Education' in *Changes in the Japanese University; A Comparative Perspective*, eds. W.K. Cummings, I. Amano & K. Kitamura (New York: Praeger, 1978) p. 38.

4. Kaigo Muneomi and Terasaki Masao, *Daigaku kyoiku (University education)* (Tokyo: University of Tokyo Press, 1969), pp. 92–96.
5. Harry C. Kelly, Interview by Charles Weiner (1975). Institute Archives and Special Collections, MIT Archives.
6. 'Reorganization of Science and Technology in Japan', Report to the American Academy of Science, issued 28 August 1947.
7. Personal communication.
8. B.R. Clark, 'The Japanese System of Higher Education in Comparative Perspective' in W. Cummings et al. eds., *Changes in the Japanese University*, p. 237 and also B. R. Clark, *The Higher Education System* (University of California Press: Berkeley, 1983) p. 143.

First published in C.G. Weeramantry (ed.) *Human Rights and Scientific and Technological Development*, pp.137–50, United Nations University Press, 1990

⑱ Human Rights and the Structure of the Scientific Enterprise

INTRODUCTION

Scientists open up many problems of human rights and lawyers attempt to solve them. So far, however, the two communities have remained isolated from each other at two extremities of a spectrum.

In a search for bibliographical information on the matter of 'human rights and science', I made some inquiries of my 'science study' community, which includes the disciplines of history, philosophy, sociology of science, and science policy studies. The typical responses of my colleagues have been that our study has reached the sociological, economic, political and even ethical aspects of science, but not the legal one as yet.

Though there have been numerous writings on human rights issues, it seems that the issue of human rights in relation to the development of science and technology has been raised only recently, and that people have developed a 'critical awareness of science'[1] only since the late 1960s. The turning-point may be found in the 'Human Rights and Scientific and Technological Developments' resolution at the Intergovernmental Conference on Human Rights held at Tehran in May 1968.

Since then, a number of possible sources of violation of human dignity have emerged, along with the rapid development of the frontiers of microelectronics and life science, as enumerated by Professor C.G. Weeramantry in his recent book *The Slumbering Sentinels*, which is, to my knowledge, the first serious attempt to tackle the issues from a legal point of view.[2] The tempo has been so rapid that the legal profession seems not to have caught up effectively. This state of affairs has been lamented by my lawyer friends who have no scientific background. Weeramantry too has pointed out that the law has been tardy in evolving concepts to deal with technology.

Scientists in their laboratories, preoccupied with their immediate goals, are not aware of the possible consequences of their research. Legalists deal with the issue of science and technology in an *ex post facto* fashion on such aspects as society and environment, aiming to fill the gap between the scientific and legal communities. I shall explore the production mechanisms of scientific knowledge and technological information in the human rights context, and then present the general pattern of attempts to discover where principles and mechanisms for technology

Table 18.1

	Academic	*Industrial*	*Defence*	*Service*
Assessors	Peer	Sponsor	Military	Public
Competition	Individual	Corporate	National	None
Expression	Open	Classified	Classified	Open

assessment can best be applied. This sort of analysis is, in my view, more essential than the piecemeal information gathered on the research front.

CLASSIFICATION OF SCIENCE BY ASSESSORS

As suggested by Dr Mushakoji in his problématique for this project, science and technology should be treated as information. There are two ends of information flow: the production of information at one end and its assessment and utilization at the other. In the post-war period, the latter has come to be overwhelmingly powerful while the former tends to occupy a subordinate position. The central concern here is who will review and assess the information that scientists and technologists produce. From the viewpoint of the sociology of science, I hold the view that the question of the audience to whom scientific research is addressed is the most important factor in shaping and defining its character. Viewed in terms of operative mechanisms, four types of scientific activities may be distinguished, as shown in Table 18.1.

From this point on, I do not in principle distinguish between science and technology, as this is only a historically meaningful distinction, which may no longer be valid, particularly in the case of industrial and defence sciences (though the classical distinction may still hold good at times in the case of academic science).

STRUCTURE OF ACADEMIC SCIENCE

Academic science is the science practised in the open scientific community, in which members present their research for debate and discussion by their colleagues, and seek recognition of their work through publication in scientific journals. Research is initiated out of personal interest and pursued for reasons of personal honour and distinction meted out through a referee system.

In academic science, researchers typically proceed by the following steps:

1. Design of research.
2. Application for research funds.
3. Research.
4. Peer review of research findings.
5. Publication.

In the above process, steps 1 and 3 are matters of highly individual concern, since it is generally believed that classical freedom of research has to be maintained.

This is the point at which academic science becomes very different from industrial and defence sciences.

In reality, however, it has been proven that nineteenth-century science can no longer compete with post-war organized mass science and hence, as the size of scientific enterprises inevitably grows bigger, it becomes difficult to maintain individual identity in formulating research problématiques, and the tendency is to conform to the norms of mass science. Sometimes the human rights of non-conforming scientists and engineers are infringed so that the immediate goal of a bureaucratic organization may be achieved, as in the case of the promotion of the SDI project. Though not always overly careful, researchers usually keep their ideas and findings secret before publication, since they are concerned about the possibility of their being stolen by competitors.

In the post-war period, practically all scientists have applied for research funds of some kind. All academic scientists are involved in the competition for securing research funds, success in which process often predetermines the winner of the game at any early stage, even before any research is commenced. At this stage the distinction between members of the scientific community and amateurs is sharply applied, placing the latter out of reach of any research funds. This situation contributes to the formation of scientific groups resembling closed-shop unions, resulting in a clear-cut demarcation between the scientific community and those outside it. In the scientific community application forms are commonly sent to an assessment committee and are subject to close scrutiny by peers. It is also not uncommon for applicants preparing their forms to conceal carefully the essential points of their ideas, so that a peer assessor cannot appropriate them.

In the 1970s, when the unforeseeable dangers of DNA were recognized, it was felt that the social evaluation of science should not be *ex post facto*, on the outcome of research, but preliminary, before the research starts. The National Institute of Health tried to use their funding mechanism for the pre-assessment of hazardous research by applying their guidelines. Since corporations could afford such research with their own funds, this standard could not be extended to industrial science.

At step 3, a general statement is hardly possible because of the diversity of topics involved in scientific endeavour. One recent trend is observable, however.

Unlike a social scientist, who clings to ideology and conscience in carrying out research, an experimental scientist is happy to change his methodology until he finally arrives at a satisfactory position. In a major scientific project, however, he cannot easily do so owing to the fear that he will be called upon to account for funds already spent, even though he might know that the project will eventually turn out to be a failure, in such cases as nuclear fusion and the SDI project.

At step 4 another opportunity for 'post-research and pre-publication' assessment ment is theoretically open. Again, on manuscripts submitted to the referee mechanism, precautions are taken to defend the researchers' prior claims to the knowledge they contain. Otherwise, a referee might hold up its publication and meanwhile appropriate the ideas and publish them elsewhere.

Along with overspecialization in science and the overproduction of research output, it is generally suspected that the old referee system is approaching bankruptcy, though people still believe that distinguished work can survive the process of selection.

In view of the recent advent of microelectronic technology, step 5 deserves to be treated in the following independent subsection.

INEFFECTUAL ACADEMIC FREEDOM OF EXPRESSION

In the old days, knowledge and information were the private monopoly of a select group of people, transmitted orally or secretly through manuscripts. The modern printing media have made knowledge publicly available and promoted the objectivity of modern science and freedom of expression for the last three centuries. With the referee system intervening between the manuscript and printing stages, science was raised to the status of public knowledge.[3]

In reality, however, academic science is defenceless in safeguarding human rights against authority, ecclesiastical or civil.

Freedom of expression is still 'academically' maintained, though the knowledge and information produced are circulated within a closed circuit, practically out of the reach of most people. Furthermore, the new microelectronic media will bypass the referee system that modern scientific tradition has so far cultivated and turn scientific knowledge back into privately circulated and monopolized information.[4] In the generality of cases, microelectronic media have been commercially developed to promote private information rather than public knowledge. There is a fear that by making knowledge private once again we will revive the exclusiveness of knowledge and information which was a feature of the manuscript age.

Moreover, as the new electronic media make it possible to disseminate small quantities of a wide range of scientific information, the trend towards specialization will be spurred on still further, beyond the reach and comprehension of the general public.

GENERAL REMARKS ON INDUSTRIAL AND DEFENCE SCIENCES

Industrial science as practised in private laboratories has had much in common with defence science and little in common with academic science. Jerry Ravetz generalized its many manifestations into a single term, 'industrialized science,'[5] emphasizing the characteristics of production of scientific information as similar to those of an industrial commodity. David Dixson would like to call them 'strategic science', with emphasis on its goal-oriented nature.[6] I prefer the Japanese term *taiseika kagaku* (Establishment Science) which stresses its characteristic of tight and rigid incorporation into the present establishment and its isolation from the general public; thus, it may be called, with Steven Rose, 'incorporated science'.[7]

Although most people still talk of science in terms of the classical academic science paradigm, this is not the dominant form of contemporary research. Major resources for research and development are now allocated to industrial and defence sciences. These incorporated sciences are vigorously promoted by the driving force of competition among corporations or nations, while profits and military strategic necessity determine how their results will be evaluated.

Industrial science is targeted to the areas that promise the greatest profits. Defence science proceeds along policy lines laid down by a handful of military strategists, despite the fact that it involves a large amount of taxpayers' money. The most problematic aspect of these sciences is that they continue to remain beyond the purview of public scrutiny or any form of public accountability. The researcher's work is assessed wholly within the organizations, and is not reviewed or evaluated by anyone from outside. This is why industrial and defence sciences are referred to more generally as 'incorporated sciences'.

Scientific works openly assessed and publicly recognized are called 'knowledge', while the findings of incorporated sciences, classified and available only to a limited circle for private monopoly, are referred to as 'information'.

THE STRUCTURE OF INDUSTRIAL SCIENCE

The sociology of science as an established discipline has developed considerably during the last two decades, but so far its major interest has continued to be the analysis of traditional academic science. We badly need the structural analysis of incorporated science but, mainly because of its very corporate nature, we have to confess that the investigation is neither thorough nor penetrating as yet. Furthermore, there is no general rule applicable to all establishments that are competing with each other. However, it is still worthwhile to conceptualize the idealized case.

The steps of commercialized industrial science may be summed up as follows:

(1) Market demand.
(2) Targeting and formulation of plans.
(3) Research and development.
(4) Model, test and production.
(5) Advertisement.
(6) Dissemination into market.

In step 1, consumer assessment is incorporated when business enterprises carry out market research. A leader in each research group is responsible for formulating his or her target at step 2, taking into account the overriding need for commercial success, and submits his or her research proposal to the upper strata of the corporation. Unlike academic science, researchers in industrial science cannot determine their targets autonomously and hence their creativity is said to be considerably reduced.

At step 3, the assessment by top managers of corporations comes in. Many research plans are abandoned at this stage. In industrial science, assessment by top managers and sponsors, rather than the wishes of consumers, tends to prevail in determining the final product at step 6.

Steps 2 to 4 are absolute corporate secrets. Confidentiality is maintained against competing companies, not against consuming customers, but its ultimate effect appears to be similar.

Old Marxists claimed that the large corporation, with its desire to maintain market stability, purchased the patents of scientists and inventors and kept them secret, resulting in the distortion and negation of healthy progress in science and technology; in other words, monopoly capitalism killed off competition in science and technology. There have been many cases in history to prove the above statement, but the present-day reality is that industrial science is ruthlessly promoted by competition between rival enterprises in the oligopolistic system.

With regard to research and development, it is cynically remarked in the scientific community that, in rapidly changing and innovative industrial science, the findings of primary importance are classified as the 'know-how' of the company and those of secondary importance are patented and sold on the market, while only insignificant results from the corporation's viewpoint are reported in scientific journals and academic meetings following the convention of academic science. In the most innovative fields, however, intellectual property rights cannot be maintained too long as such property becomes outdated quite quickly.

At the step of basic research, classification is not yet too strict, but as the production step is approached, precaution is exercised to an excessive degree. Social assessment can be applied only to steps 5 and 6. Step 1 is then returned to, thus repeating the cycle. It is, however, often pointed out that post-advertisement assessment is too late, especially on the frontier of biotechnology and life science, where past experience and simulation do not help much.

STRUCTURE OF DEFENCE SCIENCE

The structure of defence or armaments science is the most difficult for us to analyse, as it is deeply embedded in the self-perpetuating military-industrial complex. The following model is taken from the well-publicized, recently formulated SDI Project of the USA.

(1) Idea and design.
(2) Proposal to the military.
(3) Governmental approval and parliamentary hearing.
(4) Research and development.
(5) Model, test, procurement and production.
(6) Assessment by the military.

Whereas all the steps in industrial science are normally conducted within the corporate system, the common procedure in respect of defence science projects is twofold: scientists in universities and corporations initiate ideas and military procurers assess and accept them. In reality, the research sectors of all the above organizations from an exclusive group called a military-industrial-academic complex. Hence, whether research is conducted in universities, corporations or inhouse military laboratories, such an activity can be called defence science.

The market demand for industrial science noted in step 1 is absent in the case of defence science in peace-time. Ideas for new weapons development have never originated from professional career servicemen but have been initiated by scientists who have made proposals to the military; the SDI project is such a case. The military always seek quantitative expansion while qualitative leaps are made only by scientists.[8]

The professional assignment of scientists and technologists is to create something new. Thus, researchers in a military-industrial-academic complex constantly feel compelled to propose ideas for new weaponry. On the other hand, the military establishment can no longer assess the cost of innovation of weapons systems at step 2. Matters are entirely left in the hands of a handful of leading scientists and technologists.

At this point, a word is in order to clarify the nature of the 'military-industrial-academic complex'. This does not mean the weapons production departments of corporations but the complex of research and development sectors of universities, national laboratories, corporations and military establishments. In the case of the SDI project, the Lawrence Livermore Laboratory, MIT and the Lockheed Corporation are major sharers of the governmental R&D budget, where step 1 to step 4 are proceeded with and the results fed back to step 1 in a repetitive cycle. It would not necessarily be continued to step 5 of procurement and production. The complex is only happy as long as the cycle continues to be repeated and escalated. This is a point where defence or incorporated science differs from industrial science. Like a nuclear fusion project, incorporated science can exist by implementing a cyclic mechanism without reference to immediate production.

From the point of view of the social assessment of science, the problem here is obviously the exclusiveness of the project, its confinement within a very limited circle. Step 3, the only step available for social assessment, is usually avoided in the name of strategic secrecy. In step 3, the process of governmental approval and parliamentary hearing on the R&D budget proposals, public assessment is formally carried out, but in practice certain limitations on public inquiry are imposed even at the congressional hearing, under the pretext of national security. In place of market demand, the most compelling arguments are built up on the assumed effectiveness of the weapons system of a hypothetical enemy.

Research and development in step 4, in the case of defence and incorporated science, is conducted on a large scale, on the solid base of a huge national budget with which industrial science is unable to compete. The high social status of defence scientists and technologists in the American scientific and engineering community is due to the huge budget with which they can develop their ideas freely, without being disturbed by commercial market assessments.

Sometimes a rationale is provided for a high defence science budget on the basis that its findings will have a spin-off to the civil sector, thus enabling industrial science to enjoy economies in the initial cost of R&D, which private corporations cannot afford. A little thought would show these arguments to be incorrect, as defence science has to observe, except in step 2, a strictly enforced classification which prevents its findings from leaking to other social sectors. Only in those areas such as nuclear energy and space science, where the basic paradigms of research have originated in defence science but have public applicability, have spin-offs been possible. But these areas could not be called industrial science; they are, rather, incorporated science, or pseudo-civil science, as Aant Elzinga calls it.

At step 5 the position of defence science is similar to that of industrial science, but the lack of commercial assessment often leads to extravagance; the organization may have proven its efficiency for the immediate purpose of wartime crash programmes, but those scientists involved, if they work on a long enough project, can lose a sense of economy and can often be misled to corruption, which is still more prevalent in step 6.

In the final stages of step 6, peacetime assessment of new weaponry is made at practice manoeuvres, but the difference from real war is obvious. Weapons systems can best be assessed only in terms of the difference between them and those of opponents. Assessment becomes a game within the military-industrial-academic complex against a hypothetical enemy. In the last step, a new military demand and requests for further improvement will appear, as necessary for defence against a hypothetical enemy, resulting in the renewal of the cycle.

HUMAN RIGHTS OF SCIENTISTS AND ENGINEERS

Scientists' rights to publish research results are often infringed or denied because of the policy of giving priority to the interests of the business and military establishments. Especially for those scientists and engineers in industrial and defence sciences, these rights are categorically denied. As the UNESCO Recommendation on the Position of Scientific Researchers (1974) has already demanded, the right to disclose military or corporate secrets, in cases where scientific information is of a crucial nature for human existence, should be protected.

I cannot, however, be too optimistic about the ombudsman function of the scientific community. My long experience shows that only a small portion of the community (say 10 per cent) is truly qualified as guardians of human rights, the rest being prone to undertake whatever research is best funded. Most contemporary scientists are involved with the industrial and defence establishments, and thus the collective inclination of the scientific community is not necessarily sound.

Important Role of Technicians

I would like to suggest in this connection the important role to be played by technicians, or rank and file scientists and engineers, rather than leaders of projects, in affording protection against science-related hazards.

In general terms, there are three stages in the development of pollution originating from scientific research.

(1) Small-scale pioneering experiment by a leading scientist.
(2) Large scale R&D in major industrial laboratories.
(3) Industrialization and introduction of industrial products into the environment.

In stage 1 the scale of experiment is so small that its hazards are faced only by the individual scientist who designs the project. It is well known that Madame Marie Curie shortened her natural life by exposing herself to radiation at her laboratory. Early workers on the DNA recombinant experiment must also have undergone such risks. In such cases, as well as those of pioneer-adventurers, intellectual or otherwise, an individual risk may be compensated by individual success and reputation. As long as the adventurer maintains complete freedom of decision-making in exposing his life to the hazards of his experiment, no infringement of his human rights will occur. Furthermore, in the early stage the rule of the openness of academic science is usually observed.

In stage 2 large-scale R&D involves a number of technicians and rank and file scientists who are engaged in the project for their livelihood. Their risk may be compensated partially by their salary, but this is not worth the risk they are exposed to. Hence, they can assume a more cautious and detached attitude than the adventurous leader in identifying hazards. If they have no freedom to reveal and express themselves about possible hazards, there will have been an infringement of their human rights.

In the case of X-ray experiments, the International Physics Society admitted and set a threshold value of X-ray exposure only as late as 1928, when many technicians were exposed to the radiation hazard. In the case of the recent recombinant DNA experiment, labour unions such as the British Association of Scientific Workers have played a watch-dog role in formulating guidelines for the experiment through assessment committees like the GMAG (Genetic Manipulation Advisory Group). The Association of Scientific Workers has long been concerned to protect the human rights of technicians and laboratory workers.

These hazards are a sort of occupational disease that may turn into environmental pollution when scaled up.

In stage 3 the citizenry has nothing by way of compensation. They may retain the right to complain about pollution, but they have no special knowledge of it and hence they can easily be cheated in the ensuing debates. In the case of the recombinant DNA experiment, a representative of the citizenry was invited to the GMAG. Jerry Ravetz, a critical historian of science, was chosen for that work but he complained that his role was merely ornamental. It was then said that GMAG was a cosmetic exercise and Jerry Ravetz was the lipstick. None of the scientists on GMAG believed there was a real hazard.

It is rather difficult to find an expert who stands on the side of the citizenry on hazards and pollution issues. At this point, it is indispensably important for the technicians at stage 2, with their expert knowledge, to disclose and expose to the public the possible hazard before it can be diffused into the environment.

STRUCTURE OF SERVICE SCIENCE

The slogan 'science for mankind' still lingers in the popular imagination. Scientists who devote themselves to 'truth-seeking' are considered to be qualified to decide what is truth and what is not, to be providers of intellectual services to mankind and, like other professionals, to be quite independent of earthly desire for wealth and power. A classic example would be those engaged in the search for bacteria at the turn of the century.

I have defined 'service science' as science assessed by the citizenry – citizenry defined as 'those who have no direct vested interest in science and technology activities as such' and who are thus qualified to be objective, disinterested assessors. It seems that there is no such assessing mechanism by the citizenry at work at the present time, but again a classic exception may be found in medical science, in the relationship between a doctor and his client.

As we cannot expect much from academic science as a counterbalance to the menace of incorporated industrial and defence science, we must inevitably turn to another kind of enlightenment, that of service science.

For the role of protecting human rights against the aggression of industrial and defence sciences, service science needs to be promoted in place of classical academic science. Indeed, we need to elaborate the concept of service science a little further and to develop the strategy and tactics for promoting it.

Its structure is rather simple, since nothing, no bureaucracy, no interest group, intervenes between practising scientists and citizen assessors; it is as follows:

(1) Exposure of problems.
(2) Solution of problems.
(3) Assessment by the people.

All of the above processes are, in principle, kept open to the general public.

In step 1 we find outspoken messages of service science expressed by science journalism, which exposes all positive and negative aspects of scientific endeavour to the people for their own critical judgement and choice. The incident of the Asilomar Conference in 1974 marked the beginning of science journalism's assumption of this critical role in a positive sense.

Journalists present serious problems but never solve them. Journalistic provocations are relatively short-lived. Before society tires of repeated alarms, someone must take a step towards solving the problem.

It is the official duty of the non-military sector of public laboratories to undertake such service science. The daily life of taxpayers is conducted not on the basis of competition but rather compassion and co-operation, and so also should service science be conducted for the betterment of ordinary life. However, in the absence of such competition as has existed in the academic and incorporated sciences, service science for the people's sake lacks adequate financial support and motivation and remains insignificant and powerless.

In point of fact, if there is any one element that distinguishes service science from any other type of science, it is the system by which research is reviewed and evaluated. Service science addresses its findings neither to fellow scholars, as in academic science, nor to research administrators, as in incorporated science, but directly to local residents. Yet, service science still does not have any clearly defined apparatus for this purpose. In reality, works of scientists at a service science institute, like a research institute for environmental protection, are still evaluated according to the standards of academic science.

Hence, service science requires forms of communication that differ from the scientific papers of academic science or the know-how and reports of incorporated science. Since neither originality nor the accumulation of classified know-how is involved in service science, it can make its point in handbills and appeals, in a style designed to secure as wide an audience as possible. It also makes maximum use of journalism and the broadcast media.[9]

NEW ASPECTS OF SERVICE SCIENCES

So far, we have discussed service science in a positive way as a defence against the hazards of incorporated sciences. Medical and life sciences used to be primarily service sciences of the first grade and there was little intervention from industry or the military between scientists and the general public.

The service science of life is, however, increasingly industrialized now. Medical technology assessment is needed in connection with the introduction of elaborate and expensive machinery into the health-care system. Mass production of artificial organs will be the big industry of tomorrow, far surpassing the car industry of today.

The advent of the new life sciences portends future markets for the sale and hire of human organs and sale of human tissues. We do not know as yet how much these businesses will be industrialized. As they are still conducted between an individual doctor and an individual patient, people may not be aware of the possible danger of human rights infringement. In spite of the direct influence of life science on everyday life, laymen are quite at a loss in evaluating and assessing the outcomes of research.

RIGHTS OF THE IGNORANT

It is obvious that the human right of access to scientific information should be fully guaranteed, in principle. Towards this end we shall introduce a new issue hitherto not much considered nor discussed. We would like to bring to your attention an issue that perhaps makes up two sides of the same coin, namely 'the human right not to know about specific scientific information but still not to remain at a disadvantage because of ignorance' or, in brief, 'rights of the ignorant'.

It may be too early to propose such a right without fear of being misunderstood, especially when enlightened education still needs to be promoted among the people in the third world. But we can foresee that it is a logical consequence of contemporary trends that the infringement of human rights, due to the compartmentalization of scientific and technological knowledge, is destined to become serious sooner or later. This is by no means an official proposal being formally presented, but we would like to provoke our readers to consider this issue.

Suppose that maldistribution of information is modified to the extent that nobody suffers any disadvantages from inaccessibility to information resources. Still a fundamental problem remains unsolved. It is a deep-rooted issue in the contemporary intellectual environment. It is an ever-present issue in relations between technocratic specialists and the amateur citizenry.

Suppose that the technocratic establishment starts to construct an atomic power station in a region. Local residents naturally feel apprehensive of possible radioactive pollution. The technocrats then hold meetings to discuss devices to dispel residents' misgivings. They start with the premise that knowledge of all sorts concerning atomic power is held by the establishment and that residents are entirely ignorant of that knowledge. Their next move is to send out pamphlets to residents illustrating the safe nature of atomic power or to send specialists to a public forum to explain it to the residents.

The method of persuasion on the part of the scientific establishment may be as follows: the distrust of or misgivings about an atomic reactor on the part of the citizenry is due to their ignorance on scientific matters and hence they must be required to be knowledgeable in such matters to the same standard as experts. If they learn enough, they will be convinced of the safety of atomic power. It is entirely because of their lack of effort in gaining information about science and technology that they oppose governmental planning.

Resident groups may not be convinced by the persuasive arguments of the establishment, and may sometimes invite critics who are public activists in order to have a counter view.

These residents are in the main people who live average lives and have only a basic knowledge of reading, writing and arithmetic. Then, a new body of knowledge about atomic power is forced upon them. Unless they work hard to gain knowledge about atomic power and build up their ability to assess the incoming new technology, they may not be able to survive in the atomic age. Hence, whether they like it or not, they will have to devote much effort to understanding recent advances in atomic power technology, and even basic atomic physics, in order to prevent themselves being deceived. From the viewpoint of the citizenry, this is nothing but a pollution of the intellectual environment, or information pollution; namely, laymen are requested to be knowledgeable in the technicalities of atomic reactors, which is a nuisance to those who do not have any particular training in atomic science. They would like to spend their time more enjoyably. Further, the kind of technology with which citizens are incessantly called upon to familiarize themselves is not a technology of quality.

In these circumstances, how much claim can the citizenry make to the human right to remain ignorant about any particular knowledge?

In the debate on environmental issues, techno-bureaucrats often insist that they are not concealing information (say, of possible hazards of atomic reactors) from the citizenry, but the citizenry complains that, as they do not have time to specialize in atomic physics, they cannot effectively assess what the technocrat-specialists are trying to bring into their local environment. This gap of information, or more precisely this gap in the time allocated to specialized knowledge by specialists and ordinary citizens, will continue to widen so long as scientific specialization keeps advancing.

In order to prevent this situation of 'no information, but retained advantage', we shall have to introduce a human right 'not to be at a disadvantage even though one does not have training and time for mastering specialized information'. A similar problem may be detected in arguments over bilingual education in large American cities. Do Hispanic high-school students have a right to be illiterate in English? Unless such a human right is introduced, future society will be full of information pollution, from which nobody will be able to escape.

NOTES

1. Jerry Ravetz's term.
2. C.G. Weeramantry, *The Slumbering Sentinels* (Penguin, 1984), pp. 17–21.
3. John Ziman, *Public Knowledge* (Cambridge University Press, 1968).

4. Shigeru Nakayama, 'The Three Stage Development of Knowledge and the Media', *Historia scientiarum*, no. 31 (1986): 101–113.
5. Jerome R. Ravetz, *Scientific Knowledge and its Social Problems* (Oxford University Press, 1971).
6. Private communication, November 1987, Paris.
7. Shigeru Nakayama, *Shimin notameno kagakuron*. (Science Studies for the Citizenry) (Shakai Hyouronsha, Tokyo, 1984).
8. S. Zuckerman, *Nuclear Illusion and Reality* (Viking Press, New York, 1982), pp. 103, 145.
9. Shigeru Nakayama, 'The Future of Research – A Call for a Service Science', *Fundamenta scientiae*, vol. 2, no. 1 (1981): 85–97.

BIBLIOGRAPHY

Morris-Suzuki, Tessa. *Beyond Computopia: Information, Automation and Democracy in Japan*. Kegan Paul International, 1988.

Nakayama, Shigeru. The Future of Research – A Call for a Service Science. *Fundamenta scientiae*, vol. 2, no. 1 (1981): 85–97.

——. The Three Stage Development of Knowledge and the Media. *Historia scientiarum*, no. 31 (1986): 101–113.

Ravetz, Jerome R. *Scientific Knowledge and its Social Problems*. Oxford University Press, 1971.

Weeramantry, C.G. *Nuclear Weapons and Scientific Responsibility*. Longwood, 1987.

First published in *Osiris*, Vol.10, 1995

⑲ History of East Asian Science: Needs and Opportunities

Objective and value-free scholarship is no more possible in the history of science than in any other field. This essay will describe and assess assumptions and values reflected in the historiography of East Asian science. My own viewpoint is that of a historian educated in both Japan and the United States and experienced in research on the history of science, ancient and modern, in China, Japan and the West.

The vague if long-standing dichotomy between East and West has generated much shallow debate. In what follows, when I speak of Eastern and Western science, I refer to complementary, well-defined traditions. The historical centre of Western science moved from Babylonia to classical Greece to Hellenistic Egypt, India, and the Middle East, and thence to medieval Europe, thus to the modern-day Western world. In the East the centre remained in China until the period of European expansion; Korea, Japan, and Vietnam remained cultural satellites. Hence, unlike Indian and Arabic sciences, from which the main current of Western scientific thought developed, China and East Asia provide us with an independent counterculture of science.

I. A TYPOLOGY OF WESTERN APPROACHES TO EAST ASIAN SCIENCE

George Sarton and the Orientalists

When George Sarton, shortly after founding *Isis* in 1913, initiated the *Isis Critical Bibliography*, he devoted a section each to Chinese and Japanese science and invited librarians from China and Japan working in the United States to contribute. He had tried to learn East Asian languages without much success, but he remained curious about and respectful of Far Eastern scientific developments. He did not venture to make generalizations about them, believing that 'although I do not know anything about East Asia, I am sure that something important was going on there, and want to know about it'.[1] This attitude was shared by the first generation of Western specialists on premodern East Asian science, who began working in this field because they were fascinated by the unique and exotic features of ancient non-Western cultures. Of this generation of Orientalists, which included Léopold de Saussure, Willy Hartner, and to some extent Henri Maspero, Joseph Needham is the last living representative. The motivations of these scholars were of course highly individual: Saussure's interest was to a large extent antiquarian; Hartner was seeking common themes and issues in the great astronomical traditions, all of which he studied using original sources; Needham's scholarship has served a profoundly ecumenical vision of the unity of science worldwide. (Although such scholars were called 'Orientalists,' this term came to connote a low estimate of the culture studied and sympathy with imperialist goals. I therefore use the more neutral term *premodernist* below.)[2]

Professionalization and Relegation to the Periphery

After the Second World War the history of science rapidly became a professional academic field, particularly in the United States. As postgraduate education evolved, the generation of historians of science who came of age in the late 1950s and early 1960s replaced the amateurish charm of their predecessors with narrow specialization in a well-defined discipline. One result is that specialists in Western science without a strong foundation in East Asian languages avoid anything more than passing references to China and Japan in their own writings, leaving it mainly to East Asian specialists, in fear of stepping outside their own familiar field of expertise. They have been made aware that China has its own major technical traditions, thanks to Needham's monumental *Science and Civilisation in China*, the first volume of which appeared in 1954. Nonetheless, as work in the overall discipline has accumulated, few general texts provide serious coverage of Asian science, technology and medicine, although at least some acknowledge their limitations by including the word *Western* in their titles.[3]

This peripheralization of East Asian science was also institutionalized in the *Isis Critical Bibliography*. When I. Bernard Cohen took over the editorship in 1953 and 'modernized' the journal, he combined the coverage of China and Japan in a single section called 'The Far East (to ca. 1600)'. This endpoint of circa 1600 reflected the assumption that since the rise of modern science in the West, technical activity in the rest of the world has conformed to its goals. One can scarcely agree, however, with the implication that the East Asian tradition was replaced immediately. Assimilation did not begin until the late nineteenth century. The intervening seventeenth to early nineteenth centuries pose distinct research problems as Asian and occidental cultures gradually confronted each other, in the arena of science as well as elsewhere.

(Re)locating Centre and Periphery

One issue relevant to the attempt to integrate the East Asian tradition with the Western view of history of science is the issue of parochialism, more particularly, of locating periphery and centre. Parochialism can take various forms. Although I have published on the history of premodern and modern European, American, Chinese, and Japanese science, technology, and medicine, I am often treated by Western historians of science as a narrow specialist in Japanese or East Asian science. Admitting that as I reside in Japan, my work can be seen as geographically peripheral, I claim that scholars like myself usually have a wider perspective than those in Berkeley or Paris. Since scholars at the 'periphery' know what is going on at the 'centre', and few of the latter are interested in what is going on at the periphery, it is more correct to see those at the centre generally as narrow specialists in the history of Western science.[4] Furthermore, the main current of the Western scientific enterprise is now the common heritage of the history of science profession, capably summarized at the textbook level all over the world. Even in Asia the definition of the discipline draws on the Galilean-Newtonian tradition. Thus East Asian historians of science are involved in the same enterprise as Westerners.

The narrow approach of the 'centre' reveals itself in other ways. I am often invited to serve on the editorial board of international journals. Many are only nominally international, accepting foreign contributions only if written in acceptable English. *Isis*, for example, which still calls itself an 'international review', ceased some years ago to accept articles in any language but English. I often find it difficult to think of their editors, some of whom are extremely narrow in intellectual outlook and unaware of non-European cultures, as 'international'. It would be better if the editors of international journals were scholars who were able to read a non-Western language as well as being capable editors of English. Perhaps Indian scholars are particularly qualified.

The Modernizers

Although the 'Orientalists' were drawn to the special character of early Eastern science, an entirely different motivation for research soon appeared, first in the work of East Asian scholars[5] and then, in slightly altered form, among Western historians and social scientists. These scholars view acquisition of Western science and technology simply as a means of modernizing non-Western societies. Modernization implies conforming to European and American ways and changing any attitude or institution that interferes with doing so. Uncritical modernizers evaluate their subject matter according to how closely it approximates the scientific practices and institutions prevalent in the West. (In this essay I use *modernist* to refer to someone who studies the modern period, and *modernizer* for someone who, explicitly or not, takes this ideological stance towards history, described in more detail below.)[6]

Most Western historians of science unconsciously adopted this view. Until the late 1960s, hardly any questioned whether the criteria of the modernizers were valid for measuring the achievements of non-Western science. They even applied the same criteria to premodern East Asian science. In research of this kind the key question was whether Asian scientists achieved some item of modern knowledge earlier than their European counterparts. To be sure, Joseph Needham reversed the earlier tendency to use priorities as an argument for the inferiority of Asian cultures. He drew on a broad command of the Chinese literature to persuade Western readers that before modern times Eastern technologists were more innovative than Westerners. But the issue remained priorities. Needham's strategy of evaluating ancient Chinese science by modern European criteria paradoxically encouraged most of his followers around the world, including those in China, to accept the modernization perspective uncritically. It undermined the exemplary value of his own passion for comparative study.

Alternative Views of Science

The intellectual climate in the late 1960s and early 1970s, with its critiques of imperialism and the myth of value-free science, encouraged a new view of alternatives to conventional Western science. Structural anthropology provided a promising framework of analysis to account for the emerging Third World on its own terms. At the same time some observers argued that the history of science, by glorifying modern science as the quintessence of the modern European heritage, had become the last stronghold of Western chauvinism.[7]

In these new circumstances I was commissioned in 1977 to present a paper countering the assimilation thesis at the International Congress of the History of Science. Up until the 1960s, the papers I wrote in English for the Western audience were neither understood nor accepted unless they were framed to fit the assumption that the goal of East Asian science should be to assimilate modern Western science. The 1977 essay, entitled 'Alternative Science of the East', argued instead that East Asian science should be appreciated in terms of its own unique paradigms, and its development measured by its own yardstick.[8] Although Needham, for example, always maintained optimistically that in the future Eastern and Western science will converge in an ecumenical synthesis of 'modern science', as he defines it, I believe that unless external pressure had been applied, the two traditions would have developed along their own normal lines, diverging so much that easy synthesis would have been impossible.[9]

Once we accept this alternative stance, using modern Western science as a yardstick no longer appears attractive. As a leading member of the history of Western science, recently retired, put it, 'Comparison with the West is instructive to Western readers but of less importance than the intrinsic fascination of the Chinese material.'[10] This is the approach of the post-Needhan generation of Western historians of East Asian science, such as Nathan Sivin. They address their message to the Western academic community of the history of science, by now firmly established, where the thesis of modernization, once taken for granted, is now being critically reexamined. In order to obtain a unique position in the community, they are now trying to understand more deeply than before the intrinsic values and structures of Eastern science. True, traditional native scholars also sought out those intrinsic values; their perspective, however, was not comparative. The 'alternativist' generation should be able to make more informed comparisons. This may be the only way to integrate the Asian experience into the world history of science, as Needham originally suggested.[11]

We might well argue that the complementary Eastern and Western traditions defined in the opening paragraphs of this essay form a better-balanced unit for study than the Western tradition alone. This comparative and relativistic approach also has its utility in studying the evolution of Western science. The characteristics of the latter might be better defined and its contours carved more clearly when considered from the perspective of another distinct culture.

Science and Western Expansionism

Modernizers are aware that what convinced the East of Western superiority was not Newton but Newcomen. The East began to modernize not because of the Newtonian paradigm or, more generally, the outcome of the Scientific Revolution in the seventeenth century, but because of military technology and its underlying power, engineering – both products of the Industrial Revolution that made the Western military threat overwhelming.

Whether the Industrial Revolution could have eventuated without the preceding Scientific Revolution can scarcely be considered without studying the non-Western historical experience. Although mechanistic views of nature were embedded in both revolutions in the West, elsewhere, in one country after another, industrial change has swept over societies upon which modern scientific education has had practically no impact aside from the training of a small corps of technocrats. Third-world scholars have generally concluded that the most deeply felt influence of the West was the imperialistic impact of the Industrial Revolution.

As the integration of the European Community proceeds, its ideologists have tended recently to seek an historical basis for a common identity in 'European expansion' in the age of the great navigators. In order to avoid an imperialistic tone, they have tried to invent a substitute for the old Eurocentric approach by creating a new dialogue between Europe and the non-Western world, those who did the expanding and those expanded upon.[12]

It might be interesting to create a history of European expansion as seen from Asia. The wealth of native language resources would allow construction of a relatively impartial account, free of imperialistic bias. For instance, the European influence on Japanese science has already been more revealingly studied by Japanese scholars than by Europeans, who have generally made scant use of materials in the Japanese language and have often taken their sources at face value rather than scrutinizing the interests that shaped them.[13]

Non-Western scholars, however, have worked only on the one-way, single channel from Europe to their native land, and rarely have an overview of world-wide influence. European scholars, located at the hub of many radiating channels of influence, are well placed to investigate the whole enterprise of imperial expansion. Some important new approaches are emerging. Lewis Pyenson, for example, examines the spread of modern science in the context of cultural imperialism, quite a contrast to the conventional modernizers' point of view.[14] I call this new approach 'critical modernist'.

It would be premature to forecast what these new approaches will accomplish, but there is some hope that the alternativists and the critical modernists will replace the conformist and outworn orthodoxies of their predecessors.

II. OLD VERSUS NEW: SCHOLARS OF THE PREMODERN AND MODERN PERIODS

Differences in Work Habits and Training

In a number of respects the differences between the historiography of premodern and modern science suggest two distinct cultures. This dichotomy applies to students of the West as well as to historians of Japan and other non-Western countries.

Premodernists have few enough sources that they can control them, but modernists are overwhelmed by data. The former habitually browse in libraries and bookshops, looking for additional sources and manuscripts, while the latter must sift through ever-increasing source materials.

Premodernists cultivate a bibliophilic attitude, rather like a hobbyist's, but modernists now tend to use computer databases systematically in the interest of efficiency. Of the foreigners who come to work in Japan, premodernists tend to work in Kyoto, and modernists in Tokyo. In the Western world the premodernist tradition survives in Europe, as evidenced in the activity of the Needham Research Institute in Cambridge and P. E. Will's group in Paris (Groupement de Recherche 798 du CNRS), while Americans dominate the modernists.

Premodernists have an antiquated aura of cultivation and erudition, while modernists identify with the social and political sciences. The former need formidable linguistic skills, while the latter are preoccupied with scientific methodology. As a corollary, premodernists today are primarily interested in the internal aspect of traditional science, but modernists are preoccupied almost exclusively with socio-institutional aspects.

Finally, in the history of science many premodern subjects can be dealt with by those trained exclusively in the humanities, while studies of the modern period often require insight into the behavioural patterns of the scientific community. Such insight is often impossible without some training in science.

Differences in Problems

The historiography of non-Western science is further complicated by a slightly different dichotomy, interest in indigenous science versus concentration on interaction with the West. This dichotomy parallels that between premodern and modern studies, but the demarcation is even clearer. In Japan those who work on pre-Meiji history of science differ noticeably in approach as well as background from specialists on the post-Meiji era. In China one sees an analogous difference between those doing research on the periods before and after the Opium War.

China's is the dominant culture for those who study premodern science. The paradigms that inform premodern science derive from the Chinese classics, written in the last five centuries B.C. The primary sources of science were usually written in classical Chinese even by people in Korea, Vietnam and Japan, who shared the Chinese paradigms.[15] One can speak of all these sources as Chinese science in the same sense that works written in Greek by Syrians, Egyptians and others are sources for what is called Greek science.

The cultural dominance of China does not hold for the twentieth century. Modernists may suggest, for example, that the earlier Japanese experience of modernization could instead provide a model for China – and other non-Western countries as well.

Differences in Audience Concerns

We have so far dealt mainly with Western scholarship written in a Western language and addressed to the Western academic community. The issues may be further complicated when we turn to the works of Eastern scholars who write in their native language and address a native audience. I myself, when writing in Japanese, choose very different topics and write in a very different style than when writing in English. In the latter case I try to take the viewpoint of an objective outsider or even a proponent of the value of Eastern science. When writing in Japanese and to a native audience, which I can address more effectively, I can write more critically on the same subject.

Native audiences are usually interested in what helps to resolve issues of current concern to them, while international audiences tend to focus on unique local characteristics. To take an example from Japan, Confucianism in the Edo period is an important topic for native premodernists. It is less interesting to Chinese and Western premodernists, who consider Japanese neo-Confucianism a mere derivative of the Chinese orthodoxy. Similarly, in the field of the history of science non-Japanese may be interested in unique aspects of Japanese culture such as traditional mathematics (*wasan*) and native Japanese flora and fauna. These premodern topics, however, attract a Japanese audience less, for they assign them to the realm of minor arts and consider them irrelevant to later scientific development in Japan. Japanese audiences are instead more interested in, for example, the great precursors who translated Western science during the Dutch period, not because of unique features that invite international attention, but because of the topic's bearing on the practical business of assimilating Western science.

The differences of style between premodernists and modernists are amplified in the case of native scholars, because of the impact of modern nationalism. In approaching premodern subjects, natives and foreigners can keep a nearly equal historical distance. Natives working on modern subjects, however, sometimes find it difficult to avoid parochial viewpoints because of their involvement in their own societies. For instance, when comparing the modernization of Chinese and Japanese mathematics and science, Ogura Kinnosuke expressed uneasiness that so many more Japanese than Chinese sources were available to him, and concern lest his assessment might be coloured either by nationalist pride or, conversely, excessive self-criticism. The detached appraisals of outsiders would be helpful in striking the proper balance.

How to bridge the two cultures and find a synthesis is an intractable problem deeply embedded in the disciplinary structure. A hasty synthesis would have little value.[16] All I can suggest is to begin extending our research horizon by applying the methods and problems nurtured in one camp to the other.

III. LINGUISTIC AND COMMUNICATION PROBLEMS OF WESTERN SCHOLARS

In the study of East Asian science, the formidable requirement for research in primary sources sharply divides those proficient in Asian languages from those who are not. This counts far more than boundaries of nationality. In a day of professionalization and specialization, few scholars are prepared to study many foreign languages extensively, although the most successful graduate programmes in the United States require students who want to do research on Chinese science to learn classical Chinese and modern Chinese and Japanese in addition to at least one European language. Western historians who merely aim to make informed comparisons – a reasonable expectation for any historian of science – ought at least to know the European-language sources on the topic under scrutiny, and to have the overview of Asian science that might be provided by undergraduate survey courses.[17]

I have already considered the dichotomy between native and foreign audiences; in this section and the next I will discuss the options available to different kinds of researchers. I find a basic tripartite division among scholars: nonlinguists who can only read and write European languages, and are thus not normally considered researchers in the field; linguists who can read the native language and write European languages; and native scholars who speak and write only in the native language, although they may read English or another European language.

Turning 'Nonlinguists' into Researchers: Two Approaches

The trend towards increasing specialization is irreversible in any modern discipline, including the history of science: the result is an atrophy of comparative studies. Many books that claim to compare cultures or deal with worldwide topics merely collect specialist articles on various societies, with no common ground.[18] The few recent bold attempts at comparative history have had to transcend a linguistic imbalance. My *Academic and Scientific Traditions in China, Japan, and the West* compared Chinese and Greek institutions without a command of the Greek language, and Geoffrey Lloyd did the same in his recent *Demystifying Mentalities* with a limited knowledge of Chinese.[19] Such experiments would be improved by actual collaborations. One such work by Lloyd and Sivin, tentatively titled 'Tao and Logos', is now under way. Outright collaborations can be supplemented by interdisciplinary, multicultural conferences or symposia designed to overcome the deep gaps between disciplines and imbalances due to the inherently limited viewpoint of the culture one belongs to.

One systematic way to overcome the current state of compartmentalization is to create language networks in various subject areas. The members would each specialize in one language area, and the network would sponsor sessions devoted to comparative topics at international gatherings. How such a network might operate can be illustrated for the history of astronomy. The old premodernist approach was on the whole philological. Scholars such as Willy Hartner mastered many languages and compared technical terms and astronomical parameters over a large span of the Eastern and Western hemispheres. Even the most linguistically versatile historians today, such as David Pingree and Yano Michio, cannot approach that span. In fact not all the older historians did. Otto Neugebauer, despite his deep knowledge of the Western tradition, never evinced interest in Chinese astronomy.

If a network of premodern astronomy was formed, E. S. Kennedy and his pupils could be consulted for data on Islamic astronomy, Yabuuti Kiyosi and his pupils for Chinese astronomy, or I myself for Japanese astronomy. This approach is most suitable for mathematical astronomy, where a comparison of astronomical parameters can be used to trace relationships between one area and another. A network could search more widely than any scholar can do alone.

Another topic that invites the network approach is comparison of names for flora, fauna and natural products in different cultures – a favourite project of an earlier generation of Orientalists, who knew many languages. Such small-scale projects aside, a broad comparison of paradigms and characteristic approaches in each culture would be feasible on such an occasion as an international congress of the history of science or, better still, in a small specialist seminar.

Another approach turns a weakness into a strength. It is nearly impossible for a Westerner to search as widely as a native Chinese for historical sources in the Chinese language. Conversely, although the historiography of early China has been transformed by archaeological discoveries, particularly since 1970, many traditional scholars still depend almost wholly on written sources. Outsiders may open new perspectives by going beyond the written word. This is true for the modern as well as the premodern era.

Among the relevant innovations for modern science are sociological methods, such as formal surveys and statistical studies of groups, and anthropological methods, such as interviews and participant observation. We already have two successful examples in which foreign scholars studied the Japanese physics community without knowing much of the language. Male physicists found it easy to relax with them and tell inside stories to researchers who, as women and foreigners, were clearly outsiders.[20]

Bibliographies: Disciplinizing the Field

Because of new methods used to teach non-Western languages, particularly in the United States, an increasing number of Westerners are proficient in Chinese and Japanese, among them professional historians of science. They may bridge the gap between nonlinguists and native historians and play other international roles. Yet the state of affairs in Western language scholarship on Asian science still reveals some lacks.

One is for more bibliographies. For premodern subjects, Western scholarship – or, more precisely, scholarship written in Western languages – has been disciplinized. That is to say, those who work in this area know what other works are available in Western languages and are aware of the need to cite them in their own publications. The historiography of modern Asian science, however, has not been yet disciplinized, even in native-language works. In Western languages the situation is still worse. For instance, I have been asked repeatedly to write a short general survey of the history of modern Japanese science in English. That I have already written several does not release me from the obligation, since the editor always assumes that no one has read or will bother to look up the surveys already published. This repetitious activity contributes little or nothing to the normal accumulation of new knowledge. If I do not do it, whoever commissions the paper is unlikely to look into the state of the field, and may well ask someone who has not seriously studied Japanese science and will merely add to the store of misleading information. The low level of disciplinization is particularly severe in government publications and the like, where the author is often an Eastern bureaucrat who has no idea how to carry out and write up a scholarly investigation. Such literature is often more useful as a primary than as a secondary source.

We might begin disciplinizing the study of modern science by making a comprehensive collection of sources, classifying them into primary and secondary sources, and compiling a computerized bibliography that is easily kept up to date. Even the premodern period could use more such work. For premodern Chinese science Joseph Needham's *Science and Civilisation in China* gives the most comprehensive bibliography available for sources in both Western and Asian languages. Sources cited in the earlier volumes are, of course, out of date. Nathan Sivin has recently published bibliographical essays and annotated bibliographies of scholarship in Chinese, Japanese and Western European languages.[21]

Nothing analogous to these reference sources, however, is yet available for Japanese or Korean science. A bibliography of English-language works on modern Japanese science (with emphasis on physics) by Morris Low is the best available at present.[22] To supply that need is the aim of the *Bibliography of East Asian Science, Technology and Medicine* that I have compiled with the help of Ho Peng Yoke, Sivin, Low, Jeon Sangwoon and Christine Daniels, now in press at Garland Publishing.

IV. EAST ASIAN HISTORIANS OF SCIENCE: SOME PROBLEM CHOICES

Scientists who can read only English tend to think that all important writings are available in the English language. This is largely true in the community of physical science, where internationalization is most advanced today. Works in humanities and social sciences are instead still written in many languages, and the highly culture-bound concepts employed in such fields render translation extremely difficult. This is true of the history of science.

Most historians of science in East Asia never intend to write in English. They communicate in their native language, forming a local citation group, often with local paradigms. They can of course read English or another major European language. They freely cite Western works, but their own publications are practically never read in the Western world. Some authors who are not native speakers of English but have a scientific background are inclined to follow the convention of the scientific community and write in English. Even they, however, become aware of the culture-bound character of authorship, which imposes a number of limitations.

Premodernists as Textual Editors

For the premodern period, varying language skills have led to a *de facto* division of labour. While Western premodernists concentrate their energy on a limited number of representative classical works, native scholars tend to search widely for new manuscripts and editing collections of primary sources, often of second- and third-order importance. For Japanese and Korean sources, manuscript searches have nearly exhausted the supply. For Chinese sources, there are plenty of printed works available because of the larger audience, and scholars have little free time to look for unpublished sources. Yet the number of researchers is also large and can be mobilized to search for, collect and edit rather recent manuscripts and local gazetteers. Since mainland China was isolated from the rest of the world for some time, some scholars still maintain the old-fashioned approach of giving priority to the first appearance of concepts and events without much analysis of their historical environment.[23] This work can be done mechanically with the advent of computerized databases.

Modernists as Modernizers

Much as with the Eastern assimilation of modern Western science, native historians of science working on modern subjects try to make their analyses conform to Western historiography, whether positivist or Marxist. This leads them to find similarities between East and West, while Western Orientalists are interested in finding the differences. The native historians are predisposed to the modernizers' approach. They tend to take the West in general or the world's most advanced level of research, that of Germany in the past and the United States at present, as the model; to compare the state of affairs of the local scientific community with the model; and to bewail local institutional and cultural defects and to attack what they see as the indifference of their own government towards science.[24] Most of these modernist writers are ex-scientists who largely share the value system and frustrations of the international scientific community, to which the Western Orientalist's approach is unacceptable. Although modernists trained as professional historians of science are a little more sophisticated, most of them also, whether pre-war Japanese Marxists or post-war proponents of democracy, have shared the psychology, if not the ideology, of this preoccupation with catching up.[25]

The Japanese are somewhat more relaxed and freer of this psychology nowadays, but Korean, Taiwanese and mainland Chinese modernists are still obsessed with catching up with the West and, increasingly, with Japan.[26]

Peripheral Citation Groups

I mentioned in the opening of Section IV that East Asian historians may form local citation groups. Many such independent citation groups exist not only in the humanities and social sciences but in the natural sciences.[27] These last offer a special opportunity for those seeking a broader base for comparative studies of science and are an ideal research subject for those speaking the citation group's language. Imanishi Kinji's non-Darwinist ecology group is a notable example. I have reported on the post-war 'grassroots geology' research movement centred on the charismatic Ijiri Shoji.[28] Even in theoretical physics, which is almost completely international, Yukawa Hideki's elementary particle group formed such an independent citation group in the secluded circumstances of the Second World War. Some such groups boast that they often have richer resources than those who can read only English, since they can draw on native as well as Western research. Their concepts often can hardly be rendered into English.

I have cited these Japanese groups from my own knowledge; analogous Chinese and Korean groups, cited only in their native language, should certainly exist. But they are not visible to foreigners, and their existence is seldom recognized as part of modern Western science. They may constitute a nuisance for those who strive to make a universal model of science, but they provide an illuminating counterexample for sociological study.

V. A NEW DEVELOPMENT: HISTORY OF SCIENCE AND SCIENCE POLICY

Japan's experience with modern science and technology has made it a special focus of studies that combine history and science policy. First, its technical modernization predates that of most other non-Western societies, including China. Second, its post-war development, alluded to above, seems unique. These two apparent success stories have led to approaches that could be described as applied history of science.

The UNU 'Japanese Experience' Project: A Failed Approach

During the 1970s the United Nations University (UNU), with headquarters in Tokyo, launched a big project entitled 'Technology Transfer, Transformation and Development: The Japanese Experience'.[29] The motive was to derive lessons from the Japanese experience in the nineteenth century for contemporary developing countries. The underlying ideology was purely that of the modernizers.

From the outset I was critical of the validity of such an approach at a time when the more prudent science policy planners in third-world countries were concentrating on 'appropriate technology' or 'alternative technology'. It also seemed to me self-evident that the nineteenth-century context of science and technology is so totally different from the contemporary world as to invalidate simple historical analogies.

Most of the works produced for this project by Japanese historians were addressed to a Japanese academic audience. Many of these writings have scholarly merit, but they fail to share the concerns of the Third World. Only a handful of historians of science and technology, such as Hoshino Yoshiro and Nakaoka Tetsuro, who did not belong to a tight disciplinary group of historians and were accustomed to addressing wider intellectual audiences, were sensitive to third-world problems. They later visited nearby third-world countries to discuss their research findings with local scholars. Among other findings, they discovered the plain fact that third-world people are not particularly inspired by the nineteenth-century Japanese experience. These historians modified their project, working on post-war Japan with international and particularly third-world collaborators, but because of the financial drain UNU could not extend the project.

The Business School Approach

When rebuilding its industrial sector, post-war Japan followed a different path from the rest of the world. Technical development was led by the MITI-industrial complex (that is, by close cooperation between the Ministry of International Trade and Industry and various industries), but conducted mostly within the private sector. In this sense the complex differed from the publicly sponsored alliance of industry and the Department of Defense in the United States. While academic science is in an impoverished state in Japan, corporate science flourishes. The privatization or capitalization of science has now taken an extreme form, in which management experts, rather than historians or sociologists, are most interested and concerned.

Work that explores industrial science is of the highest importance, especially since most works in the sociology of science today are still largely confined to topics relating to traditional academic science. Assessment of post-war Japanese science and technology took on a new dimension in the 1980s, initiated by science policy groups. To promote the national as well as corporate interest in science and technology, a society was formed in 1985, entitled Kenkyu Gijutsu Keikaku Gakkai (the Society for Science, Technology, and Planning) with a membership that included corporate scientists, bureaucrats and some scholars from university business schools.

The focus of their attention is not the academic science that historians and sociologists of science are accustomed to dealing with, but industrialized science or corporate science practised in secrecy, the research findings of which are seldom published. Mainly because of this secretive nature, historical investigation is very difficult, even though corporate science now dominates many fields in Japan. An increased trend towards privatization is likely to occur in the post-Cold War period, with the Japanese model of a privately dominated structure of science replacing the government-sponsored military industrial complex model dominant in the immediately post-war world.

The society's aim is to focus on the efficiency and cost performance of research investment. Their language is aggressive, largely borrowed from military strategy, and intended to promote survival in the competitive marketplace rather than the pursuit of disinterested scientific concepts. Meetings of the society are taken up with the success stories of ex-engineers, which may interest us to some extent as case studies, but often no generalizations can be made, as the strategy of each corporation is different. The members do not use sociology to study laboratory workers.

Similar changes are occurring in the West. In the mid-1980s I made a trip to the United States to lecture on the post-war development of Japanese science and technology. Most of the audience consisted of business-school management specialists rather than the premodernist academics with whom I had previously been familiar.[30] I found myself unintentionally involved, not in civilized Orientalist conversation, but in an adversarial and controversial discussion of national and business interests. This made me uncomfortable, since I was not prepared to speak for 'Japan, Incorporated'.

But consciously or unconsciously, we are inevitably involved in such nation-state viewpoints, and this is likely to continue as long as data, statistics and indicators are all related primarily to the nation state as a unit. It is thus essential to make explicit our independence of nation-state interests. Unfortunately, comparative histories of modern science in East Asia have conventionally tended to be success stories about Japanese cases. We should seek to establish perspectives removed from questions of national pride.

A 'Four Sector' Approach

To counterbalance the business-school defence of national and corporate interests, I have introduced a 'four-sector' approach to analysing post-war Japanese scientific and technological activity. It incorporates not only academic, government, and private corporate viewpoints, but also that of the ordinary citizenry. A preview was published by Kegan Paul in 1991, under the title *Science, Technology and Society in Postwar Japan.* A massive Japanese-language version of this project, the product of team research by more than fifty historians of Japanese science, will soon be published by Gakuyo Shobō. I hope that it will be translated into English by the turn of the century.

Historians are not in a position to defend the viewpoint of nation-states or of business firms. Most traditional historians of science have depended on the value system of the academic community, of which they themselves are members. For most of them science is an object of observation, testing and criticism, not of promotion. The disinterested position of historians must be linked with the concerns of citizens, who by definition have no vested interest in promoting scientific activity. It is only from this vantage point that the critical examination of nationalistic or industrialized development of science and technology is possible.

This new dimension to the history of science is not confined to Japan. Newly industrialized economies in Asia, such as Korea and Taiwan, are rising quickly, and eventually continental China, with its population of a billion, will emerge into scientific and technological parity with the rest of the world. We need to ask such questions of these countries as well, taking care not to adopt positions limited to a nation-state perspective, or simplistic comparisons with the United States or Japan.

NOTES

1. Personal communications in his final days.
2. The appearance of Edward W. Said, *Orientalism* (New York: Pantheon, 1989), helped discredit the original term. On Needham see Nakayama Shigeru, 'Joseph Needham, Organic Philosopher', in *Chinese Science: Explorations of an Ancient Tradition*, ed. Nakayama and Nathan Sivin (Cambridge, Mass.: MIT Press, 1973), pp. 23–43.
3. For example, John G. Burke, ed., *Science and Culture in the Western Tradition: Sources and Interpretations* (Scottsdale, Ariz.: Gorsuch Scarisbrick, 1988), a very popular introductory text, which does not devote a single page to Islamic, Indian or East Asian science. More egregiously, a recent reference work that included a grand total of two pages on China and Japan, three on Islam and less than two on 'Hindu science' out of five hundred does not hesitate to call itself without qualification *Dictionary of the History of Science* (ed. W.F. Bynum, E. J. Browne and Roy Porter [Princeton: Princeton Univ. Press, 1981]).
4. See Nakayama Shigeru, 'The Shifting Centres of Science', *Interdisciplinary Science Reviews*, 1991, *16*(1):82–88.
5. Ogura Kinnosuke's pre-war writings on the modernization of Eastern mathematics, such as *Sūgakushi kenkyū* (Studies on the history of mathematics), two vols. (Tokyo: Iwanami, 1935–1948), are a good example.
6. For an example of the modernizers' conception of science see William Beranek, Jr., and Gustave Ranis, *Science, Technology, and Economic Development: A Historical and Comparative Study* (New York: Praeger, 1978). Modernists are not necessarily modernizers, although in China and Japan the two categories tend to coincide.
7. It can be pointed out that the history of science is largely the domain of white researchers. Few black scholars attend international conferences in the history of science, perhaps because they, unlike Asians, may find no scientific tradition in their past.
8. See Nakayama Shigeru, 'Alternative Science of the East', in *Human Implications of Scientific Advance: Proceedings of the XVth International Congress of the History of Science*, ed. Eric G. Forbes (Edinburgh: Edinburgh Univ. Press, 1978), pp. 36–44. See also Nakayama, *Academic and Scientific Traditions in China, Japan, and the West* (Tokyo: Univ. Tokyo Press, 1984).
9. Nakayama, 'Joseph Needham' (cit. n. 2); and Nakayama, 'Alternative Science' (cit. n. 8).

10. A. Rupert Hall, 'A Window on the East' (review of Joseph Needham et al., *Science and Civilisation in China*), *Notes and Records of the Royal Society of London*, 1990, *44*(1):108.
11. Nathan Sivin argues, e.g., that the premodern classification of scientific knowledge is quite different from that of Europe: see Sivin, 'Science and Medicine in Chinese History', in *Heritage of China: Contemporary Perspectives on Chinese Civilization*, ed. Paul S. Ropp (Berkeley: Univ. California Press, 1990), pp. 164–197.
12. In a little over a year I was invited to three such meetings: 'Science and Empire', Paris, April 1990; 'European Federalism', Lisbon, October 1990; and 'History of European Expansion: Technology Transfer between East and West since Vasco Da Gama', Leiden and Amsterdam, June 1991.
13. James R. Bartholomew, *The Formation of Science in Japan: Building a Research Tradition* (New Haven: Yale Univ. Press, 1989), is a rare example in Western scholarship. One of the earliest attempts to translate a Japanese source into English is Watanabe Masao, *The Japanese and Western Science*, trans. O. T. Benfey (Philadelphia: Univ. Pennsylvania Press, 1990). There are hundreds of others worth translating.
14. See, e.g., Lewis Pyenson, *Cultural Imperialism and Exact Sciences: German Expansion Overseas* (New York: Lang, 1985); and Pyenson, *Empire of Reason: Exact Sciences in Indonesia, 1840–1940* (Leiden: Brill, 1989).
15. Although Thomas S. Kuhn has abandoned the term *paradigm*, and it was never widely used by historians of science, I find it particularly applicable to classical times, when printing did not exist. Printed works guarantee that scientists in a given school will share new information, but in the days when academic and scientific traditions were handed down by copying manuscripts, paradigms were more easily maintained through many generations. Later efforts tended to be attributed to a single ancestral, paradigmatic, common figure or treatise and shared by all members of the school, as in the schools of Aristotle, Pythagoras or Confucius. See Nakayama, *Academic and Scientific Traditions* (cit. n. 8), Chs. 1–2.
16. I have tried to bridge this gap for Japan by dealing with the transitional period in Nakayama Shigeru, *Bakumatsu no yogaku* (Western learning at the end of the Tokugawa period) (Kyoto: Minerva shobō, 1984).
17. Nathan Sivin, 'Over the Borders: Technical History, Philosophy, and the Social Sciences', *Chinese Science*, 1991, No. 10, pp. 69–80.
18. A typical example is A. F. Aveni, ed., *World Archaeoastronomy: Selected Papers from the Second Oxford International Conference on Archeoastronomy* (Cambridge: Cambridge Univ. Press, 1986). The Comparative Studies of Health Systems and Medical Care monograph series published by University of California Press contains few volumes that address comparative issues.
19. Nakayama, *Academic and Scientific Traditions* (cit. n. 8); and G. E. R. Lloyd, 'A Test Case: China and Greece, Comparisons and Contrasts', *Demystifying Mentalities* (Cambridge: Cambridge Univ. Press, 1990).
20. See Lillian Hoddeson, 'Establishing KEK in Japan and Fermilabs in the United States: Internationalism, Nationalism, and High Energy Accelerators', *Social Studies of Science*, 1983, *13*:1–48; and Sharon Traweek, *Beamtimes and Lifetimes: The World of High Energy Physicists* (Cambridge, Mass.: Harvard Univ. Press, 1988).
21. Nathan Sivin, 'Science and Medicine in Imperial China: The State of the Field', *Journal of Asian Studies*, 1988, *47*(1):41–90. See also Joseph Needham et al., *Science and Civilisation in China* (Cambridge: Cambridge Univ. Press, 1954–).
22. Morris F. Low, 'The Butterfly and the Frigate: Social Studies of Science in Japan', *Soc. Stud. Sci.*, 1989, *19*:313–342.
23. See Nakayama Shigeru, 'The History of Science as Practiced in China,' in *China: Development and Challenge*, ed. Lee Ngok and Leung Ci-Keung (Hong Kong: Univ. Hong Kong Press, 1981), pp. 287–296.
24. A typical pseudohistorical example is Qian Wen-yuan, *The Great Inertia: Scientific Stagnation in Traditional China* (London: Croom Helm, 1985).
25. Nakayama Shigeru, 'The History of Science: A Subject for the Frustrated', in *Science and Society in Modern Japan*, ed. Nakayama, David L. Swain and Yagi Eri (Cambridge, Mass.: MIT Press, 1974), pp. 3–16.
26. See the publications of the Science Policy Section of Academia Sinica.
27. Examples in the social sciences are Maruyama Masao's school of political science and Otsuka Hisao's school of economic history. For Yanagita Kunio's folklore group in the humanities see Bernard Bernier, 'Yanagita Kunio's "About Our Ancestors": Is It a Model for an Indigenous Social Science?', in *International Perspectives on Yanagita Kunio and Japanese Folklore Studies*, ed. J. Victor Koschmann et al. (East Asian papers, 37) (Ithaca, N.Y.: China-Japan Program, Cornell Univ. 1985).

28. Nakayama Shigeru, 'Grass-roots Geology: Iriji Shoji and the Chidanken', in *Science and Society in Modern Japan*, ed. Nakayama, Swain, and Yagi (cit. n. 25), pp. 270–289.
29. See the Human and Social Development Programme – Japanese Experience series of reports published by UNU in 1979–1982.
30. Representative publications are Sheridan Tatsuno, *The Technopolis Strategy: Japan, High Technology, and the Control of the Twenty-first Century* (New York: Prentice Hall, 1986); and Gene Gregory, *Japanese Electronics Technology: Enterprise and Innovation* (New York: Wiley, 1986).

First published in A.Burgen, P.McLaughlin and J.Mittelstrasse, eds, *The Idea of Progress*, de Gruyter, 1997

20 The Chinese 'Cyclic' View of History vs. Japanese 'Progress'

ASTRONOMICAL BACKGROUND OF THE CHINESE CYCLIC VIEW OF NATURE

History repeats itself. When the ancients identified cyclic movements in the heavens, they naturally must have been struck with the idea that terrestrial events conform to the celestial cycles, and that everything under the heavens repeats itself. Thus in the ancient world was born the cyclic view of nature as well as of human events, which is logically incompatible with the notion of positive progress. This is particularly true for the Chinese mentality, which maintained the principle of Yin and Yang, to counterbalance positive and negative. The Chinese were quite aware that technology and material culture had become more elaborate over time, but they kept this idea so carefully segregated from their cyclical view of society and politics that there was no need to inquire into what would bother someone today as the tension between the two.

The notion of periodical cataclysmic destructions and recreations of the physical world existed in ancient Chinese civilization, as among the Indians and Greeks. Over the last three centuries B.C. Chinese thinkers made great cycles the basis of a cosmology that encompassed everything under the heavens. They gave it an astronomical basis by beginning and ending these cycles with a grand conjunction, when the sun, moon and planets are together in one division of the sky, and all the calendrical cycles begin simultaneously.

The dynastic cycles reenact this cosmic rhythm, which explains history and subsumes all social institutions at the same time as it patterns Nature. It justified the new centralized imperial order and patterned its rituals.

ASTRONOMICAL PRECISION AS PROTOTYPE OF PROGRESS

As the lengths of the solar, lunar and planetary cycles were determined more precisely by observation, the length of a great cycle, because it was the least common multiplier of the increasingly elaborate numbers, increased by many orders of magnitude. By roughly the time of Christ, it had become so large that it had lost its connection to practical measures of length and was finally abandoned. Simple cyclic methods could no longer maintain the old cosmological scheme. At this point, there were two ways out. One was to abandon the astronomical basis for purely numerological cycle-counting; the other was to abandon the cyclic view of nature.[1]

The first alternative was to replace increasingly complex and abstract numbers based, however indirectly, on empirical procedures, with simple symbolic numbers such as those associated with qualitative discourse based on the Five Phases and sexagesimal year-month-day cycles. Such a move could have led to fantastic number mysticism like that of Indian cosmology. It could have been more tempered, like the move in Greek and later European mathematical astronomy away from a luni-solar calendar (based on the actual spring equinox and new moons) and towards the Egyptian calendar with its integral number of days in a month and a year. The Chinese were too committed to their cyclic view of nature to move in either of these directions.

Greek adherence to the idea of cosmos as harmony was also challenged by the discovery that the lengths of month and year are incommensurable, so that intercalation cannot be treated as a simple harmonious relationship. This problem did not cause Europeans to move away from a rational astronomy based on geometrical models until the ideal was compromised beginning in Kepler's time.

The Chinese, on the other hand, abandoned rational astronomy based on simple postulates, preferring numerical and later algebraic methods of computation and giving priority to empirical agreement with celestial phenomena.

In reality, however, even with the demise of cyclic view of nature, the Chinese continued to solve problems dealing with the great-cycle conjunction as the origin of calendrical cycles in mathematical textbooks as late as the fourteenth century. This became a conventional set-piece using indeterminate equations, a most interesting problem for astronomers and mathematicians.

The grand conjunction ceased to play an important role in calendrical calculation two thousand years ago but was not formally abandoned until the thirteenth century when it was replaced by an epoch in the recent past. At this point cyclic numbers, expressed as fractions, were also replaced by decimal numbers. This coincided with the era of highest achievement in basic astronomical observations as the Chinese defined them, far surpassing in some respects the earlier Greek and Islamic achievements. In those days, the notion of progress in the precision of observation took firm root in astronomers' minds. They did not hesitate to say, in a society that normally put its glories in the distant past, that 'the astronomical techniques of the ancients were less precise than those of today's system'.

A result of the increasing precision of astronomy was the 'discovery' that the length of a tropical year had changed over history. Astronomers had found that it was possible to reconcile ancient observations (or records) of solstices with precise contemporary observations using an algebraic formula that made year length shorter as time went on. This would have been utterly impossible in Greco-Islamic astronomy, in which the supralunar world was believed to be invariable.

THE CHINESE NOTION OF SLOW HISTORICAL RATHER THAN SPATIAL PROGRESS

As noted above, astronomers were the first to postulate progress over time. Even as early as the time of Ptolemy, Chinese astronomers must have been aware that angle measurements of celestial positions were increasing in precision. With a continuous improvement over time, astronomers were, by the thirteenth century, proud of the radical progress they had made in determining the exact time of syzygies. This was not just a matter of meticulous observation; the precise measurement of time had also begun to improve.

Modern Western technology and its associated time consciousness have shaped the idea of progress over the past three hundred years. According to the *Oxford English Dictionary*, the word 'progress' itself changed its meaning during the seventeenth century, from a movement through space to advancement with earthly time, to a progress of advancement with the passing of earthly time. As utopia was no longer placed in unknown distant lands, dreams of perfection came to be projected into the future.[2]

Unlike the Western evolution of intellectual and cultural activities, in which the geographical centre has moved from one place to another since Babylonian times, China has remained central (and the location of its capital irrelevant to this cosmological centrality) in the East Asian tradition since the first millennium B.C. In this respect as well as in those already discussed, the Chinese sense of history and progress differs radically from that in the European tradition.

The Chinese experienced slow continuous progress over time in one geographical location, as evidenced most clearly in the increase of astronomical precision. Instead of the spatial progress characteristic of the Western tradition, the Chinese have nurtured a sense of history that has let their linear temporal progress coexist with the older cosmological notion of cyclic recurrence.

In that intellectual climate, China became the home of one of the greatest historiographical traditions, in which history provided a paradigm for scholarship.[3] This provides quite a contrast with the medieval European scholastics, who ignored history in search of eternal, invariable Truth.

THE JAPANESE VARIATION ON THE CHINESE PARADIGM

In China bureaucratic inertia played a role in maintaining a sense of continuous history, but the Japanese lacked such formal, rigid institutions. Japanese intellectuals could thus flexibly accept change originating inside or outside their national boundaries.

I have mentioned earlier in this article the thirteenth-century Chinese notion of a secular change in year length. Although it was eventually abandoned in China, Japanese astronomers in the late seventeenth century revived it, on the assumption that such a minute quantity must have been based on exceedingly precise observations. In the late eighteenth century, Japanese astronomers extended this idea of celestial change to every astronomical parameter and adopted it for some time in their official calendar.

Ogiu Sorai (1666–1728), a leading Japanese Confucian scholar, supported the idea of celestial change, claiming that 'heaven and earth, sun and moon are living bodies'. Since heavenly bodies are organic, they should be subject to incessant change. There can be nothing like eternal truth.[4]

JAPANESE RECOGNITION OF PROGRESS

During the sixteenth to nineteenth centuries, the Japanese were making their own judgements about whether, on the whole, things Chinese or Western were better. In medicine, which concerned and interested everyone, there was no obvious basis for judging the superiority of one over the other, and in clinical practice eclecticism and complementarity prevailed. On the other hand, astronomy offered objective standards for assessing the two cultures, such as precision in predicting of eclipses.

Thus, Japanese astronomers were shocked in the early eighteenth century, when they noticed that the Chinese had adopted Western parameters and observational values already in the mid-seventeenth century. This discovery came late because of the shogunate's seclusion policy. Once Japanese scientists realized what progress had been made in Western astronomical precision, they without further ado abandoned the Chinese paradigm for the Western one.

They made this fateful shift of allegiance solely on the grounds of astronomical precision. Their recognition of Western superiority over the Chinese was gradually extended to other fields of science. Some, of course, unaccustomed to the notion of rapid progress or unwilling to value a culture on such narrow criteria, demurred on the grounds that since the parameters of Western astronomy change, they cannot be reliable.

PROGRESS OF NORMAL SCIENCE DEVELOPMENT

As indicated in the above, progress is a characteristic of normal science (in Thomas Kuhn's original sense). The precision of astronomical observations is a paradigmatic example of progress. No scientific revolution is required to establish its fundamental criterion: 'the better the precision, the more valuable'. That simple value system has survived since the beginning of astronomy.

One does not have to be a scientist to conclude from Kuhn's analysis in *The Structure of Scientific Revolutions* that scientific progress cannot be permanent, for revolutions change the value system that defined progress. Only those engaged in research on a given line of normal science will find it progressive. It does not appear to be progress for those who do not share the same value system.

Laymen recognize progress in more materialistic terms, generally judging it by technological elaboration. The precision of machine tools is as basic as astronomical precision was for early scientists. In the modern world, technological precision is often derived from scientific precision. Hence educated people are persuaded that technological progress is based on scientific progress.

When we consider the intermediate technologies prevalent in nineteenth-century non-European development, we can no longer apply such a simple value system as mechanical precision. At this point, the value of linear progress has to be replaced with appropriateness to the circumstances in which a multi-valued technological system is at work.

PROLIFERATION RATHER THAN ONE-DIMENSIONAL PROGRESS

A society or community that depends on a professionalized and narrowly defined paradigm may appear from its own point of view to be progressing. Those who believe in progress come to depend on one-dimensional measures. To those who do not share their value system such a claim is nonsense. For instance, applying a simple technological yardstick, Japanese ceramics like teabowls, which did not evolve handles, remain functionally premodern, primitive and hence inferior, compared with those produced in Islamic and Western culture.

The orthogenetic notion of progress turns out to be ridiculous when brought to bear on complex matters such as arts and crafts, where the diversity of values rules out facile appraisals of superiority.

Archaeologists prove that primitive artifacts proliferate as they evolve in different climate and geographical settings. In modern market-oriented production, however, advanced technology can endlessly create demand by junking its predecessors as obsolete. Modern capitalistic production and open market economies sacrifice convenience, amenity and fit to local environment in the name of progress.

If cultural proliferation cannot be called progress, then the idea of progress remains useful only in a very limited sense, that is, to replace the preceding with the new one on some simple criterion of value. An obvious example is the measure of central processing unit speed in computers, which leads to an announcement every six or nine months that what recently became the latest and most advanced computer is now obsolete.

The increasingly predominant way of defining progress, regardless of how much effort goes into it, cannot be convincingly applied to more complex systems that depend on plural values and human needs. Although the language used in advertising fashions in clothing is larded with assertions about progress, no aspect of modern industry is more cyclical and dependent on rummaging in the past.

The modern world has seen the growth of such simplistic paradigms in a good number; many independent of each other and some contradicting each other. In sum, it may appear that a modern multi-paradigmatic society is progressing, when a particular component paradigm makes its own progress.

PROGRESS OF MILITARY TECHNOLOGY

The next example of one-dimensional progress was the development of modern military technology, in which the measurement of progress is made with a simple yardstick: that which wins. The Japanese state of mind was perhaps flexible enough to accept the Western notion of progress. But the threat of the Western subjugation in the middle of nineteenth century reduced the basis of such judgements to the simple criterion of progress and superiority in military technology. The value system of Western imperialism, rather than the philosophy of enlightenment, became the key to future.

The first impression that Western progress made on alert samurais came from news of the Opium War of the 1840s in China. For ordinary Japanese, it was the American gunboats anchored in Tokyo Bay in 1853. Japanese judged correctly that the moral and ethical progress of mankind in the mode of Condorcet was beside the point. That notion of progress was for those confident of victory. Asians, understanding that such victories would make them losers, faced it with mixed feelings.

It took until the 1870s before Japanese intellectuals, and Westernization-oriented government as well, recognized Condorcetian progress lay in the background of Western military superiority. The Chinese were slower to appreciate this process. The Japanese word for 'progress' was coined from Neo-Confucian antecedents and was soon adapted into the Chinese language.

Japanese intellectuals eventually distinguished the imperialistic interest of nation-states from individual human nature. Before that point, the early impression of military progress was so predominant, among Japanese as among other non-Western peoples, that when they contemplated the faith of the Enlightenment they were overwhelmed by a sense of its built-in contradictions.

I conclude this section with these reflections that progress can be measured only in simple matters as astronomical precision, technological precision of machine tools and military destructiveness. It has played no historic role in such matters as the ethical and moral perfection of human society as well as individuals.

CONCLUSION: DARWINIAN VS. CONDORCETIAN PROGRESS

I have pointed out that in the nineteenth century military superiority impressed the value of progress on non-Western people. In the seventeenth and eighteenth centuries Chinese and Japanese specialists and intellectuals had recognized Western superiority solely in astronomical precision and, in Japan alone, in anatomy (which is not the same as medical practice). They pointedly did not generalize from the technical domain to others they valued more, such as moral philosophy.

For the Chinese, military superiority became the measure of progress during the period of hostilities in the 1840s. People had good reason to abandon the older criterion of scientific precision as they strove, for the sake of survival, to catch up with rapid Western advances in the manufacture of firearms and the construction of warships.

The Japanese, by comparing Chinese and Western astronomy, had acknowledged Western superiority in the early eighteenth century. News in the mid-nineteenth century prompted the samurai class (which had abandoned firearms in the seventeenth century and had practically no experience of military action since then) to master the art of artillery for the defence of their country. Those early leaders must have had sleepless nights for the 'struggle for existence', a Social Darwinian term.

With the change of political regime in 1868, the new Japanese government and modernizing intellectuals began to Westernize in earnest, as a matter of state policy, and since then European enlightenment philosophy has played a certain role in the ideology that supported this effort. When these reformists coined a Japanese word for progress, what they had in mind was not the ideals of Condorcet. They were simply adapting for local purposes the nationalistic Social Darwinian version of progress that social and political propagandists in Europe had created out of Darwinism. The Japanese thus always measured progress by comparison with the West, first in military superiority in the pre-war period and in material wealth in post-war time. 'Progress or perish' and 'Progress for survival' became their state-authorized slogans. The progress of human nature, its ethical and moral character, played no part in this calculus.

Even after the high economic growth of the 1960s, the Japanese government and industry sectors seem not to be liberated from the paranoia of Darwinian progress.

NOTES

1. Nathan Sivin, *Cosmos and Computation in Early Chinese Mathematical Astronomy*, (Leiden: E. J. Brill, 1969).
2. Samuel L. Macey, 'Literary Images of Progress: The Fate of an Idea', J. T. Fraser, N. Lawrence, and F. C. Haber (eds.) *Time, Space, and Society in China and the West* (University of Massachusetts Press, 1986) p. 93ff.
3. Joseph Needham, 'Time and Knowledge in China and the West,' in J.T. Fraser (ed.), *The Voices of Time*, reprinted in Needham, *The Grand Titration: Science and Society in East and West* (London: Allen and Unwin, 1965).
4. Shigeru Nakayama, 'Japanese Scientific Thought', *Dictionary of Scientific Biography*, XV, (Scribners, 1978) pp. 731–2.

Keynote Speech at the first Conference on Redefining Twenty-first Century Chinese Culture, Taiwan, 27 October 2000 (*Harvard Asia Pacific Review*, Fall 2002: 87–88)

21 The Ideogram versus the Phonogram in the Past, Present and Future

THE EAST AND THE WEST: MY DEFINITION

I thank those who have kindly given me the chance to speak today. It provides me with the opportunity to present an idea of mine which would be impossible at a normal academic gathering of historians. I wish to make an historical comparison of East and West, drawing on my personal insights.

When I was at Harvard as a graduate student in the late 1950s, it struck me how my East Asian friends, mostly Taiwanese and Korean students, were calm and much less talkative than the Westerners and Indians. There appears to be an Eastern over-dependence on reading rather than hearing and discussion, visual rather than aural. As I carried out my studies in the discipline of the history of science, I found a remarkable cultural difference in that the Aristotelian trivium (rhetoric, logic and grammar) is largely missing in China and East Asia.

When I speak of the Eastern tradition in the history of science, I always refer to the tradition of China and its satellite countries in East Asia, Korea, Japan and Vietnam. If we trace the tradition of the exact science of astronomy, we can clearly see how the centre of cultural activity moved from one place to another simply by comparing the value of astronomical parameters, in the way that Otto Neugebauer's school has been doing for years in the history of astronomy. If we find identical astronomical parameters, which are usually six or seven effective numbers like a tropical year length of 365.2422 days, we can prove unmistakable intercourse between two cultures using the same parameters. It would otherwise be unthinkable to find the coincidental appearance of such high order numerals in two different cultures. Thus, we can trace the tradition of exact science when it first appeared in Babylonia in the fifth century B.C.[1] and when it moved its centre to Greece, Hellenistic Greece, (India), the Islamic world, and then Renaissance Europe, seventeenth-century England, eighteenth-century France, nineteenth-century Germany and twentieth-century America. This is the tradition of Western civilization. On the other hand, the centre of Eastern tradition stayed in China up until nineteenth-century.[2]

When I have previously spoken of the above definition of Western civilization, Indian people have been shocked, for as far as they are concerned, Indian culture belonged to the Eastern civilization rather than the West. But from my point of view, the Indian civilization is definitely located in the extreme West. Essentially the difference between East and West may be attributed to the style of language, ideograms vs. phonograms, an emphasis on reading vs. the spoken language, visual vs. aural communication, and a recording vs. a debating tradition. Historically, the Eastern tradition has been heavily influenced by Buddhism, which started in India, and in the East debating has been practised only in the context of the Buddhist tradition.

KUHNIAN APPROACH

I was personally very close to Thomas Kuhn since our first meeting at Harvard in 1955–56. I read Thomas Kuhn's book *The Structure of Scientific Revolutions*, the first edition,[3] and regularly corresponded with him during the 1960s. An idea came to me of applying his 'paradigm and normal science' scheme to a comparative history of science and scholarship between the East and the West as discussed earlier. It eventually appeared in Japanese as *Rekishi toshite no gakumon* (*Scholarship as History*) (Chūō kōron, 1973) and was translated into English as *Academic and Scientific Traditions in China, Japan and the West* (Tokyo Univ. Press, 1984).

PARADIGM FORMATION

I wanted to present my ideas by beginning with the formation of paradigms[4] in each culture from Han times in the East, and Hellenistic Greece in the West.

	Nine Chapters of Mathematics (mathematics)
	Mathematical Classics of Chou Gnomon (astronomy)
East	Hanshu Calendrical Chapter (calendrical science)
	Yellow Emperor's Inner Classics (medicine)
	Shan Han Lun (medicine)
	Classical Pharmacopoeia of the Heavenly Husbandman (natural history)
	Platonic and Aristotelian Corpus in doxographical tradition
West	Hipparchian and Ptolemian: Almagest (astronomy)
	Pythagoreco-Euclidean (mathematics)
	Hippocratean-Galenian (medicine)

These were the times when paradigms were consolidated and normal science started accordingly. Then, there was pre-paradigmatic time in the Hundred Schools period and pre-Socratic classical Greece. In these periods, I have noticed the symptom of bifurcation between the Eastern documentative tradition and Western argumentative scholarship. Its very paradigm must be the Shihchi 史記 and Aristotelian trivium respectively.

Originally, excited with my broad attempt at comparison, Geoffrey Lloyd and Nathan Sivin worked on the comparative history of science in the East and West in a much more detailed and scholarly way for many years. Last summer, there was a workshop held at Cambridge centring on the manuscript of their forthcoming book of 'Tao and logos'[5] by Lloyd and Sivin, which compares ancient Greece and China much more deeply than I did before. The close comparison highlights the difference between them. While in pre-Socratic Greece debate was conducted democratically among citizens, in China, Hundred School scholars mainly addressed feudal lords with the anticipation of being employed and being able to implement their theories. The difference is in the final analysis due to the difference of social base; while Greece was a small mercantile city-state, China was a big agrarian, irrigational state. I would agree with their major thesis with a minor criticism that in order to make the difference more visible, they are comparing Han China with pre-Socratic classical Greece, rather than the more bureaucratic Hellenism.

Later developments which exacerbated the difference between East and West, are shown in the following chronological table. We may contrast the difference in cultural style or academic paradigm by referring to the infrastructure of each civilization.

Period	*West*	*East*
5 cent. B.C.	pre-Socratic philosophy	
		Warring State 諸子百家
	Platonic and Aristotelian	
	Paradigms	history paradigm 史記
– 2 to+2	Euclid, Hippocrates, Ptolemy	九章算術、周碑算経、神農本草経
	doxographical traditon, Galen	黄帝内释、傷寒論、漢書律曆志
	Roman law	paper and bureaucracy
7 to 10	Islamic madrasa	civil service examination
	parchment	printing
12 to	medieval university	

There are some commentaries to be added to this table.

THE DEVELOPMENT OF THE CHINESE DOCUMENTATIVE TRADITION AND PAPER AND PRINTING

As to the origin of paper-making, there is still an on-going controversy. Whether everything is attributable to Tsai Lun 蔡倫 of the later Han is very much in doubt from the viewpoint of the history of technology.[6] We can apply to the nascent state of technology then, unlike the present-day situation of modern technology with targeted R&D activity, a Darwinian thesis that there are lots of seeds of technologies which can be later developed when the environment is favourable. This might have been the case with the idea of paper-making in the pre-Han period. Unless the seed of the idea was followed by favourable social conditions, it would be forgotten and disappear. There must have been a number of such instances for paper-making throughout history, though it is difficult to reproduce a disappeared history. Only when a continuous need existed, could the technology of paper-making be perpetuated. Bureaucracy was said to be the politics of paper. In trying to control a vast area, paper was quite an appropriate technology, indispensable in China since the Han times on. In contrast, in the oralistic argumentative democracy of the city-state in the West, paper was not much needed. This can be seen in the Platonic tradition where the most important matters were not transmitted by paper but orally.

We could say that the most favourable situation to perpetuate paper-making technology existed in the Chinese bureaucracy, dating from the Han period. This situation was further strengthened during the Tang and also the Sung periods by the written examination for the civil service. I do not know to what extent the still precious paper was employed for the competitive civil service examination in Sui and Tang China, or for that matter in Nara Japan in the eighth century, when the Chinese civil service examination was accepted and practised. Recently excavated Japanese evidence shows that wooden tablets were still used for communication within the bureaucracy. But when it comes to the Sung period, there should be no doubt that paper was employed for examination purposes.

Besides, the Sung period was the time when printing was disseminated widely. In the beginning, printing was employed to disseminate Buddhist sutra to the populace but we may ask why, at a time when the vast amount of Chinese classical encyclopedias were being printed, that we have a situation in Japan where little such activity was occurring. Such vast printing of classical collections would be unthinkable without a wide readership of candidates who were preparing for the civil service examination. In Japan, such an examination system was faded out in the ninth century and never resumed until the late nineteenth century. How could the Chinese have such a printed tradition (as evidenced in Ssuk'u ch'uan-shu 四庫全書) that they could account for more than half of the printing in existence in the eighteenth century world? Without the tradition of civil service examination, was it really conceivable?

WESTERN DEBATING TRADITION

The debating tradition of the West hardly needed paper for records. Papyrus was too fragile to preserve historical records and parchment in the medieval West was too costly to be used for printing. Thus, without paper, printing was utterly impossible.

In contrast to the dating practice of any historical record in China, the lack of dating in Indian historical sources must have been due to their debating tradition and its associated way of thinking in terms of unhistorical, logical abstraction. Paper-making technology was transmitted from China to medieval Islamic culture, which preserves the most abundant Aristotelian sources, more so than in Latin culture.[7] This promoted the debating seminars conducted at madrasas. But printing was not adopted particularly early, perhaps for some religious reason.

Finally, I would like to add the rider that I am not exactly a technological determinist. The earlier Chinese invention of paper and printing did not necessarily give them an advantageous position in history. In the West during the Renaissance, printing started with religious sources, the Bible, just as in China, but unlike China, the Western debating tradition contributed to the creation of modern critical journalism, magazines and newspapers, and modern scientific journals, none of which the Chinese developed.

Thus, the general notion of scholarship in China was about memorizing classical phrases for the ultimate purpose of repeating them at civil service examinations, while in the West it was something to defeat and persuade an opponent on a public occasion, the model being the lawsuit in a court or public debate at an open forum.

The most basic paradigm of all Western academic tradition was the Aristotelian *trivium* that was introduced to seventeenth-century China by the Jesuits. The Jesuits coined Chinese translation words for logic, rhetoric and grammar, but these were totally forgotten by the Chinese in the course of later development, and it was only in the late nineteenth century that they were reintroduced by Protestant missionaries as a part of Western curriculum. Around the same time, the Japanese were translating the Aristotelian *trivium* into their own words, also as a part of Western curriculum. Since it was the time when rhetoric ceased to be taught in the modern Western curriculum, the introduction of (oral) rhetoric was so unfamiliar in their culture that it was not properly understood by the Chinese, nor by the Japanese. Logic and grammar were treated in the modern East as a bookish way, separated from the Western origin of debating tradition.[8]

On the other hand, I believe that the most genuine Chinese invention that was imported into the West from the Renaissance on was the invention of the civil service examination system, perhaps arguably more significant than the invention of paper, ammunition and magnetic needle. It impressed the seventeenth-century Jesuits sent to China and it was imported to Collegio Romano but until the nineteenth century it did not become a widely-known practice in Western culture. The nineteenth century was characterized as 'the century of the written examination', on the basis of which meritocracy and egalitarianism took root in Western society, while prior to that nepotistic practices were customary in the recruitment of bureaucrats. The Japanese had abandoned the civil examination system as far back as the ninth century but they re-introduced the written examination system in the late nineteenth century, not from China but from England.

The above instances exemplify how different academic traditions coexisted without much intercourse in between. The difference was so deeply rooted and embedded in both cultures that the transmission of ideas from one to another was nearly impossible, unless the whole institutional or social system *in toto* underwent paradigmatic change. As I noted in the beginning of this article, a dichotomy between the written and oral tradition still continues.

IDEOGRAMS VS. PHONOGRAMS

What clearly divides the East and the West was not an historical accident. There was a deep-rooted difference, which made translation difficult or often impossible. It is due to the radically difference style of language, which can be contrasted in terms of binary opposites such as Eastern ideograms vs. Western phonograms, language for reading vs. language for talking, language for eyes vs. language for ears, the electromagnetic wave vs. the vibration of air in physics term, and finally double-byte vs. mono-byte in language code of word processors. Which is better or more promising will be discussed towards the end of this paper. We should not overlook the fact that the relationship between two language styles was not always competitive but actually complementary and cooperative through mutual translation.

Parallel to the development of written language translation, an attempt was made to translate spoken language. At one time there seems to have been a hope for spoken language translation but it requires the memory of an individual voice-print. At the moment, we are convinced that the right track of translation is written language rather than oral, as written language is subject to clearer and easier digitalization.[9] Thus, we are now in a position to contemplate the future of our digitalized ideographic society.

MY OWN EXPERIENCE WITH THE COMPUTER LANGUAGE TRANSFORMATION

In the 1970s, I had a rather critical view of the development of word-processors. The microelectronic revolution in the Western world appeared to lead to a particular kind of technological unemployment, in which three-fourths of all secretaries and typists were said to have lost their jobs, as the Western offices were, in a sense, a place to type all the documents and transactions. Since a word-processor is four times more effective than an ordinary typewriter, the simple outcome of the replacement of a typewriter with a word-processor was estimated to be the redundancy for three-quarters of the office labour force. This never happened in Japan, as the Japanese office, and I believe Chinese office as well, still involved writing office documents by hand. Only very official documents were produced using a Japanese typewriter, which was slow and inefficient. It was operated by a specially-trained female typist, who had a licence to use it.[10] Big offices usually had such a specialist typist. It follows that only a handful of Japanese typists would lose their jobs when word-processors with character conversion started to be used in 1979. Thus, the resulting unemployment was much less in Japan. Business people as well as academics all welcomed the introduction of the word-processor. For Westerners, the introduction of word-processors was merely an evolution from typing on a mechanical or electrically driven typewriter to a computer word-processor. For the Easterners, it was a dramatic transition from handwriting to word-processing, largely without the interim experience of mechanical typing in between.

In my own experience, I used to prefer to work in Western universities or research institutes, where a secretarial service was provided mostly for the writing of letters and drafts. In the absence of much typewriting work, Japanese offices on the other hand did not know how to efficiently utilize secretaries, their main job remaining to serve tea.

With the introduction of word-processors, we do not need much secretarial assistance. I can write, edit and print myself quite easily and perhaps more quickly than if I asked for the assistance of a secretary. With the advent of Internet communication – I believe that it was mainly designed to fit into the academic work style[11] – we do not need secretarial help even with correspondence.

IS MANY-BYTE TECHNOLOGY ANOTHER PARADIGM?

Up until the 1970s, I had never foreseen the time when Chinese characters could be handled on computer and utilizing an alphabetical keyboard. Unlike other phonogramic languages, Chinese characters were not to be able to deal with mono-byte letters. Then, in late 1979, the conversion from phonograms to characters that required two-bytes to express total ideograms was successfully carried out and since then, we have been enjoying the power of conversion technology. It was really fascinating. Compared to the Western experience, it meant to us, phonogram users, a double-fold revolution from handwriting jumping to digital. What can we expect in terms of further development? perhaps a shift from mono-byte to double-byte technology? Is it possible to apply two or still more bytes for purposes other than character conversion?

Although I am no expert in computer science, and lacking an understanding of technical feasibility, it appears that two-byte culture can deal with complex phenomena as complex, rather than reducing it to the mono-byte. I dare to contemplate this, even as an amateur, as we can see occasionally in the history of science, amateurish ideas generated out of the community of the established normal science tradition can lead to a scientific revolution. What can we do with two-byte culture besides using it for representing Chinese characters? Some people have suggested to me that it be used for the digitalization of illustrations and some have proposed the digitalization of handwritten scripts.

With the advent of modern science and technology, Western culture has so far swallowed the Eastern. With the advent of two-byte culture, there is the potential for Eastern culture to swallow the Western if the bigger can swallow smaller and the two-byte can swallow the mono-byte.

Considering the toil spent on realizing the use of two-byte, it is too precious to limit the use of two-bytes only to character conversion. I am claiming it from the hindsight of those who have already mastered thousands of characters. We have already memorized thousands of characters and upon it constructed the two-byte or many-byte representation of characters. With the further development of two-byte culture, we have the hope of the East overcoming the West.

At the moment, the American-made Unicode system, that requires the use of two bytes, has not been widely employed by Westerners for the obvious reason that they do not need it. But if the Unicode system proved to be useful for conversion of Chinese characters among Chinese. Taiwanese, Japanese and Korean, two-byte culture will settle down as a common vehicle of culture in East Asia.

Some people in the East would like to go beyond the Unicode system, such as Japanese TRON, in which two or many-byte could extend its code to oracle bones.[12] It is still the matter of language code. More than that, if we could extend it to illustration or pattern recognition, it could be a revolution.

This is a revolution which cannot be completed at once but requires some time to be consolidated into normal science development. If you stay with two-byte culture without attempting to reduce it to mono-byte, a new line of normal science will follow and the revolution would be completed. It is a risky play, as a revolutionary endeavour used to always be, but worth trying.

If the two-byte paradigm revolution becomes successful, the Easterner will have a certain advantage at least for a time being as we are more accustomed to it. At least we could label sixty thousands more than single byte people who can label only 256. Beyond Eastern chauvinism, however, if the two-byte paradigm proved to be successful, the Westerner would participate in it and soon catch up.

Then, we could at least appreciate historically the liberating and pioneering effect of the Han characters conversion on the progression from mono-byte reductionism to many-byte culture.

According to the Western viewpoint, whether mono-byte or two-byte is not profound revolutionary matter but the matter of system engineers to deal with and ordinary users even may not notice the difference.[13] Still, we would like to adhere at least emotionally something extraordinary would come out of double to many-byte as we could extend from our experience in Han characters conversion revolution.[14]

Or, we may stay in the present status quo of the exclusive use of two-bytes for character conversion. It is all right for us for those who have so far spent a lot of time for learning characters. For the next generation, though, they will forget the order of strokes. Or they may even forget how to write, and only learn how to convert from phonograms to characters. It is occurring even in my generation. Will this trend eventually lead ideogramic culture to be reduced to the mono-byte? Unless we go beyond the two-byte revolution, characters and its associated culture will eventually disappear or be completely swallowed by the mono-byte Western culture.

NOTES

1. Otto Neugebauer, *Exact sciences in antiquity* (2nd ed., 1966)
2. Although Western numerical values in astronomy were adopted in the seventeenth century China and eighteenth century Japan, its paradigmatic framework of Chinese exact science of of calendar-making never changed
3. Published in 1962.
4. *Academic and Scientific Traditions in China, Japan and the West* chapter 2.
5. To be published in 2001.
6. Pan Chihsing 潘吉星、 *Chungkuo tsaochih ishushih kao* (the history of paper-making techniques in China) 中国造紙技術史稿 1978
7. Hossein Nasr, personal communication (1959).
8. Shigeru Nakayama 'Educational institutions and the development of scientific thought in China and west' *Japanese studies in the history of science* no.5 (1966) pp. 172–179
9. This impression is derived from my annual visits to the ATR (Advanced Telecommunication Research institute international) at Kansai Science City.
10. Nakayama Shigeru 中山茂 'Nyu tekunoroji shimatsuki ニューフテクノロジー始末記 (Acccursed new technology)' *Kikan Kuraishisu* 季刊クライシス (Quarterly Crisis) no.6, 1980
11. Nakayama Shigeru 中山茂, Niju, *Nijuichi Seiki Kagakushi 20 • 21* 世紀科学史 (History of Science in the twentieth and twenty first centuries, NTT, 2000)p.280.
12. Sakamura Ken 坂村健, personal email (2000).
13. I owe this comment to Western mathematicians, Garry Tee of New Zealand and Douglas Roger of UK.
14. I share this feeling with Jochi Shigeru 城地茂, a historian of East Asian mathematics.

First published in *A Social History of Science and Technology in Contemporary Japan*, Vol.1, Trans Pacific Press, 2001

㉒ Preface and Historical Introduction to 'A Social History of Science and Technology in Contemporary Japan'

When I visited China as the leader of the First Japanese History of Science Visiting-China Team, the Chinese Academy of Science asked me to give a talk addressing their question 'Why did the Japanese succeed in developing science and technology despite their defeat in the Second World War?' This was in 1980, about the time that the particular style of Japanese development in science and technology began to receive worldwide recognition. Hence, this was a legitimate and important question.

My ready-made answer at the time was that, while the pre-war paradigm of Japanese science and technology was military-oriented research and development, after the war the focus was almost totally economic. The actual course of development was, of course, not so simple, consisting of a continuous series of trial and error, of ups and downs. While Japan has not produced many Nobel laureates, the post-war development was characteristically Japanese, quite in contrast to the Western path in the Cold War environment.

Thus, in the 1980s, many countries, both developed and developing, sent observation teams to Japan to find out the 'cause' of the Japanese 'miracle' – and whether or not it really was a miracle. There are no simple or easy answers to such questions, and this work makes no pretence at providing any such thing. Instead, we have attempted to chart the course of development, both its merits and its demerits – from the 'inside' the shortcomings are often more visible – to enable readers to discover this unique course, with both its positive and negative lessons, for themselves.

In the 1990s, we Japanese came to a turning point in history after half a century of this particular course of development. Thus, it is a good time to examine the course of development and to consider the possible future of development with the wisdom of hindsight, without being distracted by the short-term ups and downs of the economic cycle.

In 1999, we completed the seventh volume of this project in Japanese and turned our attention to making this work available in English, so that our lessons can be shared with those who do not read Japanese. In the process of translation we encountered numerous problems that the translators and editors struggled to make comprehensible and, hence, some further explanation became necessary. The Japanese authors had originally addressed their message to a Japanese audience, so they had included contexts and inferences that were internationally unintelligible, while at the same time omitting some information necessary for an international audience's understanding. To overcome these difficulties Professor Sugimoto, working in Australia, and I, in Japan, tried to fill the lacuna between the Japanese and world contexts.

Our thanks must therefore include email and internet technology; we have had daily exchanges of electronic files back and forth between Australian translators and Japanese authors. Without such technology, international editing like ours is virtually impossible.

I am most obliged to Professor Yoshio Sugimoto of La Trobe University, who not only organized translators and editors in Australia – which I consider to have the highest level of proficiency in Japanese language among English-speaking countries, perhaps higher than that of the US and UK – but also provided competent editorial care on this difficult, many-authored work. Thanks are also due to Karl Smith, style editor of the volume, who has worked tirelessly, paid meticulous attention to the details of each chapter and contributed to the production of a manuscript more coherent and integrated than the Japanese version. I am also grateful to a team of translators, including Francis Conlan, Mike Danaher and Naohiko Shimizu. Finally, we acknowledge the Japan Society for the Promotion of Science for their 2000–2001 Grant-in-Aid for Publication of Scientific Research Results for the translation and publication of such a large work, unprecedented in their experience.

Shigeru Nakayama
Melbourne, Australia
18 February 2001

[See Historical Introduction overleaf]

Historical Introduction

There is a widespread belief that the Second World War (WWII) was brought to an end by the atomic bombs. From such a perspective, history can be periodized according to the science and technology discovered, invented and employed during any given era. Thus, the post-war era is known as the atomic age. The atomic age was brought forth by scientists – physicists, in particular – whom many regard as possessing the only valid perspective on this new age. Hence the spokesmen of the scientific community were in great demand by the mass media.

WWII was, however, by no means terminated with the atomic bombs. Rather, the bomb marked the advent of the new and more enduring 'Cold War'. The Manhattan Project that developed and manufactured the A-Bombs was not only the prelude to the atomic age but also provided a paradigmatic model for post-war research and development (R&D) due to its strategic success; science became a 'big' project, supported by big powers and closely connected with technology. The 'little science' practised in pre-war academia became simply antiquarian. A dominant belief developed among post-war scientists that the best and easiest way to advance the research frontier was simply to expand the scale of their research facilities and organizations far beyond the pre-war standard.

The USA and the USSR created military-industrial-academic complexes that were self-perpetuating and continually growing. The 'complex' played a leading role in R&D throughout the post-war period, with minor fluctuations in its power and influence over the years. The science policy of the European nations was generally modelled after these complexes, too, although they were not always militarily oriented. The Cold War lasted for half a century, an unexpectedly long period. So too, the military-industrial complex endured and prospered, mutually excited and enlarged. It appeared to be an everlastingly perpetuated mechanism.

Now that era has passed. In the last decade of the twentieth century, the military-industrial complex began to decline and was wholly restructured. At the present point in history, it is the duty of historians of science to recapitulate the scientific activities of the Cold War era and develop a prospective for future generations.

Uniqueness of Japanese Development

Japanese science and technology (S&T), however, followed a unique course of development, considerably dissimilar to the major Western nations.

S&T in pre-war Japan was heavily inclined towards military application. Japan shares the same experience as other invaded countries in having been awakened to the necessity of modernization through the threat of the Western powers in the nineteenth century. The difference from other non-Western countries lies in the 'successful' transformation from an invaded to an invading nation in the course of development. At the time of transformation the only thing needed was more advanced military science and technology.

The departure from militant imperialism was rapid after 1945, following Japan's military defeat in WWII. The pace of change is partially explainable by the pressure from the Occupation Forces to demilitarize, but endogenous motivation provides a better explanation for the overall changes in direction. In the Japanese scientific community, as well as the political and business communities, a pacifist mood prevailed. Following their bitter lessons, they admitted the irrationality of the militaristic 'progress' that had ultimately invited the catastrophic misery of 1945 and the immediate post-war period. Consequently, they chose to forge a new path, a path that led to post-war Japan being a military-political dwarf but an economic giant.

This path was diametrically opposed to the predominant trend of the military-industrial complex established by the USA and the USSR. The only significant parallel to the Japanese experience is West Germany. Hence, it is a unique path and is worth recording for posterity.

THE STS (SCIENCE, TECHNOLOGY AND SOCIETY) PERSPECTIVE

'Science, or S&T, is not an idea but a social institution.' The truth of this statement becomes increasingly clear as history unfolds. Modern science began in the seventeenth century as an intellectual revolution. Much of the history of science appears to be a history of intellectual elites. Today, however, it is apparent that science is simply a social infrastructure.

It was the Marxist historians of science in the 1930s that first noted the social and institutional aspects of science. At the second International Conference of the History of Science in 1931, Boris Hessen of the Soviet Union presented the socio-economic basis of Newtonian mechanics. Following the Hessen paradigm, a number of materialist historians of science appeared. Hessen's work was translated into Japanese in two different versions and was very influential amongst Japanese historians of science before WWII.

The materialist historians had disputes with conventional intellectuals who believed that the history of science was driven purely by the achievements of a limited number of geniuses. The materialists argued, instead, that science is influenced by the social infrastructure. In the pre-war period, in fact, some materialist historians made an even stronger claim, arguing that science is socially determined. However, in the days of 'small science', science played only a minor role in determining the course of social events. Science was like many other cultural phenomena, remaining in the 'superstructure', with little influence on the social base. In other words, science was socially determined, but did not, in turn, determine society. The social origin of science is now accepted even amongst academics who previously maintained that science is self-determined; that is, that scientific matters (questions, validity, methods, etc.) are determined internally by the logic of each discipline. Still, this concession has not ended the disputes between the social determinists and the internalists.

During the Soviet regime – under Stalinist intimidation – it was a dangerous taboo for a historian of science to deal with the contemporary (post-revolution) history of science. On the other hand, in the capitalist regimes up to the 1950s, particularly in the US, scholars kept their distance from the social history of science under the threat of McCarthyist purges and, instead, practised the safer internalist approach, i.e. the intellectual history of science.

Western capitalist society experienced a watershed in the 1960s, culminating in widespread protests in 1968 (some people call it a 'revolution'). Previously academic historians of science focused on the classical period and/or the seventeenth century Scientific Revolution. Subsequently, the younger generations extended their interests to the eighteenth, nineteenth and twentieth centuries. It became apparent that science had been increasingly institutionalized over this period of history. Hence, rather than considering science to be an isolated activity, they regarded scientific activity as tightly integrated into the scientific community. This approach is called the sociology of science, and was highly influenced by Thomas S. Kuhn's *The Structure of Scientific Revolutions.*[1] The sociology of science is a social constructivist approach that, at its extreme manifests as social determinism. Social determinists maintain that all scientific activity is determined by the group structure of the scientific community. While the extreme is still debated, more moderate social constructivist views are now almost universally accepted.

In the 1970s a group of researchers and educators, following the constructivist trend, began to advocate the Science, Technology and Society (STS) approach. STS is not a specific method or discipline but, rather, a problem-area. In the post-war period, as S&T's power appeared everywhere, it generated a great deal of problems *vis-à-vis* society at large. Solutions would not be found within traditional disciplines such as the history or sociology of science. For instance, 'the social acceptability of nuclear reactors in Japan' is a problem area in the sociology of science. Such problem-areas need to be approached from a variety of angles, transcending disciplinary boundaries.

Japan is not the birthplace of modern science. Hence, in the historical recollections of the average Japanese, science does not appear to be as significant as it is for Europeans. Japan's first encounter with 'modernity' was when it awoke one morning in 1853 to discover Western gunboats anchored in Tokyo Bay. Thus, for the Japanese, 'modern' originally referred to the military superiority of the Western powers rather than Western science. They soon realized that behind the military threat lay the Industrial Revolutions. Thus, for modern Japanese historians it made more sense to investigate the technological achievements of the nineteenth century than the Scientific Revolution of the seventeenth century. Their historical experience generated a keen interest in the institutional and social history of technology transfer processes. The Japanese thus readily accepted the STS problematics, and members of the STS-oriented generation were keen to participate in the project of researching and recording *A Social History of Science and Technology in Contemporary Japan*.

FOUR-SECTOR APPROACH

I have elsewhere criticized the naiveté of applying Marxism's base-superstructure model to the history of science[2] and have argued for the insertion of a middle structure consisting of the institutional and sociological aspects of the scientific community between science and society.[3] Thus, I have been sympathetic towards the STS approach from the beginning. Both economic base-determinism and scientific community (social) determinism are appealing for their theoretical simplicity; problems are simply reduced to one single factor. The reality, however, is much more complex. Most things cannot be explained in such simple terms.

According to the opinion of those who were once deeply committed to base-determinism or base-reductionism and later deeply disappointed by them, complex problems must be treated as complex phenomena and approached from many viewpoints. Hence, I propose here a 'four-sector' approach – public (governmental), private (corporate), academic (university) and citizenry – to provide a historical account of the phenomena at the interstices of the interests of the different social sectors. In post-war Japan, S&T is where all of these different sectarian interests meet. The balance of power between the four sectors has determined the direction that S&T has developed.

I proposed that the four-sector approach should provide the 'official' perspective for our Social History Project, expecting it to be acceptable to all members. There seems to have been no particular objections: some participants – particularly the sociologists – were quite willing to explore these phenomena from the perspective of the citizenry. Other participants in our project are public or private sector professionals.

However, most of the contributors to our project were neither bureaucrats nor corporate executives, but rather university staff. Hence their standpoint does not represent public or private interests; it is academic. Historians of science could readily assume the perspective of the citizenry or the academic scientist while experiencing more difficulty in sharing the public sector or corporate standpoints, even where the latter is essential. In other words, it seems somehow easier to share the perspective of the citizenry who had no vested interest in the promotion of S&T.

A mass operation such as ours should not be conducted under a rigid official ideology and methodology that restricts members' personal freedom and is often alienating. Rather, a loosely outlined approach, such as the four-sector perspective, is more liberating and therefore, perhaps, more fruitful.

CLASSIFICATION OF SCIENTIFIC ACTIVITY ACCORDING TO THE SOCIOLOGY OF SCIENCE

We deem it essential that S&T be classified according to the 'social actors' who have the power to assess scientific activity, the people to whom scientists address their work. In other words, the question 'To whom is the scientific research addressed?' is the most important factor in shaping and defining the character of each scientific activity. Other distinctions such as research location, form of presentation and sources of funds are not necessarily essential and often remain derivative. The following sections explain our classification of science according to the social sector of the assessors.

Academic Sector and Academic Science

Within the scientific community, members present their research for debate, discussion and criticism by their colleagues. This form of science is based on the principle of publishing or otherwise making the results of research available to the scientific community. I refer to this sort of scientific activity as 'academic sector science' or simply 'academic science'. In the case of 'academic science', fellow scientists conduct professional assessments.

Research is initiated out of individual interest and pursued for reasons of personal honour and distinction, which is meted out through a referee system. For example, it is simply an academic concern of engineers whether Japanese technology is more advanced than the rest of the world; a matter of pride for individual engineers but nothing more.

Academic science as practised in the academic sector can be called 'science' rather than S&T (this distinction is clarified below). In academic science, the academic freedom of choosing one's own research topic is well respected. It leads to the idea of scientism (science for its own sake), but for that reason is often less directly significant for society at large.

Universities, where the traditional ideal of academic freedom is supposed to be maintained, are generally thought to be the best locations for conducting academic science. We can call it academic science when it is practised in various professional schools – such as the school of medicine, technology or agriculture – as long as the researchers have freedom to choose their own research topics and are peer-reviewed via established procedures.

This mechanism has guaranteed and strengthened the universality and objectivity of modern science. Academic science retains its pride because of its educational role. Nevertheless, with the advent of science with its big budgets and large numbers of personnel, the judgements of sponsors became more important and industrialized so they became the mainstream of the R&D community. It is here where tensions emerge between the mechanisms for assessment and sponsorship.

Public Sector and Public Science

Such academic scientists alone have not characterized the main currents of post-war S&T. Few would disagree that, apart from the academic side, the outstanding feature of post-war S&T has been the way in which the politics of bureaucrats –

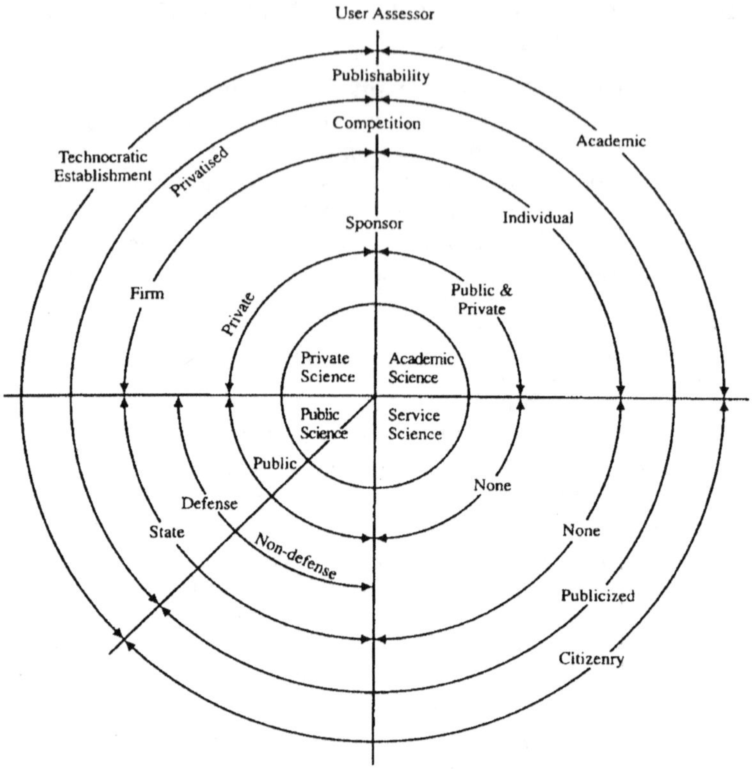

Figure 22.1 Circular diagram illustrating various scientific activities

the public sector – have largely determined the direction that S&T has progressed, e.g. for strengthening national defence and welfare, or simply strengthening bureaucratic power.

Scientists in the public sector organize and plan mission-only research. Throughout history, the government's contact point with science has been primarily focused on defence-related research. Other scientific activities can also be found in modern nation-states; in nationwide surveys of resources, the maintenance of national standards and, more recently public healthcare and managing the global environment. These activities are mostly performed in the in-house public laboratories and hence we shall define them as 'public sector science' or 'publicly-performed science'. In the post-war period, however, government sponsorship was extended to the academic and private sectors; thus it could be seen as 'publicly-sponsored science'. For example, the US system of scientific R&D since WWII is the product of an amalgamation of the government and industrial sectors – whose two largest branches are nuclear and space technologies. Because the government is the largest purchaser (consumer of the technology produced – i.e. nuclear weapons, spacecraft, commendations equipment, etc.) this approach can be called 'publicly-sponsored science' or 'public science'. The major direction of 'publicly-sponsored science' is decided by, and its output assessed and appraised by bureaucrats. Since we tend to place more emphasis on assessors and performers, we call it 'public science.'

Private Sector and Private Science

The R&D carried out in the private laboratory of a corporation is conducted in the interests of the industrialists and is hence called 'industrial science' in a narrow sense. While academic science is primarily pursued in universities, industrial science is mainly practised in private enterprises. A distinction based on the type of institution where research work is conducted might be more readily understood, in which case we would speak of 'university science' and 'corporate science'.

The broader term 'industrial science' refers to sponsored research including research at a think-tank. Similar to 'publicly-sponsored science', 'privately-sponsored science' as practised in the private sector is appraised by private enterprises in order to increase profits. As we place emphasis on sponsors and assessors rather than the site of performance, we name it 'private science'.

Depending on whether the sponsor of R&D belongs to the public or private sector, the nature of scientific activity differs. The US system of scientific R&D since WWII is the product of an amalgamation of the governmental and industrial sectors – whose two largest branches are nuclear and space technologies. In contrast to the US, post-war Japan did not develop a military-industrial complex; the military was not the major sponsor or consumer of science. The hallmark of post-war Japanese R&D is, as was internationally recognized in the 1980s, the predominance of 'private science' over 'public science', for the sake of maximum corporate profit.

Citizen Sector and Service Science

The citizens may be defined as people who do not belong to either the academic, public or private sectors described in the above categorization; in other words, ordinary people who have no vested interest in and derive no direct profit from the advance of science and technology. Hence, there is no place in a sectorial analysis based on public, private and academic for the S&T activities that aim directly at the interests of the citizenry or for the benefit of mankind. For these practises we require another science sector. Those scientific activities that directly contribute to the welfare of the citizenry, I refer to as 'citizen-sector science' or 'service science'.[4] A typical example of service science, that is, science that is directly assessed by the citizen sector, is health care. In spite of the recent bureaucratized trend, a general physician's medical service is even today directly assessed by his patients.

In spite of the devastating effects of nuclear bombs and the continual, cumulative destruction of the natural environment, it is still firmly believed that S&T will contribute to, if not actually bring about human happiness. Various questionnaires confirm that this belief is widespread.

Medical research and services are reportedly the fields most anticipated to provide benefits. We repeat the definition from the sociology of S&T that the citizenry is the social sector that does not benefit directly from the promotion of science and technology. Academic science, public and private sciences all have internally instituted promotion and assessment mechanisms that exist to protect their own interests. Therefore, the fourth sector also requires its own promotion and assessment mechanism by the citizenry. Here, I am more proscriptive than descriptive, attributing it with characteristics that it 'should have' rather than what it does have at the present time (we will discuss citizenry activism in more detail below).

Public Service Science

Narrowly defined, service science must be public service science conducted in public laboratories and financed by tax money for a return[5] to the taxpayers. In addition to medical research, prime examples of public service science are: weather forecasting, environmental protection, nutritional research and information, and library services. Such government activity constitutes a grey area between 'service science' and 'public science'.

Of the Japanese public science that is funded by tax money, defence R&D accounts for an insignificant part. Most public service science seems to be conducted in various national laboratories, which should be in the interests of the taxpaying citizenry. However, this type of institution lacks the competitive mechanisms that are integral to every other sector, academics, nation-states and corporations. It also lacks mechanisms for responding to the opinions and judgements of the citizenry. Thus, scientists in these public laboratories have no avenue for professional assessment other than through publishing articles in scientific journals, just like academic scientists.

There are also R&D services that are not considered to be public service science. For example, many public laboratories belonging to MITI (Ministry of International Trade and Industry) serve the industrial sector and often aim to patent their works. Likewise, the laboratories of each government ministry provide services to the sector of society for which their ministry is responsible.

The knowledge produced by public service science should be public knowledge that remains openly available to the general public. In contrast, military science, which is also funded by tax money, is very secretive. Nuclear research is also highly secretive as its military applications prohibit the disclosure of information even within the nuclear science community, thus creating tensions between the public and citizenry sectors.

Other Criteria for Classifying Science

SCIENCE AND TECHNOLOGY ARE INDISTINGUISHABLE

The adoption of the term 'Japanese S&T' in our project title generally does not distinguish between science and technology. Any such distinction between science and technology is a Western conception. At the beginning of Japanese modernization in the late nineteenth century, samurai bureaucrats did not understand a clear-cut distinction. For them, Western science and Western technology, combined, were the tools for civilization, industrialization and modern armament.

In Europe, there was an aristocratic tradition of 'high science', which was maintained by the 'gentlemanly class' at universities. Technology was nurtured within the apprenticeship system and later taught at polytechnics to the newly emerging lower and middle class city dwellers. In Japan there was also a class system carried over from the premodern Edo period with four categories: samurai, peasantry, artisan and tradesman. In the 1870s the Meiji government recognized technology as a legitimate subject to be studied within the university system. Yet, although from 1886 the Imperial University distinguished the Faculty of Science from the Faculty of Technology, the practitioners – mainly samurai descendants – did not recognize any clear distinction and engaged in both. The social status of engineers was therefore higher in Japan than it was in Europe.

While the samurai-class university educated engineers were responsible for disseminating Western S&T in the public sector, the traditional artisan class developed technology for private profit. While the samurai engineers often failed in their attempts to apply technology to business because of the failure to properly consider the economic aspects, success was often achieved when that technology was transferred from the public sector samurai to the private sector artisans, as in the case of Toyota, where Western technology took root before WWII. The privatization of technology continued into the post-war period when the structure of privately led S&T and R&D neared perfection.

In the interwar period, a tradition of promoting 'pure science' was founded by the second generation of modern Japanese scientists. They accepted the same distinction between science and technology as that made in the West. Outside of this particular community, though, even in the social science community, such a distinction was not so obvious. The in-distinction intensified with the general trends in the post-war world where science was no longer 'free enquiry' but a techno-nationalistic endeavour. This in-distinction still holds true today in post-war Japanese society.

This has also been true internationally since the success of the Manhattan Project, wherein science and technology were indistinguishably intertwined. In post-war Japan S&T is characterized by the national slogan 'science and technology for economic recovery' and the private dominance of R&D. Overshadowed by the private sector success of industrial science, the relative status of pure science declined. Internationally this situation is often regarded as a deficiency in Japan's basic science research, a claim 'supported' by the observation, made since the 1980s, that in contrast to the economic success, there are few Nobel laureates in Japan.

Some sectors of Japanese society maintain that in comparison to the good old European tradition, Japan is shamefully backward and is still underdeveloped in promoting science. However, from the STS perspective it is apparent that the transfer of science from one cultural environment to another invariably generates some modification to the nature of science, if not wholesale transformation. In other words, the prominent nineteenth century German *Wissenschaft* differs substantially from the 'big science' of twentieth century America. Likewise, the personal character of researchers changes from a savant or a *Wissenschaftler* to a scientific professional. Hence, it was inevitable that post-war Japanese science-and-technology would undergo transformation in an environment entirely different from the West.

It appears that what Japan has developed in the post-war period is not quite a science but rather a production technology. Japanese *kagakugijutsu* (literally translated as sciencetechnology) is somewhat different from what has developed in the West. The Chinese coined a new word for it: Ke-I (Sci-Tech is a literal translation). Thus this failure to distinguish between science and technology is a Japanese modern tradition that is shared by the Chinese.

PUBLICIZED AND PRIVATIZED SCIENCE

The most basic criterion of post-war science is whether or not it is open to the public. The former is academic science as practised in the academic sector, namely universities, including applied science fields such as medicine, engineering and agricultural study as long as the research findings are publicized. This same value of openness is shared by the citizenry sector.

Private science as practised in the private sector, however, does not share this value; its research findings are the private property of corporations. From this perspective, scientific information that has been produced by research is worthless unless it is privately owned. The information, if publicized, can be used by anyone, even competitors, and will come to have little economic value. Public science as practised by the military sector also shares the value of closedness, and the secrecy of scientific findings. Non-military public science is in a grey zone on this issue, although nominally research findings should be open to the public since it is financed by tax-money. These differences are essential to our discussion of post-war Japanese science.

COMPETITION MECHANISM

Competition is the incentive for such intellectual activities as scientific research and development. In academic science, competition is between individual researchers, but in sponsored science competition between nation-states and competition between corporations is the main motivation for funding R&D. When competition becomes intense, governments and corporations tend to place their own needs over those of the citizenry. In order to prevent such situations occurring, it may be desirable that the citizenry be empowered to make assessments of both public and private sector R&D. However, the citizenry can never become a powerful regulating force since it does not incorporate any competitive incentives.

Since the fourth sector does not possess the professional drive and profit incentive of the first three sectors and, as it is often on the receiving end of the activities of the other three sectors, citizens' movements, such as the anti-pollution campaign, have to depend on participants' motivation and must utilize the mass media if they are to influence the tide of technocratic trends so that the damage inflicted on civilians does not exceed acceptable limits.

Technocracy vs. the Viewpoint of the Citizenry Sector

In recent years, the public, private and academic sectors have become tightly integrated in a technocratic structure, in which no critical assessment function is installed. Japan is no exception. Under these conditions a critical function can only exist in the citizenry sector. Hence, the citizenry sector's positive contribution to S&T lies in its critical competence exceeding self-interest and the regulatory power arising from consensus conferences. As such, 'critical science' –

as J. Ravetz called it – is the major role of service science. But of course, to speak of this as a 'positive' contribution is to assume a particular position on the issue – it is certainly seen as a 'negative' contribution by those who are criticized, at least to the extent that they are disadvantaged in the pursuit of their self-interests. The citizenry sector activists are therefore frequently victimized by the other, generally more powerful sectors.

The citizenry sector's active avant-garde includes environmental action groups and non-government organizations (NGOs). These social actors utilize mass-communications and the mass media to speak on behalf of the citizenry sector. The voice of the citizenry is thus often represented as a consensus opinion.

TECHNOCRACY AND DEMOCRACY: POLARIZED CONCEPTS

In the following discussion, we treat technocracy and democracy as polarized concepts. We rather simplistically define technocracy – an amalgam of 'technology' and (hierarchized) 'bureaucracy' – as a top-down communication mechanism in pursuit of efficiency while democracy is a bottom-up mechanism of majority rule. 'Technocratic structure' denotes a complex consortium, the incorporated structures of the public, private and academic sectors (i.e. government, industry and universities). Historically, the dominant power in a society – usually the government – assessed, controlled and protected the S&T that emerged outside of the power structure. In recent years, however, the integration of three sectors in most developed nations gave rise to technocratic structures that incorporated and promoted S&T as its main function. Japan is one (obvious) instance of such processes.

Table 22.1 Characteristics of the four science sectors

Science	*Academic*	*Private*	*Public*	*Service*
Assessor	peer	sponsor	sponsor	citizenry
Publishability	publicized	private	private	publicized
Competition	individual	firm	nation-state	none
Performer	university	private laboratory	public laboratory	media
Sponsor	public or private	private	public	none
Utilizer	academic	technocratic establishment	technocratic establishment	citizenry

The modernization of Japan was largely samurai-spirited (samurai: warrior-bureaucrat). Under imperialist competition, modernization was encapsulated by the celebrated slogan 'Fukoku Kyōhei' (make the nation wealthy and strengthen the army). In spite of a major turn after the surrender in 1945, the same spirit has been embedded in the post-war period in a nationalistic slogan 'nation-building through science and technology' – or 'techno-nationalism'. The techno-nationalist vocabulary is rife with terms of warlike strategy and competition, such as 'the struggle for survival', which is not entirely comfortable for, or compatible with, the consumerist citizenry seeking peaceful coexistence. Consequently, the function of assessment should be transferred to the fourth sector, the citizenry.

Technocratic literature is peppered with clichés such as 'under the close cooperation of the public, private and academic sectors, the research and development of science and technology is rapidly and efficiently conducted'. Here, the citizenry sector is significantly avoided. The truth of this cannot be judged until we look closely at what science means to the citizenry sector.

In reality, the three technocratic sectors together have controlled and directed S&T so far. But something is missing: unless the citizens' perspective is integrated, the 'society' part of the 'Science, Technology and Society' approach cannot be well represented.

The Citizenry's Concern with Science Arose in the 1970s

'The pattern of industrialization in Japan and other Asian countries is technocratic developmental dictatorship.' A small number of elite bureaucrats produced a highly centralized plan of industrialization and executed it through the absolute power of the newly emerged nation-state. This approach has been called the Asian model of capitalism. Modern Japan set the precedent for this model which the other Asian nations later followed.

Langdon Winner claims that the US, West Germany and Japan attained a peak of technocratic achievement in the 1960s.[6] At the same time, though, the citizenry sector began to speak out about STS problems, the environmental (anti-pollution) and anti-war movements appeared and eventually the course of development of S&T was significantly altered. These events are impossible to explain without reference to the citizenry sector. For instance, when planning the Narita International Airport in the 1960s, technocrats did not anticipate the impact of opposition from the citizenry, but by the 1970s they realized that they had to seriously consider citizens' claims on every significant development.

Up until the 1960s, the consumer citizenry had no direct interest or concern with S&T, and there was no discussion about its social meaning. It was not until the 1970s, that a pollution-concerned citizenry began to recognize that they were directly connected with the S&T activity of corporations and began to question these activities. Subsequently, the Toyota Foundation was created, primarily to address criticisms of technocratic ways of thinking.

THE TOYOTA PROJECT

This 'Japanese Science and Technology' project began when I met Yoshinori Yamaoka, a programme officer of the Toyota Foundation, in 1983. He was looking for a new project to replace the 'Minamata Project' which was nearing completion. He was also a supporter of the 'service science' concept and had created an 'environment competition' programme to encourage the citizenry to participate in environmental research at the Toyota Foundation

Yamaoka suggested that I should coordinate a major project for the Foundation on the post-war history of Japanese science and technology. I revised my life-long research plan and decided to accept the offer. I calculated that the project would take ten years to complete.

I recalled my experience from a quarter of a century earlier when I participated in a project with the History of Science Society of Japan to produce *Nihon kagaku gijutsu-shi taikei* (Source book on the history of Japanese science and technology, 25 vols.)[7] which had taken ten years. That project was organized and completed with a tremendous contribution from Professor Mitsutomo Yuasa. I have often tried to model myself after him.

While I was enjoying the benefits of working on this large-scale, fully funded project, the younger generation of Japanese historians of science urged me to plan some similar project for their training.

When the Toyota Project began I was managing 'discipline formation' seminars with the Ministry of Education's Grant-in-aids programme and the 'Science and Society' Forum at the Toyota Foundation. The staff members of these organizations in their twenties and thirties were the most eager workers for my project. Twenty-five years earlier, Professor Yuasa was in his fifties and we active workers were in our thirties. So, for the Toyota Project I made a strategic decision to invite older historians of science to be resource persons, those in their fifties – my generation – would be organizers and the younger generation would be the real workers.

In the beginning I invited our older predecessors to the 'Science and Society' Forum and asked them: 'If you write the history of Japanese post-war science and technology, how do you write, what arrangement do you use?' and so on. We tape-recorded our discussions with Yūjirō Hayashi, Yoshirō Hoshino, Junnosuke Kishida and others. Some of these were edited and appeared, entitled *Nihon no gijutsuryoku: Sengoshi to tenbō* (Japan's technological power: Postwar history and perspective) in 1986 in the *Asahi Shimbun*. This provided a preview for the whole project.

The headquarters of our project, the 'Science and Society' Forum, is located in Tokyo. I invited Kunio Gotō (of my generation) to organize a similar forum in Osaka where we had workshops twice a year.

There were less than ten members to the core group who sat together for days to construct a preliminary outline and agree on our editorial principles. Our model, of course, was the *Nihon kagaku gijutsu-shi taikei*. Following its construction, we produced four *tsūshi* (Outline history) and twenty various *kakuron* (volumes on special topics) about the post-war period.

The Tsūshi team began by collecting and analysing major bibliographies and journal catalogues. Their aim was to systematically search not only academic journals but also sources from all sectors of society.

Besides the Tsūshi team, several Kakuron teams were created for particular projects, such as 'Nuclear Research and Reactors', 'Food and Agriculture', 'Grassroots Attitudes' and 'Scientific Manpower'. There were some overlaps in membership between the Tsūshi and Kakuron teams. At the half way point we decided to concentrate on completing the Tsūshi first and shelved the Kakuron projects for later publication.

Tekunohisutorii (Technohistory)

By this stage, we had already published the texts so far completed for the critical examination of the readers of the magazine *Shūkan ekonomisuto* (Weekly economist) published by Mainichi Shimbun. The series was entitled 'Tekunohisutorii hanseiki' (Half a century of techno-history) and printed between April

1990 and September 1991. These articles were then edited in a single volume of *Sengo kagaku gijutsu no shakai-shi* (A Social history of post-war science and technology).[8] It is an intermediary product as well as another preview of the larger project.

Compiling the Four Outline Volumes

The characteristic point of our editing is the arrangement of historical events in a topical-module system rather than the ordinary style of chapter-section subdivision. The topical-module approach was preferable to a simple time sequence of historical occurrences because our subject matters are highly specialized and each author has their own specialist topic that can stand alone and work as a module. We organized monthly meetings at which each article was subjected to a group discussion as well as being peer reviewed. During the university vacation period, the core members stayed together to examine and coordinate articles on various themes and topics to avoid overlap and fill lacuna. As a result, we often demanded rewriting up to three times. Hence, each chapter of this work is a refereed article of academic standing.

There is, as yet, no established viewpoint from which to write a contemporary history. Hence, it is nearly impossible for editors to impose a homogeneous unity on the problematics, viewpoints and writing styles of the various authors. Thus, the final responsibility for each chapter lies with the author, rather than the editors.

Periodization and Structure

The choice between the topical-module approach and the time sequence approach arises as part of a larger problem inherent to writing history. Historians are perpetually confronted with the decision of whether to adopt a synchronic view by exploring every event and aspect horizontally and arranging them in chronological order, or a diachronic view by vertically following a theme and topic from the specific perspective of each social sector, or to strike some sort of compromise between these two approaches.

We chose to, first, divide the volumes in chronological order as follows:

Volume 1: Occupation Period: 1945–52
Volume 2: Peace Treaty to the end of 1950s: 1952–59
Volume 3: 60s: 1960–1969
Volume 4: 70s: 1970–1979

It was in the 1980s when we initially planned our project to end up with four volumes of Japanese history from WWII to the 1970s. The first four volumes were published in 1995. We then spent another four years preparing Volume Five to present as much of the developments that had occurred during the writing of the first four volumes as possible; i.e. from 1980 to 1995 and, where possible, even later.

Thus, *Volume 5: Internationalization Period: 1980–95* (two volumes) was published in 1999.

Next, we prepared an 'Outline' for each chapter to provide a synchronic overview of the period covered. Each volume, however, is a diachronic presentation of topic-module-like chapters. The construction is modelled after Charles Singer (ed.), *The History of Technology* (five volumes). Of course we often found it difficult to achieve this compromise between synchronic and diachronic; the topics of each module-chapter do not necessarily end concurrently with the periodization of our volumes; for instance, some topics begin in the 1960s but extend into the 1970s. Such a module would be found in Volume 3 where the story begins chronologically.

Whilst we have adopted a four-sector approach it is not our intention to simply divide each topic and problematic into separate analyses from four independent and orthogonal social sectors. Our true intention is to look at any event from at least these four different angles to discover the points of contact and conflict between the various sectors, thereby revealing the dynamic moments of the history.

It is rather a casual analogy to use in a scholarly publication, but I shall use the quadruple rhythms of ancient Chinese poetry – the outset, the sequence, the turning and conclusion – to represent the development of post-war Japanese science and technology. From the Allied Occupation to the 1950s is the 'outset': the high growth of the 1960s is the 'sequence'; the anti-pollution movements and oil shock mark the 'turning' in the 1970s from heavy-thick-long-big business to light-thin-short-small industry. We 'conclude' this particular stage of Japanese history with Japan's 'arrival' as a 'world-class' S&T nation/society.

The 'outset' period lasts from J-day to the end of the 1950s, which has been divided into two volumes; Volume 1 discusses the Occupation Period and Volume 2 explores the development of endogenous policy. There is a turning point within Volume 1 when the Occupation policy shifts from demilitarization to economic reconstruction in the 'American way'. The period covered by Volume 2 is characterized by the revision and demolition of the Occupation Forces' legacy and establishing the direction of Japanese initiatives that would determine the high economic growth of the subsequent period. Thus there were two outsets to the history of post-war Japan; they were of different qualities and had different impacts on the subsequent period. We shall clarify these differences in Volume 2.

The 1960s – the 'sequence' phase of Volume 3 – saw the formation of technocratic structures in Japan, France and the US (amongst others). Towards the end of the 1960s, there was a watershed demarked by the anti-Vietnam-War movement, campus unrest and other social 'uprisings' that occurred in most of the developed countries and affected a 'turning' in the high economic growth paradigm. This 'turn' is the subject of Volume 4. Volume 5 is the 'conclusion', characterized by the arrival of Japanese science and technology at the international frontier, and the global demise of the Cold-War system.

SOURCE-MATERIALS

During the course of our Project, all of the Occupation Forces documents preserved at the National Recording Centre, Maryland, USA, were microfilmed, sent to Japan and compiled by archivists of the National Diet Library at Tokyo. In editing Volume 1, in particular, access to these documents provided us with a great advantage over the earlier work *Nihon kagaku gijutsu-shi taikei, Tsūshi 5.* These records were supplemented by the microfilmed diplomatic records preserved at Gaimushō, Gaikō Shiryōkan (Diplomatic Sources Archive, the Ministry of Foreign Affairs).

Compared to the US, the Japanese have a poor – in fact, almost no – archivist tradition. Hence, we suffer from a scarcity of available public documents. There is no public archive that preserves 'grey' literature such as internal sources to show the process of policy formulation and the proceedings of various committee meetings. The Japanese National Archive was created for this purpose but they have no authority to command that any administration hand over their records. The National Diet Library contains all of the published White Papers and other government publications. The libraries of the Ministry of Foreign Affairs and the Science Council of Japan were useful for our Project. The Science and Technology Agency and the Ministry of International Trade and Industry (MITI) also preserve some useful sources in their libraries. The Ministry of Education makes no such preservation effort and its library is not open to the public. Hence, most of the 'grey' literature we had access to was 'accidentally' preserved by individual bureaucrats and found in second-hand bookshops.

Sources from the private industrial sector are totally inadequate. Official company histories are readily available but with the exception of some few works that were edited by professional business historians, the quality is not satisfactory. Company secrecy policies prohibit free access to the basic source materials. Even commercial advertisements are not well preserved.

At one stage of our Project, Mr Satoshi Kōju proposed to construct a database of public relations movies and videos. There is no doubt about the value of these sources or the power of visual media over the written text. But the collection was so voluminous that our efforts remained experimental and far from completion.

Resources from the academic and citizenry sectors are all open to the public. These sectors have tried to create a database while our Project has been in progress. The situation should be improved in the future.

Oral History

Many people who were involved in the historical events recorded here were still alive during our research project, so we attempted to employ 'oral histories' in our research. This method 'creates' historical documents by interviewing participants of events and then formally authorizing transcripts of the interviews with the interviewee's signed agreement to publish. Such an interview is thereby enhanced in status as a publishable primary source. This methodology was, however, not very familiar to either the interviewers or the interviewees. Thus, in most cases, interviews were conducted informally and quoted without formal procedure.

Sources of the Citizenry Sector

The methodology of historical study should not be evaluated for its theoretical consistency *per se* but must rather be appreciated in terms of its capacity to discover a new angle and incorporate new findings. Though the viewpoint of the citizenry may seem irrelevant to S&T activity, approaching S&T from this perspective may bring something fresh into our historical picture.

In practice, however, it is virtually impossible to uncover historical sources and documents of the citizenry that might be directly related to S&T activity. The primary sources are – and will probably remain – government reports, private sector documents written from the industrialist perspective and academic scientific papers. These are all written, collected and published by individuals and institutions that are clearly committed to the promotion of S&T. In a sense, they all amount to the collected advertising of the S&T sectors. Even scientific papers must be considered to be advertising for the academic sector in the broadest sense.

Hence, to develop the citizenry view of S&T from readily available sources, we are dependent upon the sources that exclusively represent the public, private and academic perspectives. For instance, the Science and Technology Agency issues a Science and Technology White Paper each year, which lists all sources and data for each nation-state. If we rely too heavily on them, we get only the public sector's perspective on science and technology. For example, the White Papers list an abundant literature about the technological conflict and competition between Japanese and US corporations; but there are few arguments presented from the perspective of consumers or the citizenry, who's interests must be different from a nation-state's or a corporation's. In fact, the interests of the citizenry are often diametrically opposed to the competition among nation-states, corporations and scientists, e.g. governmental protection of particular industries in international trade wars is frequently disadvantageous for consumers.

Unsuccessful Non-academic Approach

A number of works have appeared in the international arena of the history of science since the 1970s that have tried to reveal something new by adopting a feminine perspective. Paralleling such attempts, we tried to establish the everyday life standpoint of consumers, from which to examine basic science and industry, a reversal of the linear approach from basic to complex reality. We invited contributions from those who are closest to the consumers' perspective such as home economists and interior designers and we organized workshops for planning our project. Regrettably, it was too early to bear fruit.

We also invited non-academic people like science journalists, think-tankers, free-lance video collectors and bureaucrats to join our team, but by the final stage most of them had dropped out and the responsibility for final drafting was left to academics who could do such work for the academic achievement rather than for their livelihood. We also tried out a variety of approaches but the only one consistently adopted was the four-sector approach.

ON THE WHIGGISH VIEW OF HISTORY

In the midst of a revolution, participants often feel that everything that is happening is worth recording. Then, as time passes, some things are lost, some are destroyed and even the things left out undergo a transformation. If, on the one hand, we only collect the surviving documents and evaluate them as historically worthwhile, we will produce a distorted history. On the other hand, if we apply contemporary criteria to historical records and write history accordingly, we are also producing a Whiggish view of history that should be considered useless among professional historians. The history of science and technology, where progress is most prominent, often falls into such a trap.

When we look to our past experience, post-war democracy for my generation, the radical student movement of the late 1960s for the next generation, we find differing assessments.

For instance, the influence of the Minshu Shugi Kagakusha Kyōkai (Association of Democratic Scientists) during the post-war period of democratic activism is now almost totally forgotten: the role that the Science Council of Japan (JSC) played around the time of its creation is under-appreciated on the basis of its present diminished status. Various ideas that appeared in the late 1960s and early 1970s – e.g. the influence of the Great Cultural Revolution in China and the radical student movement around the world – risk being totally forgotten without proper assessment.

It is the task of a historian to record forgotten events for posterity. In order to develop an historical perspective, it is essential that we attempt to move our viewpoint to the time period under consideration. To reevaluate forgotten events is meaningful in revealing alternative possibilities to younger generations, but it is almost impossible to recover the lost passion of a revolutionary movement.

Conversely, it is both the privilege and a desire of contemporary generations to approach a past event through present-day problematics. It is doubtful that contemporary history can be written in the same way as it would have been, say, ten or twenty years ago. I wrote a contemporary history of the late 1960s and 1970s in *Kagaku to shakai no gendai-shi* to record the decade after the '1968 revolution'.[9] If I were to write it now, though, the style would undoubtedly be different. The former contemporary history was a record of conspicuous events that had appeared on the journalistic surface. It was the kind of history that can be written only contemporaneously with the events – 'in the moment' as it were. If we write it later, we write it from a different historical perspective. Making judgements based on the present knowledge about a trend that, at the time, was promising to develop further is utterly unavoidable after the trend has played itself out or some resolution has been reached. However, if we neglect contemporary judgements, we are then guilty of being too caught-up in the trends of the time.

There is another way of writing history, as Benedetto Croce claimed, to describe the past events in ascending order of time from the present-day problematics. Writing contemporary history is made possible at the point where two perspectives cut across; the perspective from the past point of historical events and another perspective ascending from the present-day problematics. The historical course is selected by people at each point in history. At a historical moment, the point-of-view before and after the selection should be different. Hence in the following, we intentionally distinguish two perspectives: one descending from the past and another ascending from the present. By so doing, we can create a historical dynamism; a conversation of past and present.

Therefore, in addition to approaching our topical problematic from the four sectorial angles, and organizing our material both synchronically and diachronically, we have also attempted to reveal the dynamic moments in history by developing a dialogue between the past and present perspectives.

NOTES

1. Second edition published in 1970.
2. Shigeru Nakayama, 'Joseph Needham, an Organic Philosopher', in S. Nakayama and N. Sivin (eds.), *Chinese Science*, MIT, 1973.
3. Shigeru Nakayama, *Academic and Scientific Traditions in China, Japan and the West*, Tokyo University Press, 1984. Originally published in Japanese in 1974 as *Rekishi to shiteno gakumon* (Intellectual activity as history).
4. Shigeru Nakayama, 'The Future of Research: A Call for A Service Science', in *Fundamenta Scientiae* vol.2, no. 1, 1981, pp. 85–97.
5. This 'return' may be in the form of a 'service' or 'public benefit' – i.e. a contribution to the 'social-' or 'common-good' – rather than a narrowly conceived financial return as required by the industrial-private sectors.
6. Langdon Winner, *Autonomous Technology*, MIT Press, 1977, p. 135.
7. Daiichi Hōki, 1964–70.
8. Shigeru Nakayama and Hitoshi Yoshioka (eds.), Asahi Sensho, 1994.
9. Iwanami, 1981.

First published in *A Social History of Science and Technology in Contemporary Japan*, Vol.1, Trans Pacific Press, 2001

㉓ The Scientific Community Post-Defeat

BEFORE GHQ'S ARRIVAL

Immediately after they heard the Imperial Edict of surrender many scientists set about burning secret military documents. On the previous day, 14 August 1945, the Deputy Minister of Army issued a directive to eliminate official documents. Military related documents in particular were to be burned before the incoming Occupation Forces could confiscate them and use them as evidence of Japanese war crimes. On 15 August, the smoke of burning documents issued forth from every research institute and many universities.

Ordinance resources were released even before the surrender. The Osaka Military Arsenal started to transfer their essential resources to private sector factories on 9 August when they realized that the day of surrender was approaching. The Kantarō Suzuki Cabinet believed that when the Occupation Forces came, they would confiscate everything; hence before then all resources had to be covered up and dispersed. Covered up resources amounted to one third of the Army's total resources and 100 billion yen worth (in current prices) of Navy resources. These resources have been estimated to be the equivalent of four years supply of the peacetime economy or one year of wartime military production. GHQ (General Headquarters, Supreme Commander for the Allied Powers (SCAP)) tried to prevent this hoarding and ordered the Japanese government to stop it. The Naruhiko Higashikuni Cabinet deliberately delayed taking action for eight days, but on 28 August, the Cabinet called back dispersed military resources.[1]

Until the end of the war, the science sections of universities were engaged in mobilization research, and laboratories had been evacuated to the countryside in spite of their staff's continuing teaching commitments. The period between the surrender and the Occupation Forces' arrival was very confusing; rumours were rife that the Occupation troops would rob, rape and kill Japanese people. So a lot of people, including teachers and students, stayed home to wait and see what would happen. The secretariat of the university requested home-staying teachers and students to come back to school.[2] Professors could not foresee what sort of life awaited them and many passed the time in a state of apathy. Many professional military engineers who taught war-related disciplines in universities resigned before the Occupation Forces arrived in anticipation of a purge of positions.

When the Occupation Forces arrived in September the Japanese realized that this peacetime invasion was not so disorderly. Thus, in the autumn, students who had been evacuated to the countryside returned to school and course work resumed.

PROBLEMS WITH DEMOBILIZATION

The arrival of the Occupation Forces, though, did not alleviate the confusion. The first priority of GHQ – as discussed in Part 1 of this book – was the demilitarization of Japan. Whilst disbanding military units and closing down the weapons and ordinance industries was confusing enough, demilitarizing military-related S&T research facilities created special problems. Most of these facilities were to be reassigned to non-military institutions rather than disbanded. But the command channels were confused and where a particular laboratory was to be transferred to was not clear.

The Japanese Government tried to preserve ex-military facilities and personnel by transferring entire institutions to peacetime bureaus; for example, the facilities and personnel of the Navy were to be transferred to the Ministry of Transportation.[3]

To further complicate matters, inventories and records of facilities and equipment had been burned at the time of surrender and it was thus impossible to check them. All ex-military facilities and equipment were supposed to be confiscated by the Occupation Forces and put on reparations lists. If all of their mobile resources were to be confiscated by GHQ, though, they figured that they might as well take them home for their private use. With no adequate records, those who lived nearby and could carry munitions or equipment items could get almost anything released from their workplaces. These items were often sold and circulated on the black market. During the period of confusion a lot of police activity focused on the black market operations and criminal cases were often reported in the newspapers. Police frequently uncovered stolen goods.

The literature of science and technology suffered a similar fate. The documents and books related to wartime military action were all to be inspected by GHQ. Fearing censorship by the Occupation Forces, secret documents were either concealed or transferred to the non-military sector for preservation – e.g. the Imperial University of Tokyo or Hibiya Public Library – in order to avoid GHQ searches. Navy documents had been evacuated during the war to protect them from US air raids. After the San Francisco Peace Treaty these resources were partly returned to the Navy College (Etajima) or the University of the Coast Guard (Kaijō Hoan Daigaku) located at Kure.[4]

REPARATIONS AND RESTITUTIONS OF MILITARY-AFFILIATED S&T FACILITIES

Edwin Pauley, the head of the Reparations Mission to Japan, arrived on 13 November 1945 as a Special Ambassador. He listed machine tools, etc. as reparations items. On 6 December he decided to dismantle half of Japan's production capability. In November 1946, the final inventory of the Pauley

Reparations Mission was released; it was designed to reduce the Japanese production capacity to the 1935 level.[5]

Parallel to this programme, in January 1946 SCAP issued a directive to confiscate military arsenals, aircraft factories and affiliated research laboratories. This directive was in keeping with a decision to meet the reparations demands of the Far East Command.

The Technical Intelligence Company of the US Army conducted site evaluations and confiscated facilities. Their task was so vast that the assessment work progressed very slowly. By January 1947, 218 site surveys were completed and another 205 sites remained untouched.[6] In the meantime, some research equipment was illegally sold on the black market, some of which had to be purchased back by the institutions from which it was stolen.

GHQ's initial orders to the Eighth Army were to dislodge the parts worthy of confiscating for reparations from the ex-military equipment and then destroy the rest. Immobile facilities were listed for reparations at the site. Mobile scientific instruments were assembled at the Tokyo First Arsenal and exhibited to those nations demanding reparations. Non-functioning or obsolete equipment was all scrapped. Educational institutions that had not been involved in the production of military goods were generally not considered to be subject to reparations. The Aeronautical Research Institute (Kōkū Kenkyūsho) affiliated with Tokyo Imperial University was exceptionally listed for reparations in August 1946 because of its military affiliation and some instruments therein were destroyed.[7] Very few items, however, were considered to be for military use only; most of the equipment there was confiscated for reparations. Later, when it was decided the Institute would survive – with a name change from the Aeronautical Research Institute to the Science and Technology Research Institute – the Japanese scientists submitted petitions for some facilities to be lifted from the reparations list. G. W. Fox of ESS/ST supported them.[8] Harry C. Kelly also considered lifting them from the reparations list.[9]

Disrupted by the US-Soviet rivalry, the reparations plan was quite slow to be realized. Not until early 1947 could the Far Eastern Commission define the type of scientific facilities to be listed for interim reparations.[10] For the US scientists at ESS/ST, like Kelly and Fox, Japanese facilities did not appear to be good enough to warrant sending them to the US; hence many items were considered to be not worth listing for reparations.[11] They also thought that research facilities without military application should be removed from reparations so that Japanese researchers could use them. But the Far Eastern Command and military organizations – especially Australia represented by John W. O'Brien, head of ESS/ST,[12] and China – were dissatisfied with this assessment and demanded an 'independent' inspection of Japanese S&T facilities for themselves.[13] They claimed that they themselves badly needed such equipment and argued that, in general, in order to maximize reparations, no equipment should be returned to the Japanese except for what had already been scrapped. By the end of 1947, thirty per cent of the interim reparations plan was in effect. The detailed allocation ratio of reparation value was: three for China and one each for the Philippines, Dutch-Indonesia and Australia. Small machine tools were removed from the First Navy Arsenal and sent to the Philippines by the end of the year as reparations.[14]

It is apparent that the reparations value of scientific instruments declines as the facilities get older. Moreover GHQ sources[15] reveal that around 1947 GHQ changed their position on reparations, deciding that some scientific instruments should be left for the economic recovery of Japan. Around May 1947, a revision of the reparations lists began. After reinvestigation, equipment that had been mistakenly listed for reparations were removed from the list. The standard procedure for the Japanese to object to the listing of an item for reparations was to petition the local military commander who would then forward the petition to ESS/ST for assessment and advice; but with these changes in reparations policy, many such issues were handled directly by ESS/ST.

The Japanese scientists were incredibly frustrated by the fact that some of their research facilities that had not been physically removed remained 'confiscated' and were thus sitting idle, inaccessible for anyone to use. ESS/ST documents reveal that while these scientists maintained close contact with ESS/ST, they realized that they needed Kelly's understanding if they were to be successful in their petitions for salvaging equipment from the reparations programme in the interests of conducting the research necessary for a Japanese economic recovery. When one university succeeded in 'rescuing' its facilities from the reparations list, the news quickly spread to other universities and one after another they petitioned for the removal of their own facilities from the lists. In most cases, these petitions were successful and instruments were lent to the requesting scientists. Even Occupation Forces organizations, such as the Atomic Bomb Casualty Commission (ABCC) at Hiroshima, petitioned ESS/ST for the release of equipment from the reparations lists, such as some measuring instruments to be used at the site.[16]

There was also a proposal that certain machine tools should be struck from the reparations list and given to Japanese educational institutions purely for educational purposes. On 19 January 1948, the Education Ministry surveyed these proposals and submitted a list of this equipment to ESS/ST who, considering educational use to be a top priority, approved it.

The wind tunnels at the Aeronautical Institute of Tokyo University (renamed the Science and Technology Institute) and machine tools from the Second Aeronautical Arsenal were particularly desired. In September 1946, the Railway Technical Research Institute requested that the wind tunnels be removed from the reparations list for use in basic research for a new bullet train,[17] but O'Brien of ESS/ST refused this request[18] being eager to get hold of this equipment for Australia. No record of what subsequently transpired with the wind tunnel has been found.

Also at the Aeronautical Institute were duplications of a number of scientific instruments. In 1947 a proposal was submitted to ESS/ST to provide equal access to this equipment for universities other than Tokyo University. Initially Kelly was not very enthusiastic about the use of these instruments, considering them to be somewhat obsolete. But they were more than sufficient for the student laboratories at universities and polytechnics.

On 28 June 1948, Tatsuo Morito, the Minister of Education, petitioned GHQ to transfer non-reparations instruments from the Aeronautical Institute to the (First and Second) Engineering Schools of Tokyo University, Tokyo Kōtō Kōgei

Gakkō (Tokyo Higher Polytechnic School) at Matsudo, Chiba Prefecture, and the Engineering Schools of Keiō and Waseda Universities.[19] This petition was approved, but since these were public resources, they were provided to the institutions in the form of a long-term loan, rather than a purchase acquisition.

A number of machine tools had been transferred from university engineering schools to military arsenals as part of the wartime mobilization. University researchers, therefore, rightfully made claims for their return. Then, there was a widespread rumour among university researchers that the reparations would soon be removed. Upon hearing the rumour, Yasumasa Tani, of the Second Engineering School of Tokyo University at Chiba began negotiating to get back the items confiscated at Chiba Aeronautical Arsenal, but GHQ explained to him the difference between reparations and non-reparations and refused it. Non-reparations resources were a matter to be discussed between Japanese offices. All of this suggests that there was some confusion at the time about the Occupation policy for the removal of reparations in addition to the strife between GHQ and the other members of the Far Eastern Command.

The reparations programme effectively started on 1 July 1948 and ended in March 1950.[20] The programme was not carried out smoothly because of the US government's growing determination to mitigate reparations.

In 1949–50 when the reparations programme was relaxed, GHQ planned to lend reparations items for a short time – six months or a year – to Japanese research institutions for the economic recovery of Japan. But even some Japanese considered the wartime facilities to be outdated. Director Sakurai of the Production Engineering Institute of Tokyo University, for example, was not interested in the salvaged resources of the Tokyo First Army Arsenal,[21] although Yoshio Nishina, the President of Riken Laboratories, petitioned for reparations items for use in his laboratory.[22] In the negotiation between GHQ and Japanese scientists, the Japanese often expressed a desire for surplus US-made equipment rather than the salvaged Japanese-made equipment.[23]

THE STATE-OF-AFFAIRS OF POST-WAR RESEARCH

During the war most scientists had been engaged in war mobilization related work. After the surrender, they no longer had an objective for their research. They were forbidden from conducting research in strategic fields like aeronautics, electromagnetism and nuclear energy. The heavy-chemical industry was severely war-damaged, too, and reparations had ruined their research facilities. Thus, many Japanese scientists were in a state of apathy. Even at the celebrated Riken, still remembered as the most productive research institution of the pre-war era, the famous charismatic scientists had either died or retired, and the middle level scientists had lost their affinity for the institution because research was virtually impossible. The traditions of excellence at Riken were discontinued, as evidenced by a directive from the Riken administration to the effect that any project that could earn money should be taken up as a priority to overcome the severe economic crisis.[24]

Within a year after the surrender, Japanese scientists had completely switched their research from wartime to peacetime. A number of distinguished (physical) scientists turned to problems of agricultural production and biology. Scientists had no privilege to escape from hunger and there were many projects to test their capability for solving urgent foodstuff problems. Yoshio Nishina applied radiation to plants and researched high yielding varieties. Bunsaku Arakatsu, a Kyoto University physicist, collaborated with chemists and biologists to investigate the effects of radiation for high yielding farm products, improving species, food processing and sterilization.[25]

With no resources available for experimentation, theoretical work with paper and pencil prevailed in academia. But even then, according to Shōichi Sakata of Nagoya University, in the absence of experimental data from the US, they could do only passive work such as resolving theoretical contradications.[26]

Professor S of experimental physics at Tokyo University had studied in the US before the war and had become closely acquainted with American scientists. During the Occupation, with his fluent English, he maintained close contacts with GHQ, which apparently gained him special 'privileges' concerning reparations items. He was provided with abundant research resources that made him the envy of his fellow scientists. He was thanked for providing them with necessary resources but did not reveal how he got them. Minka researchers and students tried to disclose the truth about S and picked up some rumours about him, but the case was too big for them to deal with. However, reacting to these moves against him, S departed for the US without completing a lecture course he was giving at Tokyo University. Subsequently he worked for the Nuclear Energy Laboratory and never returned to Japan.

REALITY OF SCIENTISTS' LIFE AND EMPLOYMENT

GHQ ordered a purge of all military personnel from public positions as part of the demilitarization programme. This applied to all who had military positions even if they were scientists. With no decent work available for scientists, the university laboratories were, for a time, full of disappointed people, who had no regular salaries, but nowhere else to go. Tokyo University in particular had many of these unpaid scientists.

Overall unemployment in Japan at that time is estimated to have been about ten million people. But the unemployment rate was higher in the scientific community,[27] although, since there were no reliable statistics compiled at the time, there is no concrete evidence of this. However, it has been estimated that there were around one hundred thousand unemployed engineers in the spring of 1946, with an overwhelming representation from the field of aeronautical engineering.[28]

In the beginning of the Occupation it was estimated that the official purge would apply to one third of natural scientists in Imperial Universities, which might have dealt a fatal blow to the Japanese scientific community. At that time GHQ was afraid that secret military research would be continued in Japan, but they soon realized that the Japanese were not so inclined. The terms specified for the official purge turned out to have very little impact on scientists in teaching positions. Initially the purge was supposed to apply categorically, after close examination, to all who had participated in scientific expeditions or excavation in the territories occupied by the Japanese military from 1 January 1928. The

Education Mission, however, issued a directive on 12 August 1946 to limit the purge to those who had engaged in work with purely military purposes and those who had confiscated precious resources and items.

Some scientists, such as Toshima Araki, Professor of Astrophysics at Kyoto Imperial University and a spokesman of the extreme right wing, resigned proactively before the purge began. People were not purged due to their work as scientists but for other reasons, such as ideology or politics. Purged scientists were prohibited from teaching positions but were allowed to engage in research at universities or research institutes.[29]

Japan had sent numerous scientists and engineers to Keijō (Seoul) and Taipei Imperial Universities as well as other Institutions in their pre-war colonies. After the surrender, these professionals were repatriated, but without any anticipation of appropriate jobs. In fact, many of them must have found jobs outside of their area of specialty. Many enterprises were severely damaged without any sign of recovery. Initially the National Railway was almost the only large institution to absorb returned scientists and engineers. A little later, new universities created under the Occupation educational reform also absorbed returnees.

Governmental organizations like the National Railway Service and National Universities survived relatively unscathed, while private corporations had to convert from military to peaceful industry. Peacetime businesses struggled, however, under difficult economic conditions. A small number of private research laboratories that had existed since before the war barely managed to continue their existence.[30] In private sector industries researchers were often transferred to production lines, with few remaining in research laboratories. In general the managers of private corporations believed that they did not have the luxury of running research laboratories.[31]

According to the inaugural issue of *Kagaku Bunka Shimbun* on 29 April 1946, the monthly salary for a new graduate from primary school was 200 yen and graduates from high school received 300–400 yen, while professors of Imperial Universities only earned 200–400 yen, associate professors 120–130 yen and research/teaching assistants received only 80 yen.

According to a questionnaire the Education Ministry sent to researchers in July 1946, while natural scientists always demand more research facilities and better salaries, the demand for more food rations among research workers who lived in the cities rather than provincial areas was particularly noteworthy.[32]

University scientists had the privilege of continuing research funds allocated to the professorial chairs and the Science Research Fund but in fact these funds did not keep up with the skyrocketing inflation. Hence most research funds were spent on utilities (electricity, gas, water, etc.), not for research. While national universities continued to receive research funding from the Education Ministry, the new universities established after the war were not allocated any such funds. Since the Science Research Fund of the Education Ministry could not keep up with inflation, the actual funding for research was inferior to pre-war standards.[33]

During the war, scientists and engineers had often enjoyed slightly more generous rations than average citizens since they were considered to be key personnel for continuing the war effort. After the war these privileges were lost; they were poor and suffered in all aspects of life just like ordinary citizens. They wore military surplus clothing and foot-soldier's shoes. For nutrition they cultivated sweet potatoes in the gardens of universities and research laboratories. Chemistry professors made alcohol and saccharine – which were in short supply in the market – in their laboratories. These were marketed with the university's name on the label. Under the pseudonym of 'food social science', one professor was a broker for the barter exchange of foodstuff.[34] Scientists with nothing else to offer the black-marketeer must have survived by selling their books in exchange for food. Other scientists engaged in small-scale production for the black market as a side business.[35] They all keenly desired to sustain their livelihood by engaging in decent work as scientists.

There were housing problems for many whose houses were bombed and burned. Many scientists returned to their university posts and lived in the laboratory until they were able to find proper housing. Railway travel was also difficult, as the railroad had been bombed and damaged. Special permission was required to acquire a long distance railway ticket. Thus, for scientists to attend the meetings of JSC, Liaison Committees and Renewal Committees, they needed special permission from GHQ. Without GHQ's support it would have been impossible to have meetings with members from the provincial universities.[36]

In post-war Japanese society, where this crisis in living conditions cried out, numerous labour unions were organized. Scientists in the laboratories of both the public and private sectors were organized into labour unions. Some of them were modelled after the Association for Scientific Workers[37] but professional labour unions failed to materialize. Thus, even proud professors from Tokyo Imperial University joined the workers' unions. In 1948, parallel to the general labour union movement which organized the Economic Recovery Conference, the unions representing research laboratories were united to form the Research Recovery Conference. They lost power after the Red Purge was carried out.

RESEARCH AND PUBLICATIONS

During the war, Zenkoku Kagaku Gijutsusha Dōmei (Nation-wide Union of Scientists and Engineers) allocated paper stock for scientific journal publishing. Dōmei, however, was abolished after the war. Gakkyōkai Rengō (Federation of Learned and Scientific Societies) was established to replace it, but it was only a nominal organization without any real power. It was difficult at this time to publish research articles through regular means, because of the shortage of paper stock.[38] But old military stocks of paper became available on the black market.

During the first year after surrender, there was a publishing boom that was reliant on this wartime stock but it had little to do with the publication of scientific journals. The stock was exhausted, though, after a year. The shortage of paper produced an unimaginable obstacle to the proper functioning of scientific societies. Bulletins of national universities and research institutes were officially allocated paper stocks as well as funds for publication costs since they were part of the public sector. Scientific societies dependent upon membership fees, however, were voluntary associations, which was a disadvantage in obtaining paper for their publishing activities.

For two years after the surrender, scientific publications could barely survive, and those that did published thinner journals with much less frequency than before.[39] For example, before the war *Teikoku Gakushiin Kiji* (Journal of the Imperial Academy) functioned as a bulletin of quick notification like the Western magazines *Nature* and *Science*, but in the post-war environment it ceased to exist; hence, Japanese researchers started to make 'preprints' of their own articles and distribute them to interested parties. It has been said that preprinting, the new mode of scientific communication, originated in Japan at this time.

While academic scientific journals did not recover for a couple of years after the surrender, new popular scientific magazines flourished. Freed from the wartime censorship regime, dozens of new science magazines sprang up, driven by the appeal of (re)constructing a new society through science. The left-wing journalist affiliates of Minka were quite active in this regard. We can now judge the 'hot' climate of the post-war scientific community through the essays and commentaries that appeared in such magazines.[40]

TRANSFERS IN EDUCATION FROM SCIENCE TO HUMANITIES

During the war university students with science majors were privileged with exemptions from or postponement of compulsory military service. Hence, during the war the number of science students increased. The pre-war ratio between the humanities and sciences had been 6:4. That was more than reversed during the war. Especially in the last phase of war, as all humanities students should be drafted, students entering Kōtō Gakkō (elite higher schools to prepare for Imperial Universities) as humanities majors were sent to the battlefield without exception. Hence, humanities studies were nearly annihilated, with only thirteen per cent of the total student enrolment opting for such majors. Almost all students transferred to science majors.

After the war, those who had chosen science majors for the privileges changed their major to humanities. For 1946, the entrance quota for Kōtō Gakkō was determined by the Education Ministry to be humanities : science = 1:2 but in practice the ratio returned to approximately the pre-war ratio of 55:45.

Polytechnic students also changed to humanities subjects. Comparing student numbers before and after the war, agriculture majors increased slightly and medical students decreased.[41] The School of Engineering had been very popular during the war but because of the slump in industry in the post-war period, it did not attract any new students for a while.

The period immediately following the surrender on 15 August 1945 was the most extraordinary period of poverty, misery and confusion in Japanese history. The Japanese people had no experience before and do not wish to repeat it ever again. The Japanese scientific community experienced extraordinary circumstances during this period that were not officially recorded to any great extent. To fill this vacuum, I have tried to document the situation from my personal recollections, for it was out of this confusion that Japanese science and technology was rebuilt from the 1950s onwards.

NOTES

1. Seizaburō Shinobu, *Sengo Nihon Seiji-shi* (A history of politics in post-war Japan), vol. 1, 1965, pp. 131–3.
2. *Tokyo Daigaku Dai2 Kōgakubu-shi* (History of the Second Engineering School of the University of Tokyo), p. 76. A document dated 21 August 1945 calling for teachers and students to return to the Second Engineering School, University of Tokyo; written by the Secretariat of the School.
3. Kaigun Shisetsukei Gijutsukan no Kiroku Kankō Iinkai [6] May 1972.
4. Interviews with librarians from both libraries.
5. Hitachi Seiki Kabushiki Gaisha Shashi Henshū Iinkai (Editorial Board of Company History for Hitachi Steel), *Hitachi Seiki nijūgonen no ayumi* (Twenty-five year history of Hitachi Steel), 1963, pp. 117–21.
6. *Summary of Reparations Activities of the Scientific and Technical Divisions – December '46 to March '47, inclusive.*
7. 'Aeronautical equipment for reparations', ESS/ST, 7 August 1946. A memo of a meeting held on the previous day.
8. 'Institutions of a Scientific Nature for Reparations', 14 June 1946; 'Aeronautical Laboratories and Reparations,' Col. G. W. Fox, 12 June 1946.
9. 'Laboratory Reparations', Memorandum from H. C. Kelly, 16 January 1947.
10. 'Reparations Removal' from Lt. E. C. Allan, 13 January 1947.
11. 'Japanese reparations to the United States,' by G. W. Fox, 9 April 1946.
12. 'Removal and storage of laboratory equipment from present locations to storage', Notes by J. W. O'B. 27 September 1946. O'Brien proposed to the Pauley Reparations Committee that in order to curtail Japanese military capabilities Japanese scientists, engineers and their families should be permanently removed to the Allied Forces countries (B. Dees). On the Japanese science and technology that O'Brien wanted to import for Australia, especially in the fields of magnet and metallurgy research, see Morris S. Low 'Mosquito Coils, Ball Bearings and Batteries; The Australian Scientific Mission in Occupied Japan, 1945–1947', Paper presented at the Japanese Studies Association of Australia, Sixth Biennial Conference, 6–9 July 1989, p. 9.
13. GHQ, 'Transfer of Japanese aviation research facilities to the Chinese government', 8 May 1946. 'Application of Maj. Gen. Lee Dai-chin and Mr Z. W. Chang for inspection of some Army & Navy Arsenals', to G-2 section, GHQ, SCAP, 1 July 1946. A personal letter from the Chinese Mission in Japan (The-chang Koo) to Dr Kelly, 6 November 1947.
14. Seizaburō Shinobu, *Sengo Nihon seiji-shi* (History of politics in post-war Japan), 1974, p. 622.
15. Conference memo, 20 November 1947 by Donald J. Pletch, Fundamental Research Branch.
16. 'Use of Japanese Sanction Apparatus', from the Atomic Bomb Casualty Commission to ESS, 12 September 1947.
17. Tetsudō Gijutsu Kenkyū Shochō (Director of the Railway Technical Research Institute of the Ministry of Transportation), 'Baishō wariate yori kikai o jogaisuru ken' (On the removal of machines from reparations assignment), ESS(B) 11723–34, 6 September 1946.
18. 'Memo for record', JWO'B, 5 February 1947.
19. ESS(B) 11723.
20. 'Reparations of Science and Technology in Japan', in GHQ/SCAP, *History of the Non-military Activities of the Occupation of Japan*, 1945–51, vol. 54, pp. 22–3.
21. Memo for file, Y. E. K. 15 January 1950, ESS(B)11723, 'Laboratory equipment listed as reparations materials'.
22. Petition addressed to GHQ in March 1950.
23. K. Kazato (The Electron Optical Laboratory, Inc.) to Kelly, 12 December 1949, ESS(B)11723.
24. Yoshimi Sakayanagi, 'Sengo no ichimen: Rikagaku Kenkyūsho rokujūnen no ayumi' (One aspect of the post-war period. The footsteps of Riken for sixty years), in *Shizen* (Nature), vol. 33, no. 13, special issue, p. 117.
25. *Kagaku Bunka Shimbun*, 7 September 1946.
26. 'Kagakukai no dōkō kataru' (Speaking of the trend in the scientific community), in *Kagaku Bunka Shimbun*, 15 January 1948.
27. Yasoji Kazahaya, 'Minka sōritsu no koro' (When Minka was established), in *Kagakusha undō no shōgen* (Testimony of the Scientists' Movement), 1978, p. 26.
28. Estimated from Rinji Kokumin Tōroku (Interim Poll of Japanese People). Humitaka Nakamura, 'Sengo Nihon no gijutsu dōnyū seisaku no keisei to sono seisakuteki na kiketsu' (The formation of technology importation policy in post-war Japan and its consequence), in *Meiji Daigaku Shakai*

Kagaku Kenkyūsho Kiyō (Bulletin of the Research Institute of Social Sciences, Meiji University), October 1989, p. 118.
29. *Kagaku Bunka Shimbun*, 6 September 1946.
30. Nakamura, 1947 [11] p. 38.
31. 'Kenkyū Kikan no Seiri' (The closing down of research laboratories), in *Nissankyō geppō*, May 1949.
32. *Kagaku Bunka Shimbun*, 16 November 1946, p. 2.
33. 'Sōsetsu iraino Monbushō Kakenhi' (Science research funds of the Education Ministry since its founding), in *Gakujutsu geppō* (Science monthly), vol. 5, no. 8, November 1952, pp. 59–62. Includes annual variation with price index. Also see GHQ-CIE, *Postwar Developments in Japanese Education*, vol. 1, p. 361.
34. Kaneseki, 1947 [10] p. 32.
35. Nakamura, 1947 [11] p. 37.
36. There are a number of such permits recorded in GHQ documents.
37. J. D. Bernal's *Social Function of Science* was translated into Japanese in 1946 and provided a model for the Association for Scientific Workers.
38. 'Shiryō to Dōkō' (Resources and Trend), in *Shizen kagaku* (Natural science), no. 9, September 1947.
39. Masuo Usui, 'Shizen kagaku, jiron, gakkaishi ni tsuite' (On natural science, commentaries and scientific journals) in *Shizen kagaku* (Natural science), no. 9, September 1947, p. 1.
40. Kaneseki, 1947 [10] pp. 30–6.
41. *Kagaku Bunka Shimbun*, 3 August 1946, p. 2.

BIBLIOGRAPHIC GUIDE

(1) ESS(B)11720–40.
The military facilities confiscated by GHQ included a lot of science and engineering facilities and resources. This document primarily addresses the issues of how to deal with them. Japanese scientists petitioned for the return of confiscated goods and ESS/ST responded. Lists of equipment, facilities and resources were attached to these petitions. Petitions for the equipment at the Aeronautical Institute of Tokyo University were especially numerous. From the lists of facilities confiscated from military laboratories, we can estimate military S&T activity during the war. There are also documents recording a controversy between a scientist of ESS/ST who advocated returning confiscated items to the Japanese scientific community and the Far Eastern Command, which wanted them as reparations.

(2) Nihon Gakujutsu Kaigi (Science Council of Japan), *Kagakusha seikatsu hakusho* (White paper on the living standard of scientists), 1959.

(3) 'Shizen to tomoni nijūgonen' (Twenty-five years with *Shizen* Magazine), in *Shizen* (Nature), vol. 25, no. 5, May 1970, pp. 84–9.

(4) 'Zadankai, Kenkyū minshuka ni kansuru zenkoku butsurigakusha no kondankai: Minshu sensen ni taisuru kagakusha no taido ni tsuite' (Panel discussion, a workshop of nationwide physics community on laboratory democracy: On the attitude of scientists towards a People's Front), in *Shizen kagaku* (Natural science), inaugural issue, July 1946, pp. 22–33. The workshop was held on 30 April 1946 at Sanjō Goten of Tokyo Imperial University. One hundred members of the General Assembly of the Japan Physics Society attended it. Shoichi Sakata was the chairman. They freely discussed topics such as the democratization of laboratories and workplaces, union activity and science for 'the people'. Participants included not only left-wing but also liberal scientists in the establishment. A good source for understanding the atmosphere of that time.

(5) *Kagaku Bunka Shimbun* (Science and Culture Newspaper).

Issued weekly or every five days from 29 April 1946. It was later renamed *Kagaku Shimbun* (Science Newspaper) and continues to be published now. It was valuable for its unique ability to explain the reality of the scientific community particularly in the early Occupation period.

(6) Kaigun Shisetsu-kei Gijutsukan no Kiroku Kankō Iinkai (Committee for the Publication of 'A Record of Naval Facility Engineers and Technicians') (ed.), *Kaigun shisetsu-kei gijutsukan no kiroku* (A record of naval facility engineers and technicians), limited press, May 1972.

Valuable and interesting records on what happened to the Navy technological facilities after the surrender. The list of the Kasumi Kai (Club of ex-Navy engineers) reveals the course that Navy engineers followed in the post-war world.

(7) 'Gunkankei gakuto' (Military-affiliated students), in *Kokumin no kagaku* (Science for national people), vol. 1, no. 5, 1946, p. 18.

(8) The following are examples of short-lived magazines and articles of the post-war left-wing scientists' movement.

'Kagaku senpan tsuikyū ni tsuite' (On the search for science war criminals), in Minshushugi kagaku (Democratic science), vol. 1, no. 3, June 1946, p. 2.

Tatsuichi Hishiyama, 'Warerano gikai o tsukurō' (Let's make our own Parliament), in *Warerano kagaku* (Our science), vol. 1, no. 1, March 1946, p. 3.

Masao Nakamura, 'Kagaku Akademii no shuyō na mondai' (Main problems of the Science Academy), in *Kokumin no kagaku* (Science for national people), vol. 2, no. 1, 1947.

(9) Genkichi Hara, 'Nihon ni kagakugijutsusha wa donokurai iruka' (How many scientists and engineers in Japan?), in *Kokumin no kagaku* (Science for the people), no. 3, May 1958, p. 4.

(10) Yoshinori Kaneseki, 'Sokoku saiken to kagakugijutsusha no mondai' (Recovery of Japan and the problems of scientists and engineers), in *Shizen kagaku* (Natural science), no. 9, August 1947.

(11) Sachio Nakamura, 'Shiryō to dōkō, Kenkyu hukkō kaigi no kaidō' (Sources and trends: Conference on the way to research recovery), in *Shizen kagaku* (Natural science), no. 9, August 1947.

First published in *The Asian Journal of Social Science*, Vol.30, No.2, Brill, 2002

㉔ Overcoming the Digital Divide between Phonetic and Ideographic Languages

In the year 2000, the Mori led government in Japan adopted a 'Basic Strategy for IT' policy which involved the formation of an advisery committee. One can immediately see from even a cursory examination of the committee's recommendations that nothing particularly innovative has been proposed. Rather the recommendations are highly imitative of American and European policy several years ago. As for infrastructure-building, the most praiseworthy aspect are the proposed subsidies for IT-based training courses. For the Japanese who tend to have little experience of having used a keyboard, these courses will be highly useful.

The above title is the official title I was assigned for this conference, but I must at the outset say that the actual practice and development of information technology was much less influenced by public sector policy and more the result of the desires of the people themselves. The culture of the people, as end-users, was a more decisive factor in determining the trajectory of development. In what follows, we will discuss particular aspects of Japanese IT in the context of minimizing the domestic digital divide.

1. DIGITAL DIVIDE AND KEYBOARD ALLERGY

In my keynote address, I wish to discuss some aspects of Japanese IT in relation to overcoming the digital divide between the Roman-alphabet-based West and ideogram-based East Asia. This has been a very serious, almost insurmountable, disadvantage and handicap for East Asians in catching up computer-Internet system.

The alphabet-based keyboard was a bottleneck for Japanese seeking to access the Internet. Indeed, it was a problem for all East Asians who use ideograms, as they get used to alphabet much later than their counterparts in European countries. According to a UNESCO survey of computer literacy in the late 1990s, the Japanese people are far behind Europeans or Americans, simply because the average Japanese person is ignorant of the alphabet-based keyboard up until high school time.

Although the Japanese government has recently started teaching English (or, at least the alphabet) at primary schools, the digital divide between alphabet-friendly Western countries and ideogram-using East Asian countries still cannot be reduced to zero. The average Japanese has, until recently, little to do with the alphabet-based typewriter, due to the lack of opportunities to use English in his or her daily life. Unless students possess their own typewriters or personal computers, official school training in English or on a PC does not mean much in terms of the overall promotion of computer usage in Japan.

Rather, the Japanese have been strong in developing the non-keyboard area of the IT revolution: namely audiovisual media, cartoons, videocassette recorders, fax machines, console games, etc. They have been weak in the development of software that relied on the use of a keyboard.

2. THE DOUBLE REVOLUTION OF THE WORD-PROCESSOR

Up until the 1970s, I doubted that a word-processor capable of generating Chinese characters would be possible. The introduction of a two-byte system by East Asians made it possible for Chinese characters to be written on a word-processor and for other developments to be contemplated. The Japanese word-processor was invented in 1979 and in general use from around 1984, thanks to the installation of a phonetics-ideogram conversion system.

While those in the West experienced a more step-by-step transformation from handwritten to mechanical means (via the typewriter) and then a second move to electronic means, East Asians experienced a double revolution, jumping from handwritten to electronic communications without taking the intermediate step of typewriting, once the transformation of ideograms on word-processors became possible,

Such a double revolution is really significant in East Asian history but people have had to overcome an extraordinarily high barrier. For ordinary people, there had been no need to use typewriters or PCs in their daily life. Only small numbers of specially-trained typists (usually women) used the cumbersome Japanese typewriters for producing official documents. Unlike her Western counterpart, the average Japanese woman used neither a typewriter nor a word-processor in the office. Few could overcome the double revolution involved in learning to use a PC.

3. LATE BEGINNING OF THE INTERNET

The NTT establishment was slow to realize the importance of Internet communication. For them, it was only a vehicle for private communication amongst academic researchers. Jun Murai and his academic associates experimentally connected with the world-wide Internet in 1984 and its usage soon spread throughout the Japanese scientific community.

In 1985, Nifty Serve and other companies started provider services for the personal computer network. Japanese computer hobbyists communicated amongst themselves in the Japanese language without being connected to the international Internet, while scientists opted to use it and communicate in English. It was only in early 1992 that Japanese PC networks became connected to the Internet.

The commercial use of the Internet began around 1995. The catch-cry for PC manufacturers was 'Let us play with and enjoy the Internet!' Scientists and professionals who used the Internet as part of their day-to-day work frowned on such slogans. I personally tried to promote the use of the Internet among ordinary Japanese citizens in order to share with them what is a wonderful vehicle for free and open communication. But ordinary Japanese do not to find much use for it, with many using the Internet purely for recreation. The so-called PC boom referred to in the business literature is arguably more wishful thinking than reflective of reality. Many people who were able to afford a PC acquired one but ultimately found little use for it. Business people were late to join the bandwagon and when they did, they mostly used an Intranet, working within a corporate firewall.

4. SUCCESS OF MOBILE PHONES

NTT actually started a car mobile phone service in 1979 for elite customers, but mobile phones (cellular phones) were less popular in Japan compared with Hong Kong or Taiwan due to the non-competitive aspects of NTT's monopoly over the Japanese market. Following on from the development of cellular phones came portable, hand-held phones (hereafter abbreviated as 'keitai' in Japanese) which appealed to a much wider market in 1992.

Prior to this, NTT had started a pager service as early as 1968 which enabled company headquarters to maintain contact with travelling salesmen and worried parents to stay in touch with their children. The numerical pager then appeared with only a small display which could be used to convey the telephone numbers of senders or any numbers that needed to be transmitted. A 'pager culture' subsequently emerged in the early 1990s with school kids using telephones to send messages in the form of numbers, just for fun. For example, 39 is pronounced in Japanese as 'san kyuu' which approximates the English words of 'thank you.' It then became fashionable for young Japanese to have numerical pagers (pocket bell or what they call 'pokeberu'). These were not welcomed by the world of education for these devices facilitated cheating in exams and were disruptive in classrooms. Nevertheless, the devices were developed further and two numbers were able to be used in such as way as to denote fifty Japanese syllables.

In 1994, hardware manufacturing and marketing was privatized. What is more significant was that in 1995, the PHS (Personal Handyphone System) was launched. This competed with the keitai. In the years which followed, there has been harsh competition between the keitai and the PHS (Tadashi Aoyagi, Dai-3 Sedai Keitai Denwa Bijinesu, 2000, p. 39). Mobile phones were sold very cheaply or even given away. Young Japanese have embraced the PHS phones as fashion accessories. We thus see a trend away from pokeberu in 1995 to PHS phones, and then to the keitai. Although the hardware was almost free, the cost of wireless telephone calls has been much more than for ordinary calls. At one time, kids paid so much for PHS and keitai calls that parents complained in such large numbers that young Japanese turned to using them for the transmission of messages in the tradition of pokeberu culture.

The 'DDI Pocket' phone service began in December 1996 as primarily a closed mail service between PHS terminals. NTT DoCoMo met this challenge by providing a 'short message' service between keitai cellular phones from June 1997. Furthermore, with the PHS, NTT Personal started an open e-mail service in March 1998 and then the company NTT DoCoMo followed in February 1999 with a keitai I-mode service. (Takeshi Natsuno, I-mode Strategy, Nikkei BP, 2000, p. 33.) This mobile Internet service has now become the centre of Japanese youth culture.

Ideographic characters are, thanks to two-byte culture, displayed much more concisely than alphabetical representations of European languages. Now 80 or 90 per cent of mobile phone use by young Japanese is confined to e-mail exchange, while adults use mobile phone just like before mainly for telephone communication. The problem of cost aside, shy youths often prefer communicating with visual images rather than actual speech.

A few years ago, I organized a freshman seminar to discuss mobile phones; while students all carried a PHS or a keitai, I, the teacher, was the only person who didn't. My lectures were often interrupted by the ringing of a keitai. On the other hand, they were useful for professors wishing to contact students.

What is significant is the way young Japanese e-mail each other. Teenagers have still not gotten used to the alphabet-based keyboard. Instead, they have been quick to adjust themselves to new systems of communication using mobile phone ten-key, keypads. The Japanese language involves use of a system of phonetics, kana, which has five vowels and nine major consonants. These phonetics are perhaps better suited to the decimal keypad of mobile phones, as consonants are allocated to the decimal ten-key keypad, while vowels are represented by pushing numbers one to five times. This is perhaps more natural than the alphabet-turned-to-decimal keypad seen in the Western telephone dialling system. In this way, the decimal mobile phone has become an indispensable tool or toy for fashion-conscious Japanese youth. They are sometimes called the 'thumb-tribe' due to their ability to type on a keypad with one thumb at a rate much faster than adults using PC keyboards and typing with ten fingers.

In 1999, NTT DoCoMo commenced the I-mode service, which connects with the Internet. This service includes not only an e-mail function but also enables the transmission of various data. Its use for telephone calls has become a less important aspect of the service, so much so that it can be rightly called a mobile Internet rather than a mobile phone. This has contributed towards the strengthening of a youth subculture. NTT DoCoMo designed I-mode to fit into the already developed digital youth subculture, while the dominant Western culture of the PC-Internet system has spread much more slowly. Although there are few adults who subscribe to the I-mode, the subculture of mobile phone use will arguably come to constitute the major culture of the Japanese after a decade or two when these teenagers make up much of the workforce. Looking at the content of I-mode now, it is full of playful information that teenagers would like to have, such as horoscopes and various entertainments and events, while I, an adult, need only weather forecasts.

5. PC VS. MOBILE INTERNET

In 2000, when Bill Gates came to Japan, he was asked about his assessment of the future of mobile phones by a Tokyo University student. He responded that the screen was too small and the future of the Internet rested with the PC. Despite his warning, mobile phone sales continued to boom in Japan ('Keitai culture' by Ernest S. Johnson and Sean Odani, Via, pp. 24–25, 2001).

Considering the popularity of I-mode mobile phones not only in Japan but also in Hong Kong, there appears to be a cultural divide between East and West in terms of how the Internet is accessed. Americans think big. Japanese on the other hand are capable of dealing with smaller things. While Bill Gates and the alphabet-based peoples of the West (the US in particular and including myself) prefer to use PCs to access the Internet, average East Asians will approach it via the mobile phone. Since mobile phones are getting cheaper everywhere, they will eventually bring about the dissolution of the digital divide, especially in non-alphabet-based countries.

Still, there seems to be certain disadvantages associated with the mobile Internet, in terms of the input of letters and the display of output. For input, many of my students claim that the one-thumb operation is more convenient, easier and faster than typing on a PC. Students say that they do not require a particularly large display. Those Japanese who are already familiar with the keyboard, like myself, can never cope with one-thumb operation.

Teenagers and the average adult use the email service to exchange short messages such as where to meet and at what time. In time, they will use it to find a place to eat or shop. Today, the keitai can handle most functions of a PC, despite having only a small display. Most people never write a long article or a book. For those seeking to produce a long document, a PC is a must-have.

In Japan where there is no tradition of a keyboard-typing culture, the Internet is penetrating the life of common people through mobile terminals rather than the PC. While Japanese academics and professionals prefer to stay with the PC, ordinary people prefer mobile phones. What is more important for girls is the design of mobile phones, for the phones are a fashion accessory not unlike a necklace or a watch. The young mobile Internet generation see the Keitai as constituting part of their body and mind, a lifeline to be worn at all times and throughout life.

6. CONCLUSION

We can sum up this paper by looking at the following table which represents a digital 'trivide' rather than digital divide.

Japanese Digital Trivide

Generation	*Pre-Revolutionary*	*Double Revolutionary*	*Post-Revolutionary*
Communication	Hand	PC	Mobile phone
Age	More than 50	50–25	Less than 25
Of the cohort	100%	Less than 30%	80%
Kind of people	All	Professionals, computerholic	Girls more than boys
At school	Encouraged	Non-existing	Discouraged
Cost	None	High	Low
Writing with	Pen, brush	Alphabet Keyboard	Ten-key keypad
Writing by	One hand	Both hands	One thumb

The contrast between PCs and mobile phones can be, rather jokingly, extended in the following way:

PC	*Mobile phone*
Instrumental	Consummatory
Official	Private
Day-time	Night-time
Yang	Yin
Confucian	Taoist

Whether the PC keyboard or mobile keypad will dominate the market in future is hard to judge. In the alphabet-based West, keyboard culture remains standard, while ten-key keypads are confined to phone dialling. For the Japanese, both will coexist for the time being. There will be some devices which will serve to bridge the keyboard and keypad, but for most Japanese and East Asians, the mobile ten-key keypad will eventually dominate communications, unless their own languages give way to English.

The history of technology tells us that while new technology is often resisted by the older generation, it will survive the next generation only if it keeps the attention of the younger generation. On a Japanese commuter train, people read something, a newspaper, a book or a magazine, but you will find teenagers gazing at mobile phones for writing email or to collect information. Neither the Japanese government nor corporations anticipated the emergence of such a subculture, but they are now adjusting themselves to the newly-created digital environment. NEC and Matsushita are cooperating on R&D for the next generation keitai, and even the Japanese government has committed to providing funding support, in the hope of setting the world standard and helping Japan to become the world leader.

For contemporary Japanese, the double revolution from hand to electronic keyboard is still hardly overcome, but with the advent of the mobile Internet, the next generation of Japanese can jump from handwriting to the keypad. The second revolution from keypad to keyboard is still to be seen. Whether the second revolution will happen or not is not certain, but it is clear that a mobile phone with ten keys will dominate multimedia culture because of its convenience, low cost and digital divide-free nature.

Further words on the digital divide.
Even in a high-tech society like Japan, there exists a digital divide between social strata. Handicapped people may not be able to afford a new personal computer, which costs around 200,000 yen. Major computer manufacturers have kept discouraging the used-computer market in order to prevent pricecutting. This is despite the fact that most users do not need particularly advanced computers. For most people, the word-processing function is what is most needed. A friend of mine, Dr Akiba, the mayor of Hiroshima city, encouraged a small business to manufacture a no-frills computer for handicapped people with the price of 50,000 yen. The influential big manufacturers were not in a position to influence the pricing of this computer for 'handicapped people' lest they be criticized for being socially insensitive.

First published in *The Cambridge History of Science*, Vol.4

25 Eighteenth-Century Science: Japan

The eighteenth century was one of Western recognition of Japan against the Chinese background. During that period, Japanese thinkers became critical of the Chinese scholarship with which they had struggled to keep pace in the previous century; for the first time, Japanese intellectuals from the extreme eastern regions of Asia began to compare Chinese scholarship with the infiltrating Western science. It is extremely interesting to see what happens to a paradigm from one culture – and the scholarly traditions that have evolved around it – when it is introduced into another. In the following pages we shall examine the impact of this transplantation, mainly on three disciplines: mathematics, astronomy and medicine.[1]

The Jesuits had been evangelizing in Japan since the mid-sixteenth century. Eventually, the Japanese government, considering Christianity a threat to the cohesiveness and integrity of Japanese culture, successfully banned all Westerners from the country with the exception of Protestant Dutch traders,[2] who were restricted to the port of Nagasaki. This ban, which remained in effect until the mid-nineteenth century, was reinforced with bans on Jesuit writings in Chinese in the 1630s and further intensified in the 1680s.

The beginning of the eighteenth century was thus the nadir of access to information on all things Western. Throughout the eighteenth century, a gradual relaxation of the ban brought an awareness of East–West comparison based on limited sources of available information.[3]

SCIENCE AS AN OCCUPATION

Peace prevailed throughout the century in Japan. The economy and demographics remained stable, and the class hierarchy was tightly maintained. The samurai – around 6 per cent of a total population of approximately 25–27 million – were at the top of the class structure, and their sons learned the orthodox Confucian classics in clan schools. Commoners were much less literate until village schools evolved towards the end of the eighteenth century with the development of the rural economy.

The samurai class held hereditary stipends that only first-born males could inherit. Families without sons had to adopt heirs if they were to continue. Younger sons, on the other hand, had to find their own livelihood. Most of them were adopted by other families, and others found occupations outside the old rigid structure in medicine or Confucian scholarship.

There were no clear-cut scientific occupations. Astronomy and medicine were esteemed but offered opportunities only to a few talented men. These fields offered opportunities outside the conventional structure, allowing some to take advantage of a social mobility that was not otherwise available. But attempts were continually made to subordinate men in these fields to the hereditary tradition that governed the rest of Japanese life. It was expected, for instance, that the son of a doctor would eventually be registered as a doctor, regardless of how little aptitude or motivation he might have. The shogunal and fief governments needed talented professionals, however, and governmental authorities often resolved the conflict by advising a professional family to adopt a gifted youngster.

Elsewhere in the Chinese cultural domain, including Korea and Vietnam, scientific professionals were tightly bound to central government institutions through civil service examinations. However, in Japan after the tenth century, as the Chinese-type court bureaucracy atrophied and military power became dominant, these examinations disappeared.

Even during the peaceful Tokugawa period (1603–1869), the shogunal government had no power to impose its recruiting policy on the fief government. During the eighteenth century, the shogunate discussed reviving the examinations, but this was carried out only tentatively in the last decade of the century by testing candidates in Confucian studies from the lower samurai class.

However, an egalitarian examination system was not possible within a hereditary structure. In practice, those who passed with the highest grades received only a prize, varying with their family status, but it did not bring a permanent increase in social status.

Medical examinations began some years before those in Confucian studies. The shogunate, in need of several hundred doctors, gave written as well as oral examinations. They, too, were not intended to change the social status of graduates but to encourage the sons of medical families to study diligently rather than simply claim their sinecures.

To find experts to fill posts in technical fields such as astronomy (there were only ten or twenty such posts), personal references were sufficient.

THE BAN ON WESTERN SCIENTIFIC KNOWLEDGE

During the eighteenth century, woodcut printing flourished. A set of printing blocks could produce around two hundred clear copies, and, therefore, publishing a book with a corresponding potential readership was commercially feasible. By the end of the century, a popular culture of reading evolved to the extent that more than ten thousand copies of a bestseller might be in circulation. During the eighteenth century the centre of publishing moved from Kamigata (Kyoto and Osaka) to Edo (Tokyo). Most academic works were written in classical Chinese, whereas popular works used a native style that combined Chinese characters and phonetic kana.

Even as the book trade grew, an official ban severely restricted knowledge of the West. In the 1630s the government banned imports of Sino-Jesuit writings, thus depriving the reading public of information on Western science. In 1685 the censors defaced and destroyed what previously had been two important sources of European cosmology: *Huan yu ch'uan*, written by the Portuguese Jesuit Francisco Furtado (now preserved at the Bibliotheque Nationale, Paris) and the sequel to a Chinese work with Jesuit influence, *T'ien ching huo-wen* (Queries on the Heavens), which will be discussed later in this chapter.

A collection of treatises by Matteo Ricci, *T'ien-hsueh chu hand*, had previously been available in Japan. It consisted of two parts: *li*, catechetical and theological, and *ch'i*, scientific. The latter does not appear to have been strictly censored. The world map compiled by Ricci, and some popular astronomical books that reflected some Western influence such as *T'ien ching huo wen* (Tenkei Wakumon), had escaped the attention of the censors at the port of Nagasaki. These documents, as they spread among Japanese intellectuals, influenced their worldview as well as their cosmology.

Furtado's book *Huan yu ch'uan* is a popular treatise on Western cosmology and cosmography, with the first half covering Christian theology and the second half devoted to scientific matters, as in Ricci's collection and in contemporary popular books. This format implied that the two parts, written by the same author, were distinct but inseparable. When the authorities became aware of this, public access to scientific and overtly religious works was forbidden, and at the beginning of the eighteenth century the Sino-Jesuit treatises were still heavily censored.

The Jesuits, who arrived in China in the seventeenth century, challenged traditional astronomy with their superior parameters and methods of calculation, which became apparent in competitions to predict solar eclipses. It was clear that in astronomy – the foremost subject of a traditional exact science – the criterion of quantitative precision transcended East and West, and observation of celestial phenomena precluded human manipulation. Accordingly, Jesuit astronomers took over the Astronomical Bureau in 1644, won a decisive prediction contest, and quickly carried out a calendar reform that established a largely Western system, the Shih-hsien li. Jesuit control over the Bureau, although challenged several times, survived until the end of the empire.

The Japanese learned about this Chinese reform from imported annual almanacs. Because of the ban on Jesuit works since the 1630s, not enough information could be obtained to reform the Japanese system. Shibukawa Harumi (1639–1715), the first genuine reformer, judged from the crude values of Western parameters given in the popular *T'ien ching huo-wen* that the Shih-hsien calendar was no improvement on its predecessors. Shibukawa followed the great Shou-shih computational system of 1279, which had not been significantly improved before the arrival of the Jesuits.

T'ien ching huo-wen provided Japanese intellectual circles with a standard pre-Copernican cosmological picture, but its sequel was banned at Nagasaki for the official reason that its contents were occult and therefore unhealthy. This volume has long been unknown in China. I had an opportunity to look at the copy in the Seikado Library in Tokyo, a modern acquisition from a Chinese private collector. It contained nothing fantastic. Its history of Chinese calendrical astronomy clearly stated that the Western Jesuits carried out the recent Shih-hsien reform. If contemporary Japanese read this book, the news may have caused great concern and jeopardized Shibukawa's native Jokyo reform, which was based on a purely Chinese model.

We still do not know why the sequel was banned. It is my guess that under the seclusion policy the government and its 'Confucian' censors feared that it would convince intellectuals that Western astronomy was superior to traditional Chinese astronomy and perhaps would lead eventually to the belief that Christianity was superior – something that was, after all, the ultimate aim of the Jesuits astronomical activity in China.

In the early part of the eighteenth century, members of the scientific elite in the shogunate consulting bodies began to suspect that the Chinese approach to calendar-making had been replaced by that of the West. Because of this suspicion, in 1720 Tokugawa Yoshimune, the eighth shogun (1684–1751), who was himself eager to collect Western knowledge, ordered specialists in calendrical astronomy and mathematics to carefully examine the banned books that were stored in the shogunal library. Nakane Genkei (1661–1723), a private scholar not previously allowed to see Shibukawa's Jokyo calendar, at Yoshimune's request read the Sino-Jesuit astronomical writings and concluded that they would be useful for the next calendar reform.

The shogunate adopted his recommendations and encouraged elite scholars to study foreign languages, particularly Dutch, which was the sole language used for trade with the West during the seclusion period.

This event, a watershed in the official recognition of Western science, would have shocked those who respected the Chinese model. Because the central government demanded a monopoly on information about the West, it never publicly announced the lifting of the ban. People were still wary of becoming involved with anything related to Western learning. However, for the remainder of the eighteenth century a number of intellectuals perceived that the policy was not being rigorously enforced, and they copied and circulated Sino-Jesuit works. To escape the censors' notice, they often put false titles on the front pages and avoided direct quotation.

In 1726 an incomplete set of the *Li-suan ch'an-shu* (Complete works on calendrical mathematics, 1723), by the great mathematician Mei Wenting, reached Japan. This work was influenced by the Jesuits, but because Mei's work was purely scientific and technical it was much safer to disseminate in Japan. Mei's treatises convinced the Japanese that Western astronomy was superior, and that brought about a further moderation of the ban.

TRANSLATIONS OF WESTERN WORKS

Japanese intellectuals could read classical Chinese writings without difficulty, and Sino-Jesuit publications were likely to be widely read if available; hence the ban. However, because no astronomer had mastered European languages, there seemed little need to ban these publications.

On the other hand, at the port of Nagasaki there were about fifty official interpreters of Dutch, twenty-three of them hereditary. They had enough linguistic knowledge to communicate verbally with Dutch traders and ships' doctors. In the mid-eighteenth century, with the relaxation of the seclusion policy, these interpreters began to study Dutch books and undertake translations. They usually worked at the request of feudal lords who rewarded them well. They translated the few books they received from their foreign contacts. These materials – initially they were generally maps, seamen's almanacs and other navigators' essentials – stimulated curiosity about Western countries. Eventually, Dutch merchants were asked to import various books by way of Batavia (Indonesia), but these were expensive and limited in subject matter.

The official interpreters were bound by their official duties, and their translations were never intended for publication. Doctors, on the other hand, as intellectuals and men of culture, were more open-minded, independent of government institutions, and eager to publish books that would improve medical practice. Their knowledge of Dutch was greatly inferior to that of the Nagasaki interpreters, but, nevertheless, a few of them obtained Western medical books and began to translate them. The first publication was *Kaitai Shinsho* (New Book of Anatomy) in 1773. A pioneer of the project, Sugita Genpaku (1733–1817), managed to obtain official approval in advance for the first published translation of a Western book, and, as the title of his memoir states, this was viewed as the dawning of Dutch learning in Japan. This breakthrough encouraged other physicians and intellectuals to learn Dutch and to investigate Western science. By the turn of the century, a group of physicians and intellectuals formed a society with the aim of exchanging information on Western science.[4]

THE INDEPENDENT TRADITION OF MATHEMATICS

The Western world first became aware of the independent tradition of Japanese mathematics (*wasan*) through the publication in 1914 of *A History of Japanese Mathematics* by Eugene Smith and Mikami Yoshio. Japanese mathematicians had built this tradition on the basis of late seventeenth-century Chinese mathematics and had developed it independently in the eighteenth century. Despite the influence of Western culture on other aspects of life, Japanese algorithms and the Japanese style of writing equations are quite distinct. A number of its characteristic problems did not exist, or appeared later, in the history of Western mathematics.

Reckoners used the abacus or counting rods on a grid. When symbolic algebra appeared after Seki Takakazu (d. 1708), symbolic (as opposed to merely numerical) written calculation (on paper) became possible. *Wasan* has been compared to Newton's and Leibniz's differential and integral calculus; but Seki and his immediate successor, Takebe Katahiro (1664–1739), leading figures in the tradition, showed little interest in solving mechanical problems, the calculation of the area of a circle being a more typical preoccupation. Pure mathematics, in other words, was pursued as a hobby and was not associated with physical science.

In the seventeenth century, mathematicians established the tradition of *wasan* by the practice of 'bequeathed problems'. Anyone who solved a difficult problem for the first time would write a treatise and would then bequeath a new problem;

whoever solved this problem would bequeath another, and so on. This ongoing competition gave the tradition its momentum. Emphasis was placed on problem solving by unusual means and on presenting problems for which solutions were unknown or perhaps did not exist. Mathematicians increasingly valued complexity and emphasized the transformation of systems of simultaneous equations into higher-degree equations.

This ostentation for its own sake prompted Seki Takakazu to introduce an important innovation: the *tenzan* algebra. This was a system for expressing unknowns, previously solved only via numerical equations, in symbols. He also developed a theory of equations that recognized imaginary and negative roots (*daijutsu bengi no ho, byodai meichi no ho*) but rejected them as 'sick solutions' – thereby ruling out a theory of imaginary numbers. In *wasan*, as in Chinese arithmetic, problems were presented in the form of questions and answers and often omitted the method of derivation. Therefore, they did not encourage investigation of the basic nature of equations.

Seki was exceptional in investigating the general nature of equations. His orthodox standard for posing and solving problems fits the Kuhnian definition of 'paradigm'. Once this paradigm was established, 'normal science' could follow.

Before the time of Seki and Takebe, mathematicians were concerned with the practical problems of calendrical astronomy: surveying and so forth. The paradigmatic approach ruled out application. Subparadigms appeared in the course of development – for instance, Takebe's *enri* (circle theory) calculus. Ajima Naonobu applied this method not only to circles but also to curves and curved surfaces in general. Wada Yasushi furthered the development of mathematical analysis by compiling tables of definite integrals and applying them to the mathematically infinite and infinitesimal, and so on. All these innovations remained, however, within the *wasan* tradition. There was little discussion of fundamental theories, and practitioners continued to solve increasingly complicated geometrical figures by algebraic means.

Another stimulus for amateur mathematicians came from a somewhat different source. Artists and poets had established a tradition of offering their masterpieces, painted on wooden plaques, for display in the public gallery of Shinto shrines. Mathematicians followed suit, exhibiting a tablet with both problem and answer displayed, usually accompanied by an elegant geometrical diagram for public entertainment. Amateurs seeking acclaim often spent a great deal of money on such pursuits. Thus, mathematicians valued playful competition rather than basic research or application, and intuitive breakthroughs for elaborate problems were regarded more highly than logical consistency or rigour. Indeed, when Euclid's *Elements* first came to Japan, *wasan* experts – noting its figures only – judged it rudimentary and unchallenging.

The Sino-Jesuit treatises had introduced trigonometry, and by the end of the century it was used by astronomers and surveyors. Although *wasan* mathematicians were capable of mastering it, they continued to use their traditional methods, valuing problems arising from their own tradition and not those posed by technical practice. The eccentric mathematical genius Kurushima Yoshihiro wrote, 'In mathematics, it is more difficult to raise a problem than to answer it.

Only mathematicians incapable of inventing problems borrow them from other fields such as calendrical science.'

Because of their lack of interest in practical applications, the nucleus of Japanese mathematicians did not compete with Western mathematicians in solving practical problems until the middle of the nineteenth century. Unlike other Japanese intellectuals, *wasan* practitioners, as they moved from paradigm creation to the formation of a support group, advanced to new technical frontiers without the need to consult foreign authorities. In isolation they underwent vigorous growth in normal science, largely unaware of the developments in Chinese or Western mathematics. *Wasan* did not substantially influence scholarship in other fields or in cultural matters, although those practitioners who applied their mathematical skills to the practice of land surveying or calendrical calculation were well-versed in the algorithms, formulae and notations of traditional *wasan* mathematics.

MATHEMATICS AS AN OCCUPATION

Seki Takakazu issued a licence to teach, and it was developed by his successors into a five-stage system of degrees. This system did not guarantee employment as a teacher but was primarily considered an honour, certifying that a certain level of mathematical mastery had been achieved. Without a solid occupational basis it is difficult to estimate how many people were engaged in *wasan* or to distinguish amateurs from professionals.

There is evidence that even peasants occupied themselves with *wasan* puzzles in the agricultural off-season. Mathematicians were primarily hobbyists, and the traditions existed only in the private sector. Although the shogunate attempted to maintain the occupational hereditary system, it did not consider mathematics worthy of perpetuation through this system.

Only a handful of leading mathematicians were able to support themselves. From the late eighteenth century, a number of them travelled from village to village, visiting amateur groups and enthusiasts and conducting problem-solving competitions, thereby following the practice of other arts such as *haiku* poetry.

PUBLICATION IN MATHEMATICS

Wasan mathematicians circulated their solutions to problems by copying by hand, although the more famous published theirs. Popular mathematical works for general readership were even more widely published. The printing blocks were often cut in an informal running style of calligraphy, which made it easy for the literate public to read. They were bestsellers by Japanese standards, some selling more than several thousand copies.

The literary world did not consider mathematics to be true scholarship, and, from the end of the seventeenth century, mathematics was often classified in book catalogues as a hobby on a par with flower arrangement or the tea ceremony. *Wasan* authors, in an attempt to increase the prestige of their books, invited Confucian philosophers to write prefaces, which usually bore no relation to the technical content.[5]

ASTRONOMY WITHIN THE TRADITIONAL FRAMEWORK

The traditional Chinese approach ('calendrical astronomy') investigated the apparent motion of the sun and moon to construct a method for generating luni-solar calendars. The ultimate test was the precision with which a system could predict solar eclipses. Successive reforms refined parameters for solar and lunar motions by testing them on previous records of solar eclipses. Planetary phenomena attracted relatively little attention.[6]

Throughout the written history of Japan, the Chinese luni-solar calendar had been accepted. From the late seventeenth century onwards, the shogunate adopted its own Jokyo system, merely revising the Chinese Shou-shih system of the thirteenth century to incorporate the difference in latitude between China and Japan.

An order of the Shogun Yoshimune in 1720 assigned responsibility for a new calendar reform to Nishikawa Masayasu. He was the son of the noted Nagasaki scholar Nishikawa Joken, Japan's foremost expert on the West. Masayasu was not a professional astronomer, but, assisted by professionals, he undertook a reform based on Sino-Jesuit writings. When Yoshimune died, a family of court astronomers in Kyoto tried to restore the emperor's prerogative of issuing the calendar. This conservative backlash ignored Yoshimune's goal of reform and their Horyaku system, issued in 1754, made matters worse. Masayasu was subsequently dismissed, and his associates could only edit their records for use by a future generation.

Real reform came a generation later in the Kansei calendar revision (1797). It was undertaken by Asada Goryu (1734–1799), a physician and amateur astronomer, and his followers, who had access to most of the Sino-Jesuit treatises.

In astronomy, the new Western paradigm did not replace the traditional Chinese one. Rather, new data and mathematical techniques were simply incorporated into the old framework. This was also the case in China from the seventeenth century onwards, with the structure, style, and purpose of Chinese calendrical astronomy unchanged. As Hsu Kuang-ch'i, the high official who had collaborated with the Jesuit Matteo Ricci on several projects, remarked, 'We melted down their materials and poured them into the [old] Ta-T'ung mould.' Until the mid-nineteenth century, official Japanese astronomers adopted this attitude and even repeated Hsu's slogan in their treatises.

Throughout the eighteenth century, Japanese calendrical astronomy adopted the view that astronomical parameters varied in time. In 1684, the government adopted the Jokyo system, whose originator, Shibukawa Harumi (1639–1715), restored the variable tropical year length of the Chinese Shou-shih calendar. He reasoned that such a minute variation reflected high precision. In reality, however, it provided no gain in accuracy.

Ogiu Sorai (1666–1728), the most influential Confucian philosopher of his time, supported Shibukawa's notion on ideological grounds, commenting in his *Gakusoku Furoku* (supplement to School Rule), 'Heaven and earth, sun and moon are living bodies. According to the Chinese calendrical technique, the length of the tropical year was greater in the past and will decrease in the future. As for me, I cannot comprehend events a million years ahead.' In Ogiu's dynamic view of nature, everything was subject to change and it was therefore impossible that ancient laws could still hold. Since the heavens were imbued with vital force, the length of the year could change freely, and thus constancy was not to be expected in the heavens. Indeed, only a dead universe could be governed by law and regularity. Since it was precisely the vital aspects of nature that interested Ogiu, he remained an agnostic in physical cosmology.

Lack of interest towards the search for regularities in nature prevailed in the School of Ancient Learning (*Kogaku*), of which Ogiu was the leader. Nature was observed in the light of social and ethical concerns. This moralistic, anthropocentric, and often anthropomorphic view of nature was common among Japanese Confucian intellectuals. Few of them imagined that mathematical astronomy was deserving of attention except to provide an accurate calendar. Hence, the official astronomers' recognition of Western superiority did not immediately influence conventional intellectuals.

The next calendar reform, the Horyaku (1755), replaced Shibukawa's notion of changeable parameters. The value of yearly change was much too large and the discrepancy between observation and calculation, as it increased, was bound eventually to become apparent. This secular variation, nevertheless, was again adopted without reflection in the system that was to follow, with a predictable growth in inaccuracy.

The traditional eclipse records used as benchmarks for astronomical parameters were not supplemented by Western observational records in Jesuit writings. Asada Goryu (1734–1799) collected all the available records, traditional and Western, and tried to represent them all with a single formula of his own. He varied not only tropical year length but also other astronomical parameters in a twenty-six-thousand-year cycle of precession. His approach was purely numerical, and it was incorporated in the next Kansei calendar reform in 1798.

In Asada's time, knowledge of Western astronomy was still limited to Sino-Jesuit writings in Chinese, which made no mention of Copernican doctrines. Towards the turn of the nineteenth century, Asada's pupil Takahashi Yoshitoki (1764–1804) began to study a Dutch translation of Lalande's post-Newtonian *Astronomie*. His was a purely academic interest in the kinematics of planetary motions, although celestial mechanics were still beyond him.

Because calendrical astronomy was an official domain, advisers urged that the shogunate recognize Western superiority. However those bureaucrats employed in astronomy had no authority except in purely technical matters, and they neither intended nor had the power to speak publicly on the merits (or otherwise) of Western science. Their influence in nonastronomical fields was negligible.

An index of Western influence can be taken from the use of the Sino-Jesuit 360 degrees for coordinates as opposed to the traditional Chinese count of approximately 365.25 (the old degree, *tu*, was defined as one day's mean solar motion). Official astronomers working on the Kansei calendar reform (1798) first used the former, after which it spread gradually into general use.

ASTRONOMY AS AN OCCUPATION

Calendrical astronomy was a state monopoly. Issuing the annual ephemeris and reforming computational methods were purely a matter of prestige for the ruling government. In the eighteenth century, although dynastic legitimacy could not be removed from the imperial court in Kyoto, real political power lay entirely in the hands of the military dictator, the shogun, in Edo (Tokyo). Certain astronomical prerogatives were a hereditary right of the Tsuchimikado family in Kyoto, the imperial court astrologers, but only the shogun's astronomers had the actual power to reform the calendar and apply the science. There were eight shogunal families of astronomers, who, like their imperial predecessors, had become hereditary. They and their associates totalled between fifty and one hundred officials. The Tsuchimikado, other families of lower rank, and temporary associates brought the total in Kyoto to less than fifty. Hoping to restore their ancient authority, they successfully intervened at the time of Horyaku reform, but the Kyoto revival was short-lived.

The hereditary astronomers did not require talent or even much skill to calculate the annual ephemeris. They met the greater demands of the two eighteenth-century reforms by acquiescing in the appointment of – or even by adopting – well-qualified individuals. Some fief governments occasionally hired astronomers, usually because the ruling *daimyo* family was interested in the astrological prediction of natural disasters. Some remote areas such as Satsuma, one of the larger fiefs, appointed permanent astronomers and issued their own calendars, but these did not diverge significantly from Shogunal astronomical practice.

PUBLICATION IN ASTRONOMY

As a shogunal practice, astronomy was by no means accessible to the general public. The most important official treatise on the new calendrical system, the product of the calendar reform, was never published. Three manuscripts were submitted to the Shogunal Library, the Imperial Court Library and the Library of the Ise Grand Shrines, to each of which only a few high-ranking officials had access. The government feared that criticism from the private sector would destroy public esteem for this particular governmental function.

As a result, those who wanted to learn computational astronomy could study only the past system, in particular the Shou-shi calendrical treatise of the late thirteenth century, the highest achievement in Chinese mathematical astronomy and the model for the Japanese calendar until the middle of the eighteenth century. Many illustrated guides and commentaries on this were compiled and printed in Japan. The main source for cosmology was *T'ien ching huo wen*, the seventeenth-century treatise that incorporated some Jesuit elements. Again, many Japanese-illustrated versions and textbooks satisfied the intellectual needs of the day, and the needs of the general public were met with yearly almanacs printed and distributed by a network controlled by the hereditary imperial court astrologers.

INTRODUCTION OF COPERNICANISM AND NEWTONIANISM

Because the hereditary astronomers' interest remained confined to the traditional model of calendrical science, the introduction of the core of modern Western astronomy was left in the hands of the official interpreters. The first to become involved, perhaps, was Motoki Ryoei (1735–1794), who invented his own system of transliteration from Dutch to Japanese phonetics, using Chinese characters. We know of no similar activity in China at the time.

Ryoei was interested in translating a history of Western astronomy to add to the margins of large navigational charts. However, he was concerned to learn that Galileo had been persecuted because of his writings on Copernical cosmology. Ryoei realized that this would be a delicate subject in Japan since it was related to the strictly proscribed Christianity, and his translation, drafted in 1771, omitted discussions of the trial of Galileo. However, he found Copernicanism important and interesting, and his later translations gradually revealed details of Copernicanism, a full translation being completed in 1793. Borrowing from Ryoei, Shiba Kokan (1738–1818), an illustrator and popularizer, published many books that disseminated the theory of heliocentricism in Japan.

Shizuki Tadao (1760–1806) was also born into a family of official translators. He left his inherited profession to concentrate on the translation of Western books and was the first Newtonian in East Asia to introduce such concepts as the molecule and force. Because traditional Japanese Confucian learning was not concerned with natural philosophy and was unaware of late Chinese writing on the discipline, in translating Newtonian concepts terminology had to be borrowed from the Ten Wings of the Book of Changes, Buddhist speculation and neo-Confucian writings. Tadao translated the work of John Keill, the popularizer of Newton, adding a great many comments of his own, some quite original and going far beyond Keill. Not entirely satisfied with Newtonian laws, Tado attempted to base them on traditional Yin-yang metaphysics. He tried to introduce the inverse-cube centrifugal force or quadruple of distance to explain such phenomena as chemical affinity and plant physiology. Tadao is also known for his nebular rotation view of the solar system – a similar idea later attributed to Kant and Laplace – although, essentially, he applied neo-Confucian cosmogony to the solar system.[7]

PHYSICIANS AS INTELLECTUAL CONNOISSEURS

In the seventeenth century, mainstream Chinese medicine dominated that of Japanese with the exception of surgery, which the Jesuits had introduced in the sixteenth century to meet the needs of endemic civil war.

In the eighteenth century, a new group became critical of scholarship of the physiology and pathology that had prevailed in China since the Chin and Yuan periods. They claimed to be returning to a simpler reasoning that more directly reflected the clinical practice of the ancient *Shang han lun* (Treatise on Cold Damage Disorders, between A.D. 196 and 220) but showed little interest in the more theoretical and speculative *Huangti Neiching* (Yellow Emperor's Inner Classics). This group called itself *Koiho* ('Back to Ancient Medicine School').[8]

The school preferred simple and drastic medical prescriptions as opposed to the great variety of Chinese formulas, some simple and some complex, some strong and some mild, some formed by theory and some by direct experience of drug action, all of which neutralized effects. The Koiho defined their goals in terms of utility, which made Chinese complexity seem more of an impediment. Because they wished to tackle disease as directly as possible, they refused to view it as a microcosm. As Yoshimasu Todo (1701–1773), the foremost figure of this school, declared, 'Yin and yang are the *ch'i* of the universe, and thus have nothing to do with medicine.'

The Koiho were materialists in the sense that they rejected abstraction, trusting only that which was tangible, and thus they developed abdominal palpation, which did not exist in China.[9]

FROM THE ENERGETIC TO THE SOLIDIST VIEW OF THE HUMAN BODY

In Chinese and Japanese medicine, disease was attributed to an imbalance of *ch'i*, which circulated throughout heaven and earth and thus through the human body. It is now considered imponderable and incorporeal energy, but Ch'ing Chinese considered it the material basis of life. This view was close to that of Western humouralists, who believed disease resulted from an imbalance between the humours circulating through the body rather than to a pathological abnormality in a particular organ.[10]

Goto Gonzan (1659–1733), a precursor of the Koiho school, reduced traditional physiology and pathology to a simplistic scheme in which every disease originates in the stagnation of *ch'i* and in which the *ch'i* was a more materialistic concept than the accepted Japanese abstract and incorporeal matter. Goto's successors took a position much closer to that of the solidists than had previously been possible in Japan. Lacking abstract concepts, functional analysis lost its importance, and the Koiho physicians studied the physical organs for their own sake.[11]

In conventional *ch'i* physiology, dissection does not yield meaningful information, as the dead body contains no *ch'i*. Koiho physicians, on the other hand, showed a genuine interest in dissection and their organ-centred approach brought a recognition that the traditional anatomical charts were crude and inaccurate. In 1754 Yamawaki Toyo (1705–1762), a leader of the Koiho school, was the first to examine the corpse of a criminal for anatomical purposes. He questioned Chinese anatomical charts and wrote *Zo shi* (Chart of internal organs, 1759) on the basis of his findings. Yamawaki's achievement was, however, limited to challenging the old scheme of six yin and five yang organs (*wu-tsang liu-fu*) in preference to nine. No interest was shown in the investigation of the skull, which was regarded as a reservoir of medullary tissue.

In the East Asian tradition, it made no sense to ask in which organ thought took place because physicians did not think in solidist terms. They attributed every activity, mental or physical, to the fundamental agency of *ch'i*, which permeated the microcosm as well as the macrocosm and which circulated harmoniously in both. Thus, there was no reason to attribute thought as a function of the brain. Thought, imagination, and emotion were functions, not as a physical organ but as a bodywide system of energy circulation.

In 1771, however, Sugita Genpaku and his followers abandoned the Chinese physiological tradition, relying on Dutch anatomy charts,[12] although even they showed no interest in examining the contents of the skull after the decapitation of a criminal. Genpaku therefore found it difficult to translate Dutch writing on the brain, often resorting to guesswork and borrowing from Buddhist terminology in order to coin new words for sensory perception. His confusion created difficulties for successive generations of medical students, who used his writings as a base for the understanding of cerebral function.

Thus, the translation of Western anatomical books was significant not only as the beginning of Western learning in Japan but also for the introduction of the solidist school of thought into East Asian culture. However, it is unlikely that the average eighteenth-century Japanese doctor understood the function of the brain.[13]

Practitioners of Chinese medicine viewed disease holistically, and a given disease was not usually associated with a particular body location since pathological *ch'i* as well as life-sustaining *ch'i* usually affected the whole microcosm. For example, when doctors referred to a cardiac or hepatic dysfunction they were not referring to the physical organ but to a whole-body system of functions that the organ merely regulated and that the disease affected. They also treated the body holistically. To treat a headache, for example, needles were inserted into the foot. This approach did not require precise anatomical charts in the clinic. Unlike the physicians trained in the Chinese method, those surgeons cooperating with Sugita Genpaku discovered a remote ancestry in Western origins. This would explain their openness to a solidistic way of thinking.[14]

When the power of Western anatomical knowledge was first realized by the Japanese, it was naturally assumed that associated therapies would also be more effective, although there was no evidence for this belief. Indeed, therapeutically, there was very little choice between the systems of internal medicine that were evolved in the various advanced civilizations before the end of the nineteenth century, although European medicine was more drastic and more likely to harm the patient than most.[15]

The traditionalists naturally objected to anatomy, a common response being that anatomy and dissection were irrelevant to the improvement of therapeutic practice. Other objections were based on traditional physiology. Sano Antei, in his *Hi Zoshi* (A Refutation of the Anatomical Charts, 1760), said, 'What the *tsang* [the spheres of function and their associated viscera] truly signify is not a matter of morphology. They are constant containers that store vital energy with various function. Lacking that energy, the *tsang* became no more than empty containers.' In other words, the internal organs were characterized not by their morphology but by differences in function, defined by their proper *ch'i*, and, therefore, nothing could be learned by dissecting a cadaver, since its *ch'i* did not exist. Because they were based on dissection, the anatomical charts that had caught Toyo's imagination gave no indication of the dynamic functions of the body.

The same point emerges in another criticism of Antei: Yamawaki Toyo's anatomical charts did not demark the large and small intestines. Antei himself did not believe that they were morphologically dissimilar. A physiological difference followed. What made them different was that the large intestine was responsible for absorbing and excreting solid wastes, while the small intestine performed the fluid waste functions. This crucial difference would be undetectable in the dead body. Figure and appearance were significant only in terms of their relation to function. Antei, unlike the Koiho radicals, did not claim to be a pure empiricist. 'The observation of two obvious facts is of much less value than groping speculation . . . even a child is as good an observer as an adult'; a scholar who did not investigate the connections between form and function was no better than a child.

In spite of such reactions by conventional physicians to the radical Koiho school, the solidist tradition that it had initiated paved the way for Western anatomy. Genpaku took up the study of anatomy because it seemed the most tangible and, therefore, the most comprehensible part of Dutch medicine. A solidist breakthrough resulted from this viewpoint and, at the turn of the century, physics and chemistry were studied by Genpaku's successors. The impact of anatomy challenged the energetics and its functional beliefs not only of medicine but also of natural philosophy, and eventually led to the wholesale introduction of modern Western science.[16]

THE MEDICAL PROFESSION AS AN OCCUPATION

Medical practitioners who began to take on the challenge of Western science constituted the largest scientific profession during the Tokugawa period. Medicine, unlike astronomy, was a private concern and not subject to any form of constraint in terms of response to new ideas. Because there was no public health programme at the time, medical practice was essentially a relationship between physician and patient. Each community usually had a private physician or healer. The samurai class had its government doctors and fief doctors, and townspeople and peasants had their local practitioners. Medicine was not a profession, and practitioners did not form organizations or even common ties. They were not regulated by the central government and were not subject to the traditional expectation that physicians should be sons of physicians.[17]

Edo, as the seat of the shogunate, was a centre of professional activity. The important schools of medicine were scattered as far as Nagasaki, where a tradition of Western surgery was maintained through access to Dutch interpreters. Osaka, for instance, was famous for its number of physicians, whose patients were mainly from the merchant class. This decentralization made medicine one of the few geographically mobile professions in Japan.

Unlike the medical profession in contemporary Europe, which was well established and able to develop through the universities, Japanese doctors remained marginal. Government appointments were not the only possible source of income for physicians. A doctor hired by a fief government had no more status than other petty intellectual officials (Confucian scholars, astronomers, or interpreters), but private practice could bring a much higher income.

Doctors – unlike official astronomers – were independent of government hierarchy, and, in the period shortly before the modernization of Japan, they were among those most receptive to liberal thought.

Medical practitioners were not licenced. Even those who were able to read medical classics could advertise themselves as physicians. Often, those hoping to qualify as Confucian scholars supported themselves by practising medicine. The shogunal and fief governments appointed physicians, usually with small stipends, to take care of lords and samurai families. The government often encouraged physicians to adopt a talented young man to ensure a reliable supply of medical practitioners rather than bequeathing a first-born son. Towards the end of the century, as public living standards improved, towns and even small villages supported their own doctors, although most of the peasants found Omyoji (traditional diviners) adequate to meet their medical requirements.

Physicians were usually trained by apprenticeship. A young man wishing to embark on a medical career would become the pupil of a practitioner, living in his house for several years to gain 'hands-on' experience. The apprentice would then move from place to place to gain clinical experience, finally returning home to set up in practice. The more ambitious would seek medical training as far away as Edo, Kamigata or Nagasaki (for Dutch medicine). It is difficult to estimate the number of practitioners trained in medicine, but I would estimate it to be in the region of several tens of thousands.

The end of the century saw the emergence of therapists – practising even in villages. For example, the second son of a village chief, with aspirations of becoming a country doctor, would spend many years as an apprentice to neighbouring practitioners and finally return to his native village. He would be expected not only to provide medical services but also to undertake educational and cultural duties.

Young men who showed academic promise but with no conventional prospects were often advised to study medicine to achieve a secure livelihood. Those who studied Dutch medicine as a means to a medical career often became experts in Western learning and, much later in the mid-nineteenth century, were to have a revolutionary influence on political affairs.

Among the intellectual professions of the Tokugawa period, it was only physicians who were able to achieve an independent position: they were able to view the world from new perspectives and thus bring modern (universal) science to Japan. However, their independence was bought at the cost of alienation from the true sources of power in Japan – the samurai governments. Their role was thus limited to that of connoisseurs of cultural novelty.[18]

From the late eighteenth century on, the Rangakusha (scholars of Dutch learning) were mainly free-lance physicians.[19] The more successful tended to live in cities, often with government appointment. In Edo, particularly, doctors met and exchanged information on Dutch learning. In 1794, they started their celebrations of the Western New Year and drank European wine. Some connoisseurs wrote entertainingly on curious aspects of Western culture, and their books became best sellers. Otsuki Gentaku (1757–1827), who published a heavily edited and revised version of the *Kaitai Shinsho*, founded a school of Dutch language in 1789. Most of his students were doctors employed in the public sector, but people of any social status could attend. His school was followed by other institutions of Dutch learning. Motivated by a taste for exoticism (novelty), these scholars were not hindered by feelings of inferiority towards Western science.[20] At the turn of the nineteenth century a few recognized that it provided something that was lacking in the Eastern tradition, namely the natural philosophy that generated modern science.[21]

MATERIA MEDICA

Materia medica, a practice ancillary to medicine, included the study of substances derived from plants, animals and minerals, and writings on these subjects were indispensable to practitioners. The government often sponsored these voluminous writings, which formed a large pharmaceutical encyclopedia that followed the pattern of the Chinese classification of drugs – mainly according to symptoms – and provided a rough classification of sources. A Chinese treatise, *Pents'ao kang-mu* (Systematic materia medica, 1596, imported 1607), taxonomically arranged, provided a standard pattern in Japan as well as in China. An important Japanese concern was the comparison and identification with local species of animals and plants mentioned in the Chinese classics. This concern led to a dependence on actual observation rather than on the study of classical works. This not only furthered the trend towards morphological study but also introduced a new criterion of classification according to habitat and environment, such as distinguishing insects living on or in water, and fish living in fresh or sea water, as shown in *Yamato Honzo* (Japanese materia medica, 1708) by Kaibara Ekken (1630–1714).

Most scholars of materia medica were physicians, but in the latter half of the century their interest extended from conventional writings of materia medica towards encylopedic natural history, which added new species without proven medical properties, including materials imported from the West.

CONCLUSION

From the seventeenth century on, when Western knowledge began to produce claims distinct from Chinese learning, Japanese thinkers were critically attentive. The conviction that European technical knowledge was superior brought about a switch to the new model. Yoshimune and his astronomer mathematicians clearly recognized the superiority of Western over Chinese astronomy. As bureaucrats or technicians, their interest in Western science was limited to the precision of astronomical data and methods of calculation, and they did not jeopardize their hereditary posts by entering into the mechanistic philosophy of early modern Western science.

Professional interpreters in Nagasaki, well versed in the Dutch language, became acquainted with the concepts of Western science. They too, as hereditary officials, remained within the boundaries of their duty to translate faithfully. Neither official astronomers nor interpreters wrote for a general readership.

From the late eighteenth century on, many Dutch works on *Natuukunde* (the study of nature) found their way into Japan. They aroused the interest of independent scholars, who set about translating them even though their language skills were inferior to those of the Nagasaki interpreters. The majority of these 'Dutch scholars' were medical practitioners who were not necessarily occupationally motivated and were thus able to indulge in dilettantism.[22] By the end of the century their interest extended to anything Western.

Astronomy was the first discipline to bring about the conviction of the superiority of Western learning. The idea that this was also the case in other fields of scientific endeavour first spread among the independent physicians. Although only a few realized the power of mechanistic Western science, and fewer still knew of the Enlightenment, many were professionally interested in Western medicine. Unfavourable comparisons were not usually drawn; the thinking was that East is East and West is West and that curious things were going on in the West that brought interesting comparisons with the Eastern tradition. It was believed that Japanese intellectuals could gain advantages from both.

Later in the century, Western aggression towards East Asia – especially that of Russia – was to become prominent. It was not yet on a scale that prompted a radical reevaluation of the need to change political and technological institutions.[23] This took place after the 1840s, when Western aggression increased, leaving the shogunate face-to-face with advanced Western military technology.

NOTES

1. Shigeru Nakayama, *Characteristics of Scientific Development in Japan* (New Delhi: The Centre for the Study of Science, Technology, and Development, 1977).
2. C. R. Boxer, *Jan Compagnie in Japan, 1600–1817* (The Hague: M. Nijhoff, 1936).
3. Masayoshi Sugimoto and David L. Swain, *Science and Culture in Traditional Japan* (Rutland, VT: C. E. Tuttle, 1989).
4. Donald Keene, *The Japanese Discovery of Europe, 1720–1830*, revised edition (Stanford, CA: Stanford University Press, 1969).
5. Shigeru Jochi, *The Influence of Chinese Mathematical Arts on Seki Kowa*. Unpublished Ph.D. dissertation (University of London, 1993).
6. Shigeru Nakayama, *A History of Japanese Astronomy: Chinese Background and Western Influence* (Cambridge, MA: Harvard University Press, 1969).
7. Tadashi Yoshida, *The Rangaku of Shizuki Tadao: The Introduction of Western Science in Tokugawa Japan*. Unpublished Ph.D. dissertation (Princeton University, 1974).
8. Yu Fujikawa, *Kurze Geschichte der Medizin in Japan* (Tokyo: Kaiserliches Unterrichtsministerium, 1911).
9. Yu Fujikawa, *Japanese Medicine*, translated from the German by John Ruhrah, P.B. Hoeber (1934).
10. C. Leslie (ed.), *Asian Medical Systems: A Comparative Study* (Berkeley: University of California Press, 1976).
11. Norman Takeshi Ozaki, *Conceptual Changes in Japanese Medicine during the Tokugawa Period*. Ph.D. dissertation (University of California, San Franciso, 1979).
12. Sugita Genpaku, *Dawn of Western Science in Japan* (Rangaku Kotohajime, 1815). Translated by Rytz Matsumoto (Tokyo: The Hokuseido Press, 1969).
13. John Z. Bowers, *Western Medical Pioneers in Feudal Japan* (Baltimore, MD: Johns Hopkins University Press, 1970).
14. Harm Beukers et al. (eds.), *Red-Hair Medicine: Dutch-Japanese Medical Relations*. Rodopi (Publication of the Netherlands Association for Japanese Studies, no. 5, 1991).
15. Mieko Mace, *L'anatomie occidentale et l'experience clinique dans la mecine japonaise du XVIe au XVIIIe siècle*. In Isabelle Ang. Pierre-Etienne Will (eds.), *Nombres, astres, plantes et visceles. Sept essais sur l'historie des sciences et des techniques en Asie orientale*. Memoires, 35 (Paris: Collège de France, Institut des Hautes Études Chinoises, 1994).
16. Wolfgang Michel, *Hermann Buschof – Das genau untersuchte und auserfundene Podagra, Vermittelst selbst sicher-eigenen Genaesung und erloesenden Huelff-Mittels* (Heidelberg: Haug Verlag, 1993).
17. Erhard Rosner, *Medizingeschichte Japans* (Leiden: E. J. Brill, 1989).

18. Takeo Nagayo, *History of Japanese Medicine in the Edo Era – Its Social and Cultural Backgrounds* (Nagoya: University of Nagoya Press, 1991).
19. *Japan: The Dutch Experience* (London: Althlone Press, 1986).
20. Yoshio Kanamaru, *The Development of a Scientific Community in Pre-Modern Japan*. Ph.D. dissertation (New York: Columbia University, 1981).
21. Togo Tsukahara, *Affiniti and Shinwa Ryoku: Introduction of Western Chemical Concepts in Early Nineteenth-Century Japan* (Amsterdam: Gieben, 1993).
22. Herman Heinrich Vianden, *Die Ein fhrung der deutschen Medizin im Japan der Meiji-Zeit* (Dusseldorf: Triltsch Verlag, 1985).
23. Shigeru Nakayama, 'Japanese Scientific Thought,' in Charles Coulston Gillispie (ed.), *Dictionary of Scientific Biography*, vol. 15, supplement 1 (New York: Scribner's, 1978).

First published in *Science Technology and Society*, Oxford University Press, 2003

26 Technology in History: Japan

Broadly speaking, modern Japanese technology started with the American gunboat anchored in Tokyo Bay in 1853, characterized by a clear preference and priority placed on military technology and this phase ended up with the American Occupation of Japan in 1945. Japan, then, switched their technological paradigm from defence-oriented to market-oriented. The second paradigm has lasted and culminated in the 1980s in its international competitiveness. In the following, we shall trace the development of two sorts of technology.

1. DEFENCE-ORIENTED TECHNOLOGY

1.1 Prehistory

There is an issue of Japanese imported technology from Korea and China vs. endogenous in archeological history. Imported bronze swords are found in tumuli while Japanese original steel sword was manufactured out of endogenous iron sand. The latter is evaluated as the world's best-ever quality cutting weapon made with premodern technology.

The sixteenth century Japan saw a warring-states period, in which matchlock guns were introduced from a shipwrecked Portuguese boat in 1543. Matching with the need of warring-states battles, guns were widely spread without much delay on top of the traditional technology of the swordsmith.

The following seventeenth century found the establishment of the Tokugawa Shogunate regime, which lasted for more than two hundred years of peace. Very few gunsmiths could earn a living exclusively by forging guns. Some smiths now lost their customers and turned to be firework pyrotechnicians.

At the time of Commodore Perry's arrival in Japan in 1853, Japan was ruled by the samurai class, traditional worriers, of the top 5% of population. Even though they have forgotten martial arts during the long-lasting peace, they were at least nominally worriers wearing two swords all the time, and professionally responded quickly to the Western threat. Their first reaction was to fortify Japan against foreign threat by deploying cannons on the coastline. They had quickly realized that their seventeenth century weaponry of matchlock could not compete with the rapidly developing Western military technology of the post-Napoleonic era. The best brains of Western S&T were gathered at Saga clan to protect Nagasaki port and tried to smelt steel cannon in Western way, depending on the translation of Western military books.

These attempts proved to be fruitless when some of the 'advanced' clans fought against Western gunboats and were terribly defeated in 1863, simply because Western cannons had greater range than Japanese ones. These events led to the opening of the country to Western imperialists and eventually to the change of regime in 1868 giving way to the Meiji oligarchy government.

1.2 Imbalanced Development of Public Defence Technology

Exposing Japan to imperialistic environment of the time, the most urgent concern, or even paranoia, of the Meiji oligarchy leaders was how to defend Japan from foreign military aggression. But at the same time, they did not fail to recognize the industrial background behind the superiority of Western military powers. In order to survive, their priority naturally went to the matter represented by their slogan 'Rich state and strong military'. They established the College of Technology within the Ministry of Technology in 1870. At the college education level the emphasis on S&T was evident in the high per centage of graduates in scientific disciplines from Tokyo University (85% in the 1880s, compared with 40% in the 1820s).

They established modern bureaucracy, modern institutions such as military works, installation of telephone and telegraph lines, and railways. The defence-priority was outspokenly set forth in the structure of the Imperial University founded in 1886. Alongside ordinary departments like mechanical engineering and civil engineering, they started departments of military work and ammunition, no precedents of which were found in Western universities. Furthermore, the department of shipbuilding technology was inaugurated by the demands of the Navy.

Modern Japanese scientific and technological professions were the artificial creation of the new, Western-oriented government. The main practitioners of these new professions were former samurai, who were warriors by tradition but who, during the Edo period, became primarily administrative bureaucrats. In the past they had received hereditary family stipends in exchange for their loyalty to the Shogunate or local feudatories. They were the class long accustomed to thinking in terms of public affairs and to holding public office.

In the 1870s, efforts were made by the Meiji government to abolish class differences between samurai and commoners and curtailed the inherited family stipends of the samurai class. While other classes, farmers, artisans and merchants, could continue their inherited vocations, the samurai lost their traditional source of revenue. They were forced to find new livelihoods. Science and engineering was one of the promising careers for them, as it was an entirely new occupation that was needed for the building of a modern nation like telegram network and defence production, while medical school was occupied by sons of traditional doctors and agricultural school by sons of hereditary landowners. Thus, modern Japanese scientific and technological professions were, at the beginning of their development, very 'samurai-spirited'.

Samurai graduates found their posts in the government, as their fathers found positions in officialdom in Shogunate or clan bureaucracy, in which they, as technocrats, planned the design of the technological infrastructure of the modern nation by translating Western books into Japanese practice. Civil and military engineering were their natural calling. Thus, since they were traditionally the ruling class, the status of S&T was quite high, incomparable to the European pattern, in which S&T attracted the rising middle or lower classes. We may call their technology 'public technology' as distinguished from traditional artisans' 'private technology'.

1.3 Dual Structure and Privatization of Technology

With hindsight, we notice that, although the scholars in public technology were college graduates and literary people who self-congratulated their achievement to introduce modern technology, many of their attempts turned out to be failures, as they made no economical consideration to fit their work to local and capitalistic environments. In private market-oriented technologies such as modern textile industry and other enterprises, the government started experimental governmental factories but the experiment turned out to be economically disastrous. After 1981, these were transferred to the private sector. Finally, these early attempts ended up with the dissolution of the Ministry of Technology in 1885.

Modern engineering and technology may be said to have two entirely different origins. One is the 'public-centred' engineering service practised in the public sector, best exemplified by the military engineering taught and practised at the French Ecole Polytechnique. The other is profit-making, capitalistic engineering practised in the private sector, electric engineering and pharmaceutical technology, and so on.

In those countries that have most recently attained the modern stage of economic development, S&T have usually been introduced by the public sector, often under development dictatorship oligarchy and transferred to the private sector when they have become productive and profit-making. The Japanese experience in the late nineteenth century was typical and one of the earliest examples of this process. In the process of transferring a technology to the private sector, there are critical points at which it must be determined whether imported S&T will become established in the private sector and flourish in an indigenous, self-perpetuating form, or whether their continued importation into the public sector will be necessary.

The next question is who was instrumental in initiating and transplanting S&T and who supported this effort. The technology practised in the private sector (to be called private technology for abbreviation) had been maintained by traditional artisan class, carpentry and fishery. Their motivation was economical money-making, as contrasted to samurai public technology.

In the early Meiji period, the government encouraged invention by sponsoring Domestic Industry-fostering Fairs and conferring prizes on best inventors. However, industrial designs and inventions were not legally protected in Japan and the ideas of inventors were easily stolen. It was also the time when International Industrial Property Regulations were formulated. Westerners claimed that the Japanese, who so far had freely appropriated Western inventions, should subscribe to the international patent regulations. But those who welcomed the enforcement of international regulations were, of course, the Westerners rather than the Japanese. Around 1885 a major political and diplomatic issue in negotiation with major Western powers was the adjustment of unequal treaties. The

Japanese government intended to give patents to foreigners in exchange for Western denunciation of unequal treaty terms, and made a preliminary draft of a domestic patent system. The treaty amendment was postponed to 1899, when Japan subscribed to the International Industrial Property Regulations. Until then, Japanese inventors could obtain domestic patents by imitating Western inventions. Such was the case of Sakichi Toyota, of artisan origin, the founder of the company that became Toyota Motor Corporation. During that time, the patent assessors were college graduate engineers, while inventors were mostly traditional artisans.

The contrast may well be explained by supposing that while elite graduate engineers were busy introducing and translating the technology imported from the West, inventive skill in the private sector was being demonstrated in local technological adaptation, the product of 'intermediate technology'. College graduates were engaged in public works, and thus it was perhaps not proper for them to apply for profit-making patent rights, whereas private-sector inventors were eagerly engaging in profit-making practice. Only in the twentieth century, after Japan had subscribed to the International Industrial Property Regulations, were high-level engineers concerned with patent rights.

When Japan joined the International Industrial Property Regulations at the end of the nineteenth century, most needed technology was electricity generation and allied equipment. The Department of Electrical Communication had been created within the university already in the 1870s but it was public science of telegram and thus fitted to the national plan of a nation-wide (police) telegram network, which proved to be quite useful at the time of the Japanese Civil War in 1877. At the end of the century, however, the electricity needed was quite different; it was private technology of electric lamp and power generation. Some of the early graduates of the Imperial University tried to learn such a new technology by visiting and staying in the USA, not as regular students at universities but as workmen, as this new private technology was not as yet taught and established at the institutions of higher education. Thus electric engineers were the first among the Engineering School, Imperial University who transferred themselves from public to private sector, while civil and military engineers remained in public sector. Thereafter, samurai engineers moved gradually into the private sector.

Thus, if we characterize the pre-war effort of Japanese industrial development by one phrase, it was a constant process of transfer from the public to the private sector, although often interrupted by the effects of the government arms industry. It was a rather slow process of privatization, to be completed only at the post-war time.

2. MARKET-ORIENTED PRIVATE TECHNOLOGY PARADIGM IN THE POST-WAR PERIOD

Japan's surrender in 1945 was the second big turning point of Japanese S&T, in which a transformation from military-oriented to market-oriented occurred. In the beginning of the Occupation that followed, the demilitarization and demobilization of Japanese wartime S&T was the assignment of the Occupation Army. Accordingly, the two Japanese cyclotrons were thrown into the sea. Some strategic areas, such as nuclear and aeronautical research, were forbidden by the Occupation forces.

The civilian officers of the S&T Division of Economic and Scientific Section (ESS/ST) of the Occupation Forces soon realized that the Japanese scientific community had no intention to recover military-oriented technology; instead, S&T officers intentionally promoted S&T for the sake of the economic recovery of an impoverished Japan. For promotion of export trade, they encouraged quality control technology within Japanese industry.

After the Occupation was over in 1952, the Japanese were free to engage in nuclear and space researches, two major areas of Cold-War public science, but because of the late start the gap created from other Western nations was not easily caught up even to date. The Japanese government, notably the Ministry of International Trade and Industry (MITI), pursued a sort of industrial policy to protect and encourage domestic industry, first for the domestic market and later for the export market. They did not have a comprehensive science policy at all, as often believed, but remained an adviser as to what industry should be promoted in the 1950s and early 1960s, with the power of allocation of then scarce exchange currency for the purchase of scientific machinery and devices needed for take off.

2.1 Japanese Disproval of Effective R&D Public Policy

On the other hand, the S&T Agency (STA), created in 1956, sought big national projects of public science, such as nuclear technology and oceanography, but remained a small office. STA has complained in its *White Paper on S&T* (annual publication) that the Japanese government does not allocate as much to S&T as other advanced countries do, thus indirectly criticizing the Ministry of Finance, the body in charge of allocating the national budget. The Ministry of Finance has persistently responded that the major reason for Japan's relatively small S&T budget is because Japan does not conduct military research and emphasized only R&D for economic development. Then, the *White Paper* listed international comparisons of the ratio of public/private expenditure excluding military research. On that basis, Japan's level of government expenditure on R&D is similar to the R&D budgets of other nations.

The correlation between economic growth and R&D investment remains a popular belief. But when it comes to an international comparison, the differences in basic concepts according to the countries under investigation differ greatly. Still, however, we could believe a historical trend of one country, as it employed the same concept. Then, it becomes apparent by examining Japanese funding levels, that the Japanese case disproves the validity of the assumption, that there is a positive relationship between investment and economic growth. High growth was realized in the sixties when R&D investment was low, and as R&D investment was steadily increased, growth has slowed. Japanese high growth, then, was attributable to other factors entirely.

Within the world-wide science policy community, Japan in the economic high growth period of the late 1950s and throughout the 1960s showed unexplainable and embarrassing examples. The discovery of such negative correlations in various settings frustrated the whole science policymaking community and led to policy changes and drastic budget cuts in the USA, England and France in the latter part of the 1960s.

In the mid-sixties, USA, Britain and France spent a considerable portion (more than 6%) of government expenditure for R&D. USA spent more than 11% in 1965 and France spent 11%. In US, Britain and France, large-scale budget cuts in R&D have occurred since the late sixties, perhaps influenced by Japanese counter-evidence. Funding in R&D in Japan was much lower than major Western countries at the beginning of the post-war period and had remained between 2 and 3% throughout the sixties. Unlike England or France, Japan has had no grand R&D plan like SST (supersonic transport), yet despite this Japan has been able to realize rapid economic growth. Thus, OECD, the stronghold of science policy makers, in *Conditions for Success in Technological Innovation* (Paris, 1971) concluded that 'there is no observed correlation between the proportion of national resources devoted to R&D and the rate of growth in productivity'.

Since then and throughout the 1970s and 1980s, all of the advanced nations took lessons from the Japanese experience implicitly, and gradually reduced the public sector share of total R&D. USA's non-defence public obligation was reduced from 47.2% in 1965 to 28.9% in 1987, though its defence R&D was not significantly reduced. The French non-defence commitment was reduced from 62.3% in 1965 to 32.0% in 1986, though its defence R&D increased.

This poses the problem of disputing the belief of technocrats in Western advanced countries concerning their theory of R&D investment. It invalidated the belief in the eventual accountability of basic science, the idea being entertained by post-war science policy makers as 'linear programme' that encouragement of basic science eventually leads to technological innovation, since Vannevar Bush's *Science, the endless frontier* published immediately after the Second World War.

The observation of Third World science policy makers may be summarized as follows: Wisely staying out of involvement in costly basic science development, which does not guarantee economic returns to society, Japan borrowed the basic ideas and research results from the United States and other countries, using them only in an economically viable line of development, without expending large amounts of money. Japan has, as a result, been able to reconstruct its economy and attain high economic growth.

2.2 Secret of Japanese Technological Success

During the seventies, Japanese industry had to face nation-wide anti-pollution movement and the oil crisis in late 1973 stopped economic growth. But after only one year it started growing again and survived by gradually changing priority of the industrial structure from heavy-chemical industry to pollution-free light-microelectronic industry, or as the popular slogan goes from 'heavy-thick-long-large to light-thin-short-small', so as to minimize both pollution and energy use.

1980s is the decade when Japanese market-oriented technology was seen as the producer of the best quality products. Its quality control received considerable attention in the world search for the secret of Japan's technological success. They found that what made Japanese private industry so successful was, thus, not the outcome of large R&D budget but the improvement of the quality of the production process, namely process innovation rather than product innovation, such as the introduction of robotics and the quality circle movement in the seventies. It is perhaps more a clever labour management measure to reduce work alienation within the workplace.

In the late eighties, Japan has been involved in S&T conflict on the matter of intellectual property ownership with the US. American science policy bureaucrats were internally blamed that they spend more public money on R&D and less competitive in promoting quality technology than Japan. In turn, they attacked Japan as a free-rider on American basic science. Responding to these charges and more pressingly for competing with American market-oriented technology in post-cold-war period, such as IT and biotechnology, Japan started to support R&D after 1995 with their S&T Basic Plan, contrary to the general trend of post-Cold-War shrinking of R&D. Ironically, however, Japanese science policy community now makes the same complaint as the late 1980s American science policy makers that we spend more on R&D but gain less competitive, less economic growth.

REFERENCES

Shigeru Nakayama, *Science, Technology and Society in Postwar Japan* (Kegan Paul International, 1991)

Morris Low, Shigeru Nakayama and Hitoshi Yoshioka, *Science, Technology and Society in Contemporary Japan* (Cambridge U. P., 1999)

First published in *Historia Scientiarum*, Vol.15-2, 2005

27 Colonial Science: An Introduction

GEORGE BASALLA

Those who speak about 'colonial science' start with George Basalla's paper in 1967 ('The Spread of Western Science,' *Science*, 1967 vol. 156, pp. 611–22). Being a Harvard graduate student at the same time, I was heavily consulted at the time of his writing. I provided him with the Japanese steps of modernization in the early Meiji experience. Also I read galley proofs for him. Hence, I am partly responsible for his argument of colonial science. Thus, I set up a symposium on colonial science for the better understanding of it during the 22nd International Congress of History of Science, held in Beijing, China, on 28 July 2005.

George had no background in this issue and hence his paper is superficial. So, I did not take his paper so seriously and in the meantime I had forgotten about his paper. During the 1970s I heard nothing about it at all. Some time in the 1980s when I attended an international symposium, I was astonished to find that everybody talked about Basalla's model, very very critically. Poor George, who had never worked on the same topic of colonial science since then, was a target of academic ridicule.

The major criticism was addressed to George's developmentalism, without recourse to the political and economical differences of colonial areas. There are conspicuous differences between colonial sciences of the same race, same language (first category) and difference races, different languages (second category), as exemplified between the Australian and Indian cases. Nevertheless, George followed, more or less, the Japanese case as a developmental model, but Japanese science is never colonial. In order to modernize their own science, they hired and fired foreign scientists and advisers for their own need with their own money. I mentioned the 'independence' of Japanese science; it means that science was translated into the Japanese language so that they founded scientific societies and journals and their own criteria of assessment. It had nothing to do with colonial science.

MY DEFINITION

Instead of George's developmental model, I would like to propose a functional definition of colonial science as the following: my definition of colonial science is that subjectively, or internally speaking, it is a science conducted by scientists who envisaged achieving whatever they could not accomplish at home but only in a colonial setting. Objectively, or externally speaking, it is primarily addressed to the home scientific community, without primary concern about the influence on colonial community. The above-stated definition has been so far applied nicely to the Western colonial science which sought flora, fauna and natural resources in geographical frontiers, exemplified by the activity of the Royal Asiatic Society in the nineteenth century. Such state-of-the-art colonial science can be applied to Australian cases, as reported by Rod Home in our symposium. It was mainly the science of settler or temporary visitors. Thus, they were interested in the centre-periphery problem, and not of colonial control and power problems

SUPPLY-SIDE VS. DEMAND-SIDE

From the modernization point of view, Japanese historians of science worked hard on the mechanism of modernization from the demand-side point-of-view. There was not so much work done from the European supply-side. In this regard, I appreciate Lewis Pyenson's approach from the supply or Imperialistic side. After all, we have to recognize that the supply-side has the initiative in the historical context. Encouraged by the possibility of colonial science from the supply-side, I would like to treat the Japanese case at this symposium from the supply and Imperialistic point of view to nearby Asian countries and areas since the early twentieth century.

INDIAN CASE AND JOSEPH NEEDHAM

The Indian case is the typical and the most interesting of the second category of colonial science where the power relationship between ruler and ruled constitutes more complex problems than the first category.

The Indian experience reminds me of Joseph Needham who was invited in 1950 by Indian historians of science with their idea to start a 'Science and Civilization' in India modelled after Needham's *Science and Civilisation in China*. Needham, usually sympathetic to those ruled and depressed, however, commented that while most of the Indians were too preoccupied with nationalism, saying that British colonists disfigured Indian chronology, while Indian civilization must be much older than the colonists claimed it to be. I presume that only when those who fought against British colonialism for gaining independence may not be able to stand objectively, while only the next generation who had no direct suffering from Imperialists could start evaluating the scientific legacy of colonialism. Zaheer Baber must belong to such a generation, different from the generation who had been colonized.

REVISIONIST APPROACH: MODERNIZATION VS. COLONIALISM

Whether Baber would consider himself a revisionist is a moot point; however, it is clear that the revisionist approach to colonial science has been much discussed of late. Even in my generation, things began to appear. During the post-war period right after the end of the Second World War, Marxist historiography of wartime science was dominant. According to it, nothing of wartime scientific effort was worthwhile as it was dominated and directed by ultra-nationalistic military governments. Just after the war, we looked at wartime activity reflectively because the wartime nationalistic ideology meant the Japanese science was very retarded. But the next revisionist generation claims that wartime was a most happy time for Japanese scientists, who were treated well by the government. They also say the wartime generation contributed to post-war Japanese high economic growth.

My expectation for the second generation started to appear lately in the case of Korean. For instance, the recent generation of economists started to international cooperation (Gi-Wook Shin and Michael Robinson, editors, *Colonial Modernity in Korea*, Harvard University Press, 1999). They claim that up until the 1980s, the field was dominated by the Marxist view point, in which Imperialist Japan was severely blamed. It made the Korean generation who suffered during colonization pleased and consoled. Then, the new generation, however, who called themselves 'revisionists', tried to treat colonial modernity more objectively.

Modern science is taken advantage of as meritorious from both supply-side rulers and demand-side ruled but in a different way. We cannot neglect those Korean efforts to modernize themselves under colonial rule. Revisionists even say that the high economic growth since the 1960s was carried out by native capitalists and scientists, who started to work under harsh colonial rule.

COLONIAL INDUSTRIALIZATION

In contrast to the above-stated classical colonial science, we would like to consider the Sino-Japanese relationship in our Beijing Conference. The twentieth century Japanese experience, however, shows a slightly different form from previous Imperialistic powers. As the latest attempt at colonial science, Japan envisaged the industrialization of the colony, in spite of the inherent risk. I remember that it was symbolized by the Asian bullet train railway network. The hub of its activity was located at a colonial technological university, whose graduates mostly worked in colonial enterprises. I belong to the last generation who was fascinated by such an idea of colonial science and would like to come to terms with it by organizing a symposium.

Japanese colonial industrialization began in north Korea, where Noguchi Shitagau started the chemical industry in 1926. This topic is now tackled by an American graduate student in the history of science.

I would like to compare the preceding Western attempt and ascertain the degree and kind of influence of colonial science on the colonial community. There are mixed blessings with the extreme example of atrocities in human experiments for biological warfare. Nevertheless, how was the achievement of colonial science taken over by the post-war independent government and people? My tentative conclusion from the Imperialistic side: a lesson from the Japanese experience is that colonial industrialization does not pay.

This example can also be contrasted with endogenous efforts in the colonial context as appeared in the Tata and Birla cases in Indian industrialization during the interwar period.

CASES IN CHINA

I had an earlier and bitter experience of a cooperative attempt at writing of a synergetic history between China and Japan. I contributed an article using intentionally pre-war and wartime Japanese sources. My intention was to expose literatures in Imperialistic days to Chinese colleagues as a historical reality. It embarrassed our Chinese colleagues and the translation of my article into Chinese was cancelled. As a style of writing, I was cautious not to disturb the Chinese mind by restoring past experiences of the Japanese Imperialistic aggression. My guess is, however, that the mentioning of Kojinkai, the Japanese medical organization, the counterpart of the American Rockefeller Foundation, was categorically forbidden to be put into print. I was premature.

Revisionist or not, I presume that some scholars are now seriously working on the colonial science of Imperialistic Japan. Another symposium on the Science and Empire in our Conference concentrated mainly on the Japan-Asia relationship. While they are more or less working on Japanese colonial science and East Asian problems, I try to bring our issues in the world-wide context by taking advantage of the occasion of this International Congress. Their papers are to be published in a separate volume of this journal.

PLEA FOR SYNERGISTIC APPROACH

In a recent political issue, the Japanese have been accused by the Koreans and Chinese as having a lack of historical reflection on the former's past colonial rule and invasion. In order to clear up the source of misunderstanding, there started a scholarly attempt to write our recent history synergistically from both ruler and ruled viewpoints. In this regard, the history of science provides us an ideal starting point, in which the subject matter is less political, less ideological and more transparent than other areas of history. Such a synergy could be done most easily in the cases of Taiwan, to be followed by Korea, Manchuria and China in the order of maturity of preparedness on the part of historians of science.

First published in *Historia Scientiarum*, Vol.17-1, 2007

㉘ Thomas Kuhn: A Historian's Personal Recollections

NERVOUS BREAKDOWN

Immediately after Thomas S. Kuhn (1922–1996) published the first edition of his *The Structure of Scientific Revolutions* (hereafter abbreviated as *Structure*) in 1962, philosophers of science attacked him. Karl R. Popper, for example, blamed Kuhn with his famous comment, 'down with the normal science'.[1] Kuhn's pride was so damaged by such ridicule from philosophers (especially on the points of his misuse of philosophical jargon) that he suffered from a sort of nervous breakdown for some years. Its symptom appeared in his oft-spoken expression: 'Please do not quote me. Everybody misunderstands me. Unless you have my previous permission, please do not refer to my saying!'

As a former student of Kuhn, I witnessed how he coped with philosophers' criticisms. In the late 1960s, I advised Kuhn to add a section 'postscript 1969'[2] for a Japanese edition[3] (on which I was working) and thereby end the controversy. I even said to him, 'You know, philosophers are professional debaters, who found in your thesis an attractive target that provides them the opportunities for endless attack.'

From the outset of his quarrel with science philosophers, it was quite clear to me that the difference of their views (or images) of science might never be resolved. While Kuhn primarily wrote the *Structure* as a historiography of science for historians of science,[4] philosophers of science, like the Popperians, took it to be a study on how scientists ought to be. While Kuhn described the way that ordinary scientists tend to proceed in their problem-solving activities, philosophers argue from the naively idealistic image of science that they have constructed with a logical consistency but without consideration of the manner in which scientists actually perform. Kuhn often complained that no philosopher has ever read a technical paper of a scientist to know what he is normally doing. Though Kuhn spent the rest of his life trying to explain and persuade philosophers concerning the actual mechanism of scientific research, it was a series of fruitless argumentation.

There was an even more deep-rooted difference between Popper and Kuhn. During the interwar period, regardless of specialty there appeared a generation that was eager to defend the tradition of modern Western rational science as opposed to Nazi and Stalinist sciences. Popper, the Vienna school, Robert Merton, and even Marxists like Joseph Needham and J. D. Bernal shared such a consensus view. Professionally, the philosophy of science was grounded as an academic discipline that aimed to teach the philosophical basis of Western science at colleges and universities. This perspective lasted well into the post-war period. They took Kuhn's attitude towards science as a challenge to their established authority and criticized him as a 'relativist' because he admitted the existence of plural ways of scientific development rather than extolling the absolute value of modern science. I remember that even Kuhn's contemporaries, like Robert Cohen, accused him of doing serious damage to the philosophy of science, though the latter changed his attitude later.

On the other hand, Kuhn could not understand what philosophers meant by 'relativism'. What's wrong with relativism? Kuhn belongs to the wartime generation, and was mobilized in war as a rank-and-file scientist and also saw the rise of Cold-War big sciences during the post-war period. Thus, Kuhn could not maintain such an ideal and absolutely valid picture of science. A little later, by the post-war generation of Jerry Ravetz and me, conventional science had been completely relativised.

KUHN BACK TO PHYSICS

Kuhn himself was by no means a conservative, but an ambitious liberal as his youthful days testified. When he and I were together at Harvard in 1955–56, we discussed Marxian works with Marxian terminology, though he could never be an orthodox dogmatic follower. Under the McCarthy purge that was still in the air, he once told me, 'In this country if I were to say that I was "working on a social history of science", I should loose my university job.'

With his inclination for social history, I had expected him to move from philosophical debate (or rather quarrel) to the sociology of science. He did follow my advice with thanks on the point of not repeating interminable philosophical arguments, which in my view did not lead anywhere. Everybody seemed to have expected Kuhn to work in the sociology of science to become the founder of the then-called Kuhnian sociology of science. But he said to me, 'I was not trained in sociology and so I shall go back to my old field of physics.'

Then, for nearly a decade, Kuhn devoted himself to the completion of his earlier study of the Copenhagen group of physics, although he occasionally talked about philosophical problems upon request. When I told people like Barry Barnes (who wanted to follow the line of Kuhnian sociology of science), they were deeply disappointed and resented his decision that he had moved away from sociological problems.

Why and how has Kuhn made such a thematic change in his late forties? Due to his earlier experience with philosophers of science, he was so frightened to step into the new field of sociology without mastering customary use of sociological jargon. As a physics PhD, the language of physics was the only one familiar to him.

KUHN BACK TO PHILOSOPHY

In the late sixties, along with the rise of the student movement, the Kuhnian paradigmatic shift gave support to their anti-establishment mentality in their fundamental questioning of the conventional course of studies of science. He was at one time made a Marxian charisma together with Herbert Marcuse. But, misunderstood and intimidated, Kuhn again had not responded to the call of student radicals. In his later days, I made the following remark to him: 'At the turning point of industrial civilization in the 1970s, if you had been bold enough to step outside of your philosophical outlook, you would have risen to the rank of superstar, like Levi Strauss.' He thought about this overnight and then replied, 'One cannot be happy in an unintended high place.'

Then, he tried to devote himself to completing his history of quantum physics with a very careful, if not conservative, attitude in handling the source materials. When I met him next in 1977 at Princeton, he had piled up a huge amount of manuscripts next to his desk and asked me, 'I am ready to publish my quantum physics book but how many people, do you think, would really read it?' I said, 'Well, one hundred or so.' He looked disappointed. The book appeared two years later with the title, *Black-body theory and the quantum discontinuity, 1894–1912*, which disappointed reviewers and readers due to the fact that it did not mention the 'paradigm' at all.

In his last years, Kuhn came back to philosophy. In the meantime, the community of philosophers of science became much friendlier towards him; Kuhn himself learned how to use philosophical terms more carefully, so that he was fully accepted in their circle.

The main issue was his 'Incommensurability' thesis. Kuhn's generation had no intention to make commensurable out of incommensurables or converge the aims of different sets of paradigmatic works, such as the Newtonian and the Einsteinian works, whereas those of the earlier generation (especially philosophers), such as Ernst Cassirer, Emile Meyerson, and even Joseph Needham, could appreciate a unified view of nature. Einstein appeared unhappy during his last interview (by I. Bernard Cohen) that logical positivists had tried,

Figure 28.1 Thomas and Jahane R. Kuhn at the Itsukushoma Shrine, Hiroshima, Japan, spring 1986. (a photo taken by K. Narisada)

against Einstein's own intention, to make his work commensurable with that of Newton.[5] Practising scientists think that Newton and Einstein are different paradigms; so, why should not a third generation appear in future? There is no intrinsic barrier that deems the work of others entirely incommensurable; namely, while philosophers of science welcome commensurability, practising scientists do not. Scientists (including Einstein and Kuhn himself[6]) who do not care about commensurability, usually valued a brand new solution incommensurable with the preceding one.

Towards post-normal science Kuhn was unique in promoting 'scientific community determinism' that could not be thought of by philosophers, but he himself was not satisfied with the argument of 'negotiation' among sociologists of science, in which social science was the model rather than natural science. He believed in an ontological element in natural science, but could not express it well in terms of the sociology of the scientific community.

Turning to the original direction of a social or sociological issue, Kuhn's scheme must have certainly led, as a matter of necessary course, to the analysis of the scientific community. That is what I expected Kuhn to do. We are not satisfied directly with the old Marxist method of explaining phenomena on the history of science dogmatically in terms of only the socio-economical base. In a day when science-based technology has overwhelmingly influenced world history, the old Marxist socio-economic determinism of the nineteenth-century type could not be held; but instead, big scientific paradigms like computers, internet and life science are rapidly changing our lives in the twentieth to twenty-first centuries, and we have to take into account another explanatory route that starts with the Kuhnian internal paradigm, then goes to the external academic community, and then still further to non-academic social sectors. I have done further study on the scientists' immediate institutional background and the value system of a scientific community, to the extent that appeared in my *Academic and Scientific Traditions*.[7] But this is the area where normal academic science in the Kuhnian sense prevails.

While the Kuhnian scheme of 'paradigm-normal science' can explain nicely the development of academic science, Ravetz has further extended towards a socio-economical base by coining a term 'post-normal science'. We are now increasingly faced with problems that academic normal science cannot solve. Inevitably, stepping further out into social sectors, in which people other than members of scientific community like citizenry, bureaucrats and enterprises participate. They play the role of sponsor and assessors in a place of academic peer-reviewers. Badly needed is a critical structural analysis of this post-normal science, in which various factors other than academic infrastructure play more important roles at the civilizational turning-point that we are now in.[8]

NOTES

1. Thomas S. Kuhn, *The Structure of Scientific Revolutions* (Chicago: University of Chicago Press, 1962); and Karl R. Popper, 'Normal Science and Its Dangers', in Lakatos. I. and Musgrave, A. (eds.), *Criticism and the Growth of Knowledge* (Cambridge: Cambridge University Press, 1970), pp. 51–58.
2. Subsequently, the postscript was attached to the celebrated second edition of the *Structure* published in 1970.

3. Thomas S. Kuhn, *Kagaku Kakumei no kozo* (The Structure of Scientific Revolutions), translated by Shigeru Nakayama (Tokyo: Misuzu-shobo, 1971). See also Shigeru Nakayama, ed., *Paradaimu Saiko* (Rethinking of Paradigms) (Kyoto: Minerva shobo, 1984).
4. For me, 'Normal science' is structurally an essential and indispensable part of scientific activity.
5. Cohen's April 1955 interview with Albert Einstein was the last one Einstein gave before his death. It was published in the July 1955 issue of *Scientific American*.
6. Since his early days, he wanted to prove 'incommensurability' between Newton and Einstein, as evidenced by his projected but cancelled course at Harvard in 1955–1956: 'The Rise of Scientific Cosmology: Newton to Einstein'.
7. Shigeru Nakayama, *Academic and Scientific Traditions in China, Japan and the West* (Tokyo: University of Tokyo Press, 1984) I also worked and edited Kuhnian influence on various academic disciplines in Japan in my *Paradaimu Saiko* (op. cit.), and the same subject was later more fully and systematically investigated by the Japan Science Council.
8. This last paragraph has been read and corrected by Jerry Ravetz.

Bibliography

The following is a list of published books and articles by Nakayama Shigeru, in the order of the oldest first classified as A=Books in English; B=Articles in English; C=Books in Japanese. The year of publication appearing in bold type identifies those articles which are reproduced in this collection.

A. BOOKS (in English)

1969 *A history of Japanese astronomy: Chinese background and Western impact*, Cambridge MA: Harvard University Press.

1973 (Co-editor) *Chinese science: explorations of an ancient tradition*. Cambridge MA: MIT Press.

1974 (Co-editor) *Science and society in modern Japan: selected historical sources*, Cambridge MA: MIT & Tokyo: University of Tokyo Press.

1977 *Characteristics of scientific development in Japan*, New Delhi: CSIR.

1984 *Academic and scientific traditions in China, Japan and the West*, Tokyo: University of Tokyo Press.

1991 *Science, Technology and Society in Postwar Japan*, London: Kegan Paul International.

1999 (Co-editor with Morris Low and Yoshioka Hitoshi) *Science, Technology and Society in Contemporary Japan*, Cambridge: Cambridge University Press.

2001 *A Social History of Science and Technology in Contemporary Japan*, Vol. 1, Rosanna, Melbourne: Trans Pacific Press.

2005 *A Social History of Science and Technology in Contemporary Japan*, Vol. 2, Rosanna, Melbourne: Trans Pacific Press.

2006 (Co-editor with Yoshioka Hitoshi) *A Social History of Science and Technology in Contemporary Japan*, Vol. 3, Rosanna, Melbourne: Trans Pacific Press.

2006 (Co-editor with Gotō Kunio) *A Social History of Science and Technology in Contemporary Japan*, Vol. 4, Rosanna, Melbourne: Trans Pacific Press.

B. ARTICLES IN JOURNALS OR AS BOOK CHAPTERS (in English)

Articles selected for the Global Oriental republication in this volume are specially emphasized (bold face).

The figures following the title of the book or the journal indicate volume, number and pages (in this order).

1959 The First Appearance of Aristotelian Cosmology in Japan, *Kenkon Bensetsu*. In *Actes du IXe Congres International d'Histoire des Sciences Barcelona–Madrid, 1959*, 544–546.

1961 On the introduction of heliocentric system in Japan. In *Scientific papers of the College of General Education, University of Tokyo*, 11: 1: 163–176.

1962 Galileo and Newton's problem of world–formation. In *Japanese studies in the history of science* 1: 76–82.

1962 Did astrology contribute to the development of astronomy? In *Japanese studies in the history of science* 1: 137–138.

1962 Japanese studies in the history of astronomy. In *Japanese Studies in the History of Science* 1: 14–22.

1963 Accuracy of pre-modern determination of tropical year length. In *Japanese studies in the history of science* 2: 101–118.

1964 'Hsiao-chang' method (variation of tropical year length) in Chinese astronomy and its observational basis. In *Actes du dixieme congress international d'histoire des sciences* 2: 737–741 (Paris: Hermann).

1964 Abhorrence of 'god' in the introduction of Copernicanism into Japan. In *Japanese studies in the history of science* 3: 60–67.

1964 Cyclic variation of astronomical parameters and the revival of trepidation in Japan. In *Japanese studies in the history of science* 3: 68–80.

1965 Japanese activities in the history of astronomy during the early half of 1960s. In *Japanese studies in the history of science* 4: 1–19.

1965 Characteristics of Chinese calendrical science. In *Japanese studies in the history of science* 4: 124–131.

1965 The role played by universities in scientific and technological development in Japan. In *Cahiers d'histoire mondiale* 9: 2: 340–362.

1966 Characteristics of Chinese astrology. In *ISIS* 57:192: 442–454.

1966 Educational institutions and the development of scientific thought in China and west. In *Japanese studies in the history of science* 5: 172–179.

1968 A century's progress in Japan: science and technology. In *Technical Japan* 1: 1: 76–89.

1968 On the alleged independent discovery of Kepler's third law by Asada Goryu. In *Japanese studies in the history of science* 7: 55–59.

1969 Synopsis of the history of Chinese science. In *Japanese studies in the history of science* 8: 21–31.

1969 Are Japanese only 'good imitators'? Treatise: a study of development of Japanese technology. *Technical Japan* 1: 4: 76–81.

1970 Kyoto group of the history of Chinese science. In *Japanese studies in the history of science* 9: 1–4.

1971 Survey of 'science in colonialism'. In *Japanese studies in the history of science* 10: 17–22.

1972 Externalist approach of Japanese historians of science. In *Japanese studies in the history of science* 11: 1–10.

1973 The empirical tradition. In *Half the World* edited by Arnold Toynbee, London: Thames and Hudson, 131–150.

1973 Diffusion of Copernicanism in Japan. In *Studia Copernicana* V: 153–188.

1973 Joseph Needham, organic philosopher. In *Chinese Science* edited by S. Nakayama & N. Sivin, Cambridge MA: MIT Press, 23–44

1974 Science – a national endeavour. In *Bulletin of the visiting scholars association* (Harvard-Yenching Institute, Japan branch) 4: 13–19.

1974 Grass-roots geology: Ijiri Shōji and the Chidanken. In *Science and society in modern Japan* edited by Shigeru Nakayama et al., Cambridge MA: MIT & Tokyo: University of Tokyo Press 253–269.

1974 A history of universities: an overview – from the view point of science history. In *Science and society in modern Japan* edied by Shigeru Nakayama et al., Cambridge MA: MIT & Tokyo: University of Tokyo Press 57–71.

1974 Problems of the professionalization of science in late-nineteenth-century Japan. In *XIVth international congress of the history of science, texts of symposia* (Tokyo) 83–89.

1974 History of science: A subject for the frustrated. Recent Japanese experience. In *Boston studies in the philosophy of science* 15: 213–224.

1977 Confucianism, religious influences in higher education. In *International encyclopedia of higher education* VIII: 3516–3523 (San Francisco: Jossey-Bass, Inc.).

1977 Alternative science of the East. In *Human implications of scientific advance* edited by E. G. Forbes, 36–44 (Edinburgh: Edinburgh University Press).

1978 Science and technology in modern Japanese development. In *Science, technology and economic development: a historical and comparative study* edited by William Beranek Jr. & Gustav Ranis, New York: Praeger Publishers, 202–232.

1978 Public science in the modernization of Japan. In *Science, technology & development* edited by K.D. Sharma & M.A. Qureshi, 435–448 (Sterling: New Delhi).

1978 Japanese scientific thought. In Gillispie, Charles C. (editor in chief), *Dictionary of Scientific Biography*, Vol. XV, pp.728–758 (New York, NY: Charles Scribner and Sons)

1979 Japanese view of scientific cosmology. In *Michi: a journal for cultural exchange* 2: 1: 14–24.

1980 Two academic traditions in comparison – Western and Chinese. In *Historia scientiarum* 19: 49–75.

1980 Scientization of medicine. In *Japanese Journal for the History of Medicine* 26: 2: 243–256.

1981 The history of science as practised in China. In *Occasional Paper & Monographs* (Centre of Asian Studies) 32: vol. IV, pp.287–296.

1981 The future of research — a call for a service science. In *Fundamenta scientiae* 2: 1: 85–97.

1981 The transplantation of modern science to Japan. In *Occasional Paper* (Center for studies in higher education, the University of California) 23 (26pp.).

1983 PSSC, STS and Scientific Literacy – A historian's view. In *STSA-EASST Conference Paper* 211–214.

1983 Robotics and Human Creativity. In *The Cincinnati Post*, 17 May 1983.

1983 The Historiography of Astronomy with Particular Reference to Chinese Astronomy. In *Journal of Oriental Studies* xxi: 1: 32–39.

1984 The American Occupation and the Science Council of Japan. In *Transformation and Tradition in the Sciences* edited by Everett Mendelsohn, 353–369 (Cambridge: Cambridge University Press).

1985 The Japanese response to incoming Western mathematics (unpubl.)

1986 Ways of Thinking of Japanese Physicians. In *History of Traditional Medicine* 3–19.

1986 Greenwich list of observatories: Japan. In *Journal for the history of astronomy* 17: Part 4: 42–44.

1986 The transplantation of modern science in Japan. In *UNESCO Working document: comparative philosophical studies on changes in relations between science and society NISTADS* 103–108.

1986 The Three Stage Development of Knowledge and the Media. In *Historia Scientiarum* 31: 101–113.

1987 Periodization of the East Asian history of science. In *Revue de synthese* IVe S. : 3–4e (Juillet–Decembre 1987): 30–35.

1987 The position of Futian calendar on the history of East-West intercourse of astronomy. In *History of Oriental Astronomy* edited by G. Swarup, A. K. Bag, & K. S. Schukla, 135–138 (Cambridge: Cambridge University Press).

1987 On the T'ien-ching huo-wen by Yu I. In *Bulletin of Chinese Studies* 1: 2: 33–40.

1988 Privatization of Scientific Information. In *Todai Symposium '86: Information and its Functions Institute of Journalism and Communication Studies*, 283–300 (Tokyo: The University of Tokyo).

1989 Independence and Choice: Western Impacts on Japanese Higher Education. In *Higher Education* 18:31–48.

1989 J.D. Bernal – The Founder of Science Policy Studies and the Later Development of the Discipline". In Helmut Steiner (ed.), *1939 J.D. Bernal's The Social Function of Science 1989*, pp. 392–397 (Berlin: Akademie Verlag).

1990 Human Rights and the Structure of the Scientific Enterprise. In *Human Rights and Scientific and Technological Development* edited by C.G. Weeramantry, 137–150 (Tokyo: United Nations University Press).

1990 The Four Social Sectors: An Approach to the History of Post-War Japanese Science and Technology. In *Organization and Sociology of Research* edited by P. S. Nagpaul 150–155 (New Delhi: Wiley Eastern).

1991 Human Resources and the Shifting Centres in Industrial Science. In *Perspectives of Advanced-Technology Society 1990 – Prospect for Resource–Environment preserving Civilization in 21st Century* edited by K. Takeuchi and H. Matsuoka, 216–221 (Tokyo: The Institute of Statistical Research).

1990 The Shifting Centres of Science. In *Interdisciplinary Science Review* 16: 1 (March 1991): 82–87.

1990 The Impact of New Technology on Technical and Vocational Education. In *UNESCO Studies in Technocal and Vocational Education* 33: 98.

1991 Translation of Modern Scientific Terms into Chinese Characters – the Chinese and Japanese Behavior in Comparison. In *Modern Chinese Science and Technology Review* (Taiwan) 295–305.

1991 STS Activities in Japan. In *Science, Technology and Society* 85: 12 (1991.9), (Lehigh University).

1991 Grading Chinese Science Studies: A Report on the Kyoto Conference on Chinese Science. In *Chinese Science* 10: 57–65.

1992 Study on the Impact of New Technology on Technical and Vocational Education with a Special Reference to Japanese Vocational Education. In *Studies in Technical and Vocational Education Series* 33 (Paris: UNESCO, 1992.7).

1992 Post–Communist Science to be Privatized? (Guest Editorial). *4S–EASST Newsletter* 3: 2.

1993 Defining an Appropriateness in the Technological Environment. In *AI & Society: The Journal of Human-Centred Systems and Machine Intelligence* London: Springer 163–168.

1995 History of East Asian Science: Needs and Opportunities. In *Osiris* 10: 80–94.

1995 Working with Nathan Sivin: four decades. In *Chinese Science* 12:6–9.

1995 Science, Technology and Modernization: Introduction. In *East Asian Science: Tradition and Beyond* edited by Hashimoto Keizo, Catherine Jami and Lowell Skar,103–104 (Osaka: Kansai University Press).

1995 Cosmological Outlooks and Technological Transfers: A Comparative View from Eastern Periphery. In *Physics, Philosophy and the Scientific Community* (Festschrift for Robert Cohen, Boston studies in Philosophy of Science, vol.163) edited by K. Gavoglu et al., 225–233 (Amsterdam: Kluwer Academic Publishers).

1996 Personal Recollection. In *Erdem* Vol. I:25 (published May, 1996, in honour of Aydin Sayili).

1997 (with Morris Low) The research function of Universities in Japan. In *Higher Education* 34: 245–258.

1997 The Chinese 'Cyclic' View of History vs. Japanese 'Progress'. In *The Idea of Progress* edited by A. Burgen, P. McLaughlin and J. Mittelstraß, 65–75 (Berlin-New York: de Gruyter).

1997 Morales Orientales, Sciences Occidentales. In *Les Cahiers de Science & Vie* 2–96.

1998 How to Assess the Recent Brain Drain Phenomena: A Report of a Visit to Taiwan by a Japanese Research Team. In *The Science City in a Global Context* edited by Kunio Goto, David V. Gibson, Heidi Lopez-Cepero & Graham Steward, 221–230 (Chapter 20).

1998 The Spread of Chinese Science into East Asia. In *Current Perspectives in the History of Science in East Asia* edited by Y.S. Kim and F. Bray, 13–20 (Seoul: Seoul National University Press).

1998 History of Astronomy before and after the computer. In *Proceedings of the Third International Conference on Oriental Astronomy* (27–30 October 1998), 35–38 (Fukuoka: Fukuoka University of Education).

1999 Four–stage development of academic communication. In *International Management* (Kanagawa University) 16/17 (March 1999): 7–24.

2000 (with Yong S. Lee and Koichi Kitazawa) A Radical Restructuring of Japan's Postwar S&T Policy and Institutions: The Politics and Rationality of the New Century, *Report* (Department of Political Science, Iowa State University of Science and Technology, Ames, Iowa), April.

2001 The Digital Revolution and East Asian Science. In *Historical Perspectives on East Asian Science, Technology and Medicine* (International Conference on the History of Science in East Asia 1999) edited by Alan K. L. Chan, Gregory K. Clancey & Hui–Chieh Loy, 3–13 (Singapore: Singapore University Press).

2001 Preface. In *A Social History of Science and Technology in Contemporary Japan* edited by Nakayama Shigeru, 1: 1–22 (Melbourne: Trans Pacific Press).

2001 Preface to the English Edition. In *A Social History of Science and Technology in Contemporary Japan* edited by Nakayama Shigeru, 1: ix–x (Melbourne: Trans Pacific Press).

2001 Introduction: Occupation Period 1945–52. In *A Social History of Science and Technology in Contemporary Japan* edited by Nakayama Shigeru, 1: 23–56 (Melbourne: Trans Pacific Press).

2001 Destruction Of Cyclotrons. In *A Social History of Science and Technology in Contemporary Japan* edited by Nakayama Shigeru, 1: 108–118 (Melbourne: Trans Pacific Press).

2001 The Role of Advisory Missions. In *A Social History of Science and Technology in Contemporary Japan* edited by Nakayama Shigeru, 1: 179–192 (Melbourne: Trans Pacific Press).

2001 The Reorganization of Research Structures. In *A Social History of Science and Technology in Contemporary Japan* edited by Nakayama Shigeru, 1: 193–205 (Melbourne: Trans Pacific Press).

2001 The Science Council of Japan and the Scientific and Technical Administration Committee. In *A Social History of Science and Technology in Contemporary Japan* edited by Nakayama Shigeru, 1: 224–236 (Melbourne: Trans Pacific Press).

2001 The International Exchange of Scientific Information. In *A Social History of Science and Technology in Contemporary Japan* edited by Nakayama Shigeru, 1: 237–248 (Melbourne: Trans Pacific Press).

2001 Sending Scientists Overseas. In *A Social History of Science and Technology in Contemporary Japan* edited by Nakayama Shigeru, 1: 249–260 (Melbourne: Trans Pacific Press).

2001 The Scientific Community Post-Defeat. In *A Social History of Science and Technology in Contemporary Japan* edited by Nakayama Shigeru, 1: 261–275 (Melbourne: Trans Pacific Press).

2001 The Development of Japanese Style Quality Control. In *A Social History of Science and Technology in Contemporary Japan* edited by Nakayama Shigeru, 1: 411–421 (Melbourne: Trans Pacific Press).

2001 The Association of Democratic Scientists (Minka). In *A Social History of*

Science and Technology in Contemporary Japan edited by Nakayama Shigeru, 1: 470–481 (Melbourne: Trans Pacific Press).

2002 From PC to Mobile Internet – Overcoming the Digital Divide in Japan. In *Asian Journal of Social Science* 30: 2: 239–247.

2002 The ideogram versus the phonogram. In *Harvard Asia Pacific Review* (The Scientific Revolution in Asia) Fall 2002: 87–88.

2002 The Yabuuti Paradigm in the History of Chinese Science. In *East Asian Science* 19.

2002 Oriental Astronomy: History of East Asian Mathematical Astronomy; Homage to K. Yabuuti and A.Sayili. In *History of Oriental Astronomy* edited by S. M. Razaullah Ansari. 209–212 (Amsterdam: Kluwer).

2003 STS and Environment: 1970s and Now. In *Year Book 2003 of the Institute for Advanced Studies on Science, Technology & Society* edited by Arno Bamme, Peter Baumgartner, Guenter Getzinger and Bernhard Wieser), 101–110 (Graz: Institute for Advanced Studies on Science, Technology & Society).

2003 *Japan*. In *The Cambridge History of Science, Eighteenth–century Science* edited by Roy Porter, 4: 698–717 (Cambridge: Cambridge University Press).

2003 Technology in History: Japan. In *Science, Technology, and Society* edited by Sal Restivo, pp. 620–624 (Oxford University Press (USA))

2005 Towards a Dialogue between a Japanese Historian of Science and Arabs. In *Dialogue, Proceedings of the International Symposium: Cultural Diversity and Globalization, The Arab–Japanese Experience, A Cross–Regional Dialogue*, held 6–7 May 2004, edited by UNESCO, 33–38 (Paris: UNESCO).

2005 American graduate school: professional training vs. research originality. In *Reviews in Higher Education* (The Idea of a University in Historical Perspective, Germany, Britain, USA, and Japan) 84: 49–52 (Research Institute for Higher Education, Hiroshima: Hiroshima University).

2005 The Scientist Led Peace Movement. In *A Social History of Science and Technology in Contemporary Japan* edited by Nakayama Shigeru, 2: 334–344 (Melbourne: Trans Pacific Press).

2005 Introduction: The Road to High Economic Growth. In *A Social History of Science and Technology in Contemporary Japan Japan* edited by Nakayama Shigeru, 2: 1–43 (Melbourne: Trans Pacific Press).

2005 Overseas Study Leave and Participation in International Conference. In *A Social History of Science and Technology in Contemporary Japan* edited by Nakayama Shigeru, 2: 271–280 (Melbourne: Trans Pacific Press).

2005 'Colonial Science: An Introduction. In *Historia Scientiarum* 15: 2: 93–96.

2005 Obituary Mitsutomo Yuasa (1909–2005). In *Historia Scientiarum* 15: 2: 201–204.

2006 The Central Laboratory Boom and the Rise of Corporate R&D. In *A Social History of Science and Technology in Contemporary Japan* edited by Nakayama Shigeru and Gotō Kunio, 3: 67–77 (Melbourne: Trans Pacific Press).

2006 The Restructuring of Academic Research Administration. In *A Social History of Science and Technology in Contemporary Japan Japan* edited by Nakayama Shigeru and Gotō Kunio, 3: 145–156 (Melbourne: Trans Pacific Press).

2006 US-Japan Scientific Cooperation: Japan's Changing Role in the International Scientific Community. In *A Social History of Science and Technology in Contemporary Japan Japan* edited by Nakayama Shigeru and Gotō Kunio, 3: 157–166 (Melbourne: Trans Pacific Press).

2006 Scientific and Technological Cooperation with Developing Countries. In *A Social History of Science and Technology in Contemporary Japan* edited by Nakayama Shigeru and Yoshioka Hitoshi, 4: 289–299 (Melbourne: Trans Pacific Press).

2007 Thomas Kuhn: A Historian's Personal Recollections. In *Historia Scientiarum* 17: 1: 49–53

C. BOOKS (in Japanese)

1964 *Sensei-jutsu – sono kagaku-shi no ichi* (Astrology – its position in the history of science), (Kinokuniya shinsho). Tokyo: Kinokuniya.

1965 (Co-author) *Nihon kagaku gijutsu-shi taikei 14 – Chikyū uchū kagaku* (Source books for the History of modern Science and Technology in Japan, Vol. 14. – Earth and space sciences), Tokyo: Dai-ichi hōki.

1967 (Co-author) *Taikei Nihon-shi sōsho – Kagaku-shi* (Japanese History Series – Science History). Tokyo: Yamakawa shuppan-sha.

1968 (Co-editor) *Nihon kagaku gijutsu-shi taikei 7 – Kokusai* (Source Books for the History of modern Science and Technology in Japan, vol. 7 – International Relations). Tokyo: Dai-ichi hōki.

1970 (Co-editor) *Daigaku ni kansuru Ô-bun bunken sōgō mokuroku* (A comprehensive Catalogue of Books and Articles in European Languages for University Students). Tokyo: Gakujutsu-sho shuppan-kai.

1971 (Co-editor) *Nihon shisō taikei – Kinsei kagaku shisō – ge* (Series on Japanese Thinking – Scientific Thinking in early Modern Period, vol. 2). Tokyo: Iwanami shoten.

1972 (Co-editor) *Nihon shisō taikei – Yō-gaku – ge* (Series on Japanese Thinking – Western Learning, vol. 2). Tokyo: Iwanami shoten.

1972 *Nihon no tenmon-gaku – seiyō ninshiki no senpei* (Japanese Astronomy – Advance guard for the Recognition of the West). Tokyo: Iwanami shoten.

1974 *Rekishi toshite no gakumon* (Learning as history), (Chūkō sōsho). Tokyo: Chūō-kōron-sha.

1977 *Nihon-jin no kagaku-kan* (The perception of Science of the Japanese People), (Sōgen shinsho). Tokyo: Sōgen-sha

1978 (Co-author) *Omoi chigai no kagaku-shi* (A Misunderstanding of History of Science), (Asahi sensho). Tokyo: Asahi shinbun-sha.

1978 *Noguchi Hideyo* (Biography of Noguchi Hideyo), (Asahi hoyden-sen). Tokyo: Asahi shinbun-sha.

1978 *Teikoku daigaku no tanjō – kokusai hikaku no naka no Tōdai* (The Birth of the Imperial University – The Tokyo University in International Comparison), (Chūkō shinsho). Tokyo: Chūō kōron-sha.

1980 *Uchū no hate ni nani ga aru ka – mono no mikata-kangaekata o kaeru tame ni* (What is at the End of the Universe? For a Change in Looking and Thinking Things). Tokyo: Goma shobō.

1980 *Tenkanki no kagaku-kan* (Scientific Views in the Transition Period). Tokyo: Nihon keizai shinbun-sha.

1981 *Kagaku to shakai no gendai-shi* (Contemporary History of Science and Society), (Iwanami gendai sōsho NS). Tokyo: Iwanami shoten.

1982 (ed. & author) *Tenmon-gaku-shi* (History of Astronomy). Tokyo: Kōsei-sha.

1983 (ed.) *Tenmon-gaku jinmei jiten* (Bibliographical Dictionary of Astronomy). Tokyo: Kōsei-sha.

1983 (Co-author) *Jitsugaku no susume* (An Encouragement of Practical Science). Tokyo: Yūhikaku.

1984 (ed.) *Bakumatsu no yōgaku* (Western Learning in the Bakumatsu Period). Tokyo: Minerva shobō.

1984 (ed.) *Paradaimu saikō* (Reflection on Paradigms). Tokyo: Minerva shobō.

1984 *Shimin no tame no kagaku-ron* (Science for the People). Tokyo: Shakai hyōron-sha.

1984 *Ten no kagaku-shi* (History of Scientific Heaven), (Asahi sensho). Tokyo: Asahi shinbun-sha.

1986 (ed.) *Nihon no gijutsu-ryoku – sengo-shi to tenbō* (The Power of Japanese Technology – Postwar Period and Outlook), (Asahi sensho), Tokyo: Asahi shinbun-sha.

1986 (Co-author) *21seiki e no shisaku* (Considerations for the 21st century). Tokyo: Shinchi shobō.

1986 (Co-author) *Mirai sangyō no kōzu* (Design of Future Industry). Tokyo: Aki shobō.

1987 (Co-author) *Kagaku-shi kenkyū nyūmon* (An Introduction into the Study of the History of Science). Tokyo: Tokyo daigaku shuppan-kai.

1988 (ed.) *Jozefu-niidamu no sekai: meiyo dōshi no sei to shisō* (The world of Joseph Needham: Life and Thought of a Honourable Fellow). Tokyo: Nihon chiiki shakai kenkyū-sho.

1988 *Amerika daigaku e no tabi* (A Journey to Universities in the USA). Tokyo: Rikuruuto shuppan.

1989 *Ichinoe Naozō – ya ni sagatta shi no hito* (Ichinohe (Ichinoe) Naozō – Ein Mann mit Ambitionen, für die er seine Position aufgegeben hatte). Tokyo: Riburopooto (LibroPort).

1992 *Seiyō sensei-jutsu – Kagaku to majutsu no aida* (Western Astrology – between Science and Magic), (Kōdan-sha gendai shinsho). Tokyo: Kōdan-sha.

1993 *Kinsei Nihon no kagaku shisō* (Japanese Scientific Thinking in early Modern Period), (Kōdan-sha gakujutsu bunko). Tokyo: Kōdansha.

1994 *Daigaku to Amerika shakai: Nihonjin no genten kara* (The Universities and the American Society – from the Viewpoint of a Japanese), (Asahi sensho). Tokyo: Asahi shinbun-sha.

1994 (Co-ed. & co-author) *Sengo kagaku gijutsu no shakai-shi* (A social History of Science and Technology in the Postwar Era), (Asahi sensho). Tokyo: Asahi shinbun-sha.

1995 (ed.) *Kagaku to ekorojii – Komentaaru sengo 50nen,* 7 (Science and Ecology – a Comment on 50 Years after the War, vol. 7). Tokyo: Shakai hyōron-sha.

1995 (ed.) *[Tsū-shi] Nihon no kagaku gijutsu* 1–5 (A social History of Japanese Science and Technology vol. 1–5). Tokyo: Gakuyō shobō.

1995 *Kagaku gijutsu no sengo-shi* (Postwar History of Science and Technology), (Iwanami shinsho). Tokyo; Iwanami shoten.

1996 (Co-ed.) *Minkan-gaku jiten* (Dictionary on the everyday Science of the People), Tokyo: Sanseidō.

1999 (ed.) *[Tsū-shi] Nihon no kagaku gijutsu – Kokusai-ki* 5–1, 5–2 (A Social History of Japanese Science and Technology, sep. vol. 5–1, 5–2). Tokyo: Gakuyō shobō.

2000 *20–21 seiki kagaku-shi* (History of Science in 20th and 21st century). Tokyo: NTT shuppan.

2002 (Co-ed & co-author) *Kagaku kakumei no genzai-shi* (A Contemporary History of Science Revolution). Tokyo: Gakuyō shobō.

2003 *Daigakusei ni naru kimi e: Chiteki kūkan nyūmon* (For those who want to become University Students: An Introduction into Mental Space), (Iwanami junia shinsho). Tokyo: Iwanami shoten.

2006 *Kagaku gijutsu no kokusai kyōsō-ryoku – Nichi-Bei sōkoku no han-seiki* (International Competition Power of Science and Technology – A half Century of Rivalry between Japan and the USA), (Asahi sensho). Tokyo: Asahi shinbun-sha.

2006 (Co-author) *Jujireki – Yakuchū to kenkyū* (Jujireki (Chin. Calendar): Annotated Translation and Research). Tokyo: Ai-kei-cooporeeshon.

Note: A full and comprehensive list of book translations, book reviews and further works is to be found at: http://homepage3.nifty.com/shigeru-histsci/

Index

www.ingramcontent.com/pod-product-compliance
Lightning Source LLC
LaVergne TN
LVHW022323150926
848178LV00047B/936